TOXIC METALS AND
THEIR ANALYSIS

HEYDEN INTERNATIONAL TOPICS IN
SCIENCE

Editor: L. C. Thomas

TOXIC METALS AND THEIR ANALYSIS

ELEANOR BERMAN

Cook County Hospital
Chicago, Illinois

LONDON · PHILADELPHIA · RHEINE

Heyden & Son Ltd., Spectrum House, Hillview Gardens, London NW4 2JQ, UK

Heyden & Son Inc., 247 South 41st Street, Philadelphia, PA 19104, USA

Heyden & Son GmbH, Münsterstrasse 22, 4440 Rheine, West Germany

British Library Cataloguing in Publication Data

Berman, Eleanor
  Toxic Metals and their Analysis—(Heyden International Topics in Science).
  1. Heavy metals—Toxicology
  2. Heavy metals—Analysis
  I. Title
  615.9'25'3   RA1231.M52   79-41781

ISBN 0-85501-468-7

ISBN 0 85501 468 7

Typeset in Great Britain by Santype International, London and Salisbury
Printed in Great Britain by Cambridge University Press, Cambridge

CONTENTS

FOREWORD

The object of this series of monographs is the timely dissemination of essential information about topics of current scientific interest. The subject matter is treated in sufficient detail to enable those who are not specialists in the particular field to appreciate the applicability of the text to their own work and, with the aid of the bibliographic details included in the monographs, to extend their knowledge of the subject to any desired depth.

The present volume—*Toxic Metals and their Analysis*—admirably fulfils the objective of the series. It is hard to imagine a more topical subject than toxic metals in view of the almost daily reference to the dangers of one or other of them in the environment. Once a toxic threat has been postulated the analytical requirement inevitably follows.

In this book Dr Berman first discusses the general history of trace metals and their analysis. The following thirty-one chapters deal with the individual metals and metalloids, from aluminium to zirconium, to which toxic properties have been ascribed. The book concludes with an Appendix detailing the routine analytical procedures of atomic absorption spectrometry which are used in the author's laboratory.

The chapters dealing with the individual elements give a history of each metal and its uses, discuss its biochemical role, its toxicology, its distribution in the body, its normal concentration in the diet and in the body, and finally the methods available for its analysis. Those discussed here include colorimetry, fluorimetry, chromatography, polarography, and spectroscopy. Over 1750 references are quoted. These, together with the numerous tables of normal and abnormal concentrations, and the detailed analytical and toxicological information, will make this a valuable reference volume, not only for toxicologists but also for all analysts concerned with the determination of these metals.

L. C. Thomas

PREFACE

Advances in analytical instrumentation and their subsequent application in developing refined, sensitive, specific, and accurate techniques for trace metal determinations have made yesterday's esoteric investigations today's routine analyses. As a consequence, greater understanding concerning the roles of different metals in health and disease states is being acquired. What may be normal concentrations in biological materials and what may be toxic are becoming more clearly delineated.

The literatures, both technical and clinical, generated over the last two decades are voluminous. Also, the efforts of the more distant past must not be ignored. For, therein lie much of merit that can be applicable as the era of the 'trace metal screen' approaches. All these 'riches' tend to be somewhat overwhelming. How to separate the 'chaff from the wheat'?

Hence our purpose in compiling this volume—an attempt to gain a perspective of toxic metals and their analysis.

Chicago   E. Berman
November 1979

CHAPTER 1

INTRODUCTION TO TRACE METALS AND THEIR ANALYSIS

Toxic manifestations of metals of antiquity, namely lead, mercury, antimony, silver, and the metalloid arsenic, were known and described by the ancients. Attempts commensurate with the expertise of the period were made to minimize excessive exposure in the work environments. All ancient cultures employed these metals in their therapeutic regimens, a practice continuing, more or less, to the present.

Spectroscopic and chemical methods of analysis of metals and other substances marked their beginnings before the eighteenth century. L'Abbé Marie[1] presented the first ideas about photometry by 1700. Colors imparted to flames by metallic salts were described subsequently by Geoffrey, Melville, and Marygraf.[2] By 1858, a century later, David Alter,[3] an American physician, attempted using the flame in qualitative spectrochemical analysis. Prior to 1863, Valentin[4] employed the spectroscope in the diagnosis of metal poisons.

In 1876, Emil Rosenberg,[5] a Philadelphia physician, published a monograph, an absolute gem, entitled *The Use of the Spectroscope: Its Application to Scientific and Practical Medicine.* With his relatively crude instrument and limited analytical techniques, he managed to acquire an amazing amount of information concerning the distribution of potassium, sodium, calcium, and lithium in tissues and other biological materials.

Chemical techniques for metal analysis, as well as, the art-science of toxicology were developing concurrently. For example Orfila,[6] considered to be the father of modern toxicology, discovered that the organic and inorganic poisons then known were absorbed and accumulated in certain tissues. His experimental design consisted of administering definite quantities of a substance to animals, observing the effects, and upon the death of the animal, attempting the isolation and recovery of the substance by chemical means.

Tanquerel des Planches,[7] the Parisian physician who in the mid-nineteenth century systematically studied lead poisoning in both the various industrial workers exposed and in experimental animals, was aided in his investigations by his chemist colleagues Chevalier, Devergie, and Guibort.

From 1900 until approximately 1940 a number of fine studies concerning trace metal deposition in health and disease were made by employing spectrographic methods of analysis. Furthermore, many of the sample preparative techniques then devised are used, with slight modification, today, prior to determination of metal constituents by emission and atomic absorption spectrometry. Sheldon and Ramage[8] using a Hilger quartz spectrograph and an oxyacetylene flame were able to identify and determine the lead content, cadmium, nickel, and other elements in small pieces of tissue and 100 μl aliquots of blood. Tissues were washed, dried to constant weight, and then ground into powder. Blood samples were soaked into filter paper and dried before analysis.

Most investigators of the period preferred arc-spark methods to the flame for the analysis of metals. The Gerlachs,[9,10] Benoit,[11] and Policard[12,13] examined mineral deposits by means of histospectroscopy, a technique which consists of passing a high frequency spark through a tissue section (fresh, fixed, or dried). The radiation emitted is analyzed by a quartz spectrograph. Included among the elements identified in this manner are boron, carbon, gold, iron, magnesium, manganese, phosphorus, potassium, silicon, silver, sodium, and zinc. Spectrographic methods for the determination of lead,[14,15] and thallium[14] in blood after wet ashing and dry ashing were also described.

Scott and Williams[16,17] studied the inorganic constituents of protoplasm in depth and stated a number of caveats to be applied when attempting interpretations of spectrograms of biological materials.

Gaul and Straud[18-20] analyzed tissue biopsy specimens obtained from arthritic patients receiving gold thiosulfate and from syphilitics on silver arsphenamine therapy.

Chemical means of separation, followed by emission spectrographic analysis, were employed by Boyd and De[21] in their extensive investigations of metallic constituents in animal tissue and various vegetables.

Cholak,[22] one of the very versatile investigators of this era, developed methods for the analysis of lead, bismuth, cadmium, antimony, manganese, tin, copper, and silver in biological materials. The techniques utilized included arc-spark spectrography, and colorimetry of the metal dithizonates.

Kehoe and his co-workers[23] are probably the first group of modern investigators to attempt to establish what may be 'normal' concentrations of various trace metals in biological materials. Tipton[24] determined the concentrations of 18 heavy metals in tissues obtained from accident victims throughout the US.

Developments in analytical instrumentation and techniques have resulted in a greater understanding of the mode of action of trace metals in living organisms. Elements shown to act as catalysts in enzyme systems in the cell were classed as essential. Further, the essentiality of a metal was inferred when deficiency states could be induced experimentally or when requirements for growth were established. The term 'toxic metal' was applied to those elements which were disruptive of enzymatic processes, e.g. lead, arsenic, mercury.

Though the essential nature of iron was intimated by peoples of all ancient civilizations, the role of iron in blood formation was not demonstrated, logically, until about 1832 when the iron content of chlorotics (anemics) was found to be lower than that of healthy individuals.[25,26]

McHargue's[27,28] finding that copper is a necessary element in rat diets established its essentiality. Hart and his co-workers[29] also demonstrated the role of copper together with iron in blood formation.

The essential nature of manganese became clearly evident in 1931 through the work of Kemmerer et al.,[30] and Waddell and his co-workers.[31] Toxic effects of manganese, due to excessive industrial exposures, were also reviewed at about this time by von Oettingen.[32]

Zinc has been considered to be an essential element for lower plant forms for more than a century.[33] Its essentiality for animals was not shown until 1934.[34] The physiological role of zinc in plant and animal metabolism was demonstrated subsequently.[35-37] Monitoring of the zinc and copper levels in different disease states is now a routine clinical laboratory procedure.

Bortels' report[38] in 1930 that molybdenum was an essential nutrient for *azobacter* was the first evidence of a possible biological role for molybdenum. Two groups working independently established the essentiality of the element in animals. They demonstrated[39-41] that the enzyme xanthine oxidase contains molybdenum. To date, however, the metal has not been investigated as regards its importance for the health of man.

A biological role for chromium in cholesterol metabolism[42] was revealed 25 years ago. The essentiality of chromium to normal glucose metabolism was established somewhat later.[43,44] Blood serum levels of this element are a routine analysis in certain clinical areas.

An essential role in animal nutrition for the toxic metalloid selenium was demonstrated in 1957 by Schwartz and Foltz[45] and subsequently substantiated by others.[46-48]

The development of materials with which to create environments and diets free from specific trace metals of interest has enabled the successful completion of studies designed to establish a biological role for certain metals, shown to be prevalent in the environment, but with no known function. For example, within the present decade, investigators have been able to demonstrate that tin,[49] vanadium,[50-52] silicon,[53,54] and nickel[55-57] are essential elements.

The classification of trace metals is in a state of change. The toxic and essential elements are no longer separated into rigid categories. It is now recognized that all metals will exert toxic effects when present in excess. For example, the very essential elements iron, copper, or cobalt can be deadly poisons. Furthermore, certain toxic metals, so classed because of their adverse effects at relatively small doses, may be fulfilling some essential function at more minute concentrations. Essential physiological roles for the very toxic elements cadmium, arsenic, and lead have been inferred recently.[58]

Much yet remains to learned, however. Advances in analytical instrumentation and their judicious application, in light of the characteristics of the analyte and the chemistry of the various biological matrices, will provide the key to future progress.

REFERENCES

1. L'Abbé Marie, *Nouvelle découverte sur la lumière pour en mesures et compter les degrés*, Paris 1700, cited in E. Rosenberg, *The Use of the Spectroscope: Its Application to Scientific and Practical Medicine*, G. P. Putnam's Sons, New York, 1876.
2. R. Hermann and C. T. J. Alkemade, *Chemical Analysis by Flame Photometry*, Wiley-Interscience, 1963.
3. C. A. MacMunn, *Spectrum Analysis Applied to Biology and Medicine*, Longmans, Green and Co. Ltd., London, 1914.
4. G. Valentin, *Der Gebrauch des Spektroscopes zu Physiologischen und Ärtzlichen Zwecken*, Winter'sche Buchhandlung, Leipzig 1863, cited in E. Rosenberg, *The Use of the Spectroscope: Its Application to Scientific and Practical Medicine*, G. P. Putnam's Sons, New York, 1876.
5. E. Rosenberg, *The Use of the Spectroscope: Its Application to Scientific and Practical Medicine*, G. P. Putnam's Sons, New York, 1876.
6. A. W. Blyth, *Poisons, Their Effects and Detection*, William Wood and Company, New York 1885.
7. S. L. Dana, *Lead Disease: A Treatise from the French of L. Tanquerel des Planches*, Daniel Bixby and Company, Lowell, Massachusetts, (1848).
8. J. H. Sheldon and H. Ramage, *Biochem. J.*, **25**, 1608 (1931).
9. W. Gerlach and W. Gerlach, *Virchow's Arch. Pathol. Anat. Physiol.*, **28**, 209 (1931).
10. W. Gerlach and W. Gerlach, *Die Chemische Emissions-Spektralanalyse. II, Anwendung in Medizin, Chemie, und Mineralogie*, Leopold Voss, Leipzig (1933).
11. W. Benoit, *Z. Ges. Exp. Med. Biol.*, **90**, S.421 (1933).
12. A. Policard, *Protoplasma*, **19**, 602 (1933).
13. A. Policard and A. Morel, *Bull. Histol. Appl.*, **1** (9), 57 (1932).
14. W. Gerlach, W. Rollwaggen and R. Intonti, *Virchow's Arch. Path. Anat. Physiol.*, **301**, 588 (1938).
15. P. G. Shipley, F. T. M. Scott and H. Blumberg, *Bull. Johns Hopkins Hosp.*, **51**, 327 (1932).
16. G. H. Scott and P. S. Williams, *Proc. Soc. Exp. Biol. Med.*, **32**, 505 (1934).
17. G. H. Scott and P. S. Williams, *Anat. Record*, **64**, 107 (1935).
18. L. E. Gaul and A. H. Staud, *Arch. Dermatol. Syph.*, **28**, 790 (1933).
19. L. E. Gaul and A. H. Staud, *Arch. Dermatol. Syph.*, **30**, 433 (1934).
20. L. E. Gaul and A. H. Staud, *J. Amer. Med. Assoc.*, **104**, 1387 (1935).
21. T. C. Boyd and N. K. De, *Indian J. Med. Res.*, **20**, 789 (1933).

22. J. Cholak, *Indust. Eng. Chem. (Anal. Ed.)*, **7**, 287 (1935).
23. R. A. Kehoe, J. Cholak, and R. V. Storey, *J. Nutr.*, **19**, 579 (1940).
24. I. H. Tipton in *Metal Binding in Medicine*, (ed. M. J. Seven), J. B. Lippincott Company, Philadelphia, 1960.
25. P. F. Hahn, *Medicine*, **16**, 249 (1937).
26. R. L. Haden, *J. Amer. Med. Assoc.* **111**, 1059 (1938).
27. J. S. McHargue, *Amer. J. Physiol.*, **72**, 583 (1925).
28. J. S. McHargue, *Amer. J. Physiol.*, **77**, 245 (1926).
29. E. B. Hart, H. Steenbock, J. Waddell and C. A. Elvehjem, *J. Biol. Chem.*, **77**, 797 (1928).
30. A. R. Kemmerer, C. A. Elvehjem and E. B. Hart, *J. Biol. Chem.*, **92**, 623 (1931).
31. J. Waddell, H. Steenbock and E. B. Hart, *J. Nutr.*, **4**, 53 (1931).
32. W. F. von Oettingen, *Physiol. Rev.*, **15**, 175 (1935).
33. J. R. Raulin, *Ann. Sci. Nat. Bot. Biol. Vegetale*, **11**, 93 (1869). cited in E. J. Underwood, *Trace Metals in Human and Animal Nutrition*, 3rd edn., Academic Press, New York, 1971.
34. W. R. Todd, C. A. Elvehjelm, and E. B. Hart, *Amer. J. Physiol.*, **107**, 146 (1934).
35. H. R. Marston, *J. Counc. Sci. Ind. Res. (Austral.)*, **8**, 111 (1935).
36. E. W. Lines, *J. Counc. Sci. Ind. Res. (Austral.)*, **8**, 117 (1935).
37. E. J. Underwood and J. F. Filmer, *Austral. Vet. J.*, **11**, 84 (1935).
38. H. Bortels, *Arch. Mikrobiol.*, **1**, 333 (1930).
39. E. C. de Renzo, E. Kaleta, P. Heytler, J. J. Oleson, B. L. Hutchings and J. H. Williams, *J. Amer. Chem. Soc.*, **75**, 753 (1953).
40. E. C. de Renzo, E. Kaleta, P. Heytler, J. J. Oleson, B. L. Hutchings, and J. H. Williams, *Arch. Biochem. Biophys.*, **45**, 247 (1953).
41. D. A. Reichert and W. W. Westerfeld, *J. Biol. Chem.*, **203**, 915 (1953).
42. G. L. Curran, *J. Biol. Chem.*, **210**, 765 (1954).
43. K. Schwarz and W. M. Mertz, *Arch. Biochem. Biophys.*, **72**, 515 (1957).
44. K. Schwarz and W. M. Mertz, *Arch. Biochem. Biophys.*, **85**, 292 (1959).
45. K. Schwarz and C. M. Foltz, *J. Amer. Chem. Soc.*, **79**, 3293 (1957).
46. E. L. Patterson, R. Milstrey, and E. L. R. Stokstad, *Proc. Soc. Exptl. Biol. Med.*, **95**, 621 (1957).
47. J. N. Thompson and M. L. Scott, *J. Nutr.* **97**, 335 (1969).
48. K. E. M. McCoy and P. H. Wesweg, *J. Nutr.* **98**, 383 (1969).
49. K. Schwarz, D. B. Milne, and E. Vinyard, *Biochem. Biophys. Res. Commun.*, **40**, 22 (1970).
50. L. L. Hopkins, Jr., and H. E. Mohr in *Newer Trace Elements in Nutrition*, (eds W. Mertz and W. E. Cornatzer), Marcel Dekker, New York, 1971.
51. K. Schwarz and D. B. Milne, *Science*, **174**, 426 (1971).
52. F. H. Nielsen and D. A. Ollerich, *Fed. Proc. Fed. Amer. Soc. Expt. Biol.*, **32**, 329 (1973).
53. E. Carlisle, *Science*, **167**, 279 (1970).
54. K. Schwarz and D. B. Milne, *Nature (London)*, **239**, 333 (1972).
55. F. H. Nielsen and D. A. Ollerich, *Fed. Proc. Fed. Amer. Soc. Exptl. Biol.*, **33**, 1767 (1974).
56. M. Anke, M. Grun, G. Dittrich, B. Broppel, and A. Henning, in *Trace Element Metabolism in Animals*, (ed. W. G. Hoekstra), University Park Press, Baltimore, Md., 1974.
57. A. Schnegg and M. Kirchgessner, *Inter. J. Vitamin Nutr. Res.*, **49**, 96 (1976).
58. K. Schwarz in *Clinical Chemistry and Chemical Toxicology*, (ed. S. S. Brown) Elsevier, Amsterdam, The Netherlands 1977.

CHAPTER 2

ALUMINUM

Though aluminum has long been a metal of interest, its roles in the metabolism of living animals, plants, or microorganisms, remains to be elucidated. As with other areas in science and technology there is disagreement. There are indications that aluminum may have certain essential functions. It can also induce toxic manifestations.

Myers and Mull[1] in 1928 reported that rats fed an aluminum-supplemented diet showed greater initial growth as compared with groups fed either an unsupplemented control diet or an apparent aluminum-free diet. Early studies on plants indicating the element's beneficial effects upon growth were questioned by Hutchinson[2] in 1945. However, the water culture experiments of Hackett[3] seventeen years later did show that aluminum stimulated the growth of oats, rye, and pasture plants.

In vitro experiments of Horecker *et al.*,[4] and later Stitch,[5] revealed that the metal promotes the reaction between succinic dehydrogenase and cytochrome *c*. Harrison and co-workers[6], reporting on the inhibition of hexokinase by aluminum, suggest hexokinase or all phosphate-transferring systems involving ATP and magnesium ion are possible biological target sites for aluminum.

To further establish the essential function of a trace element, it is thought necessary to produce a deficiency disease experimentally in some species or to show that a specific anomaly or disease has resulted from its lack in the diet. Because of the ubiquitous nature of aluminum in the environment, this has been difficult. Hove *et al.*[7] in 1938 did attempt valiantly, though unsuccessfully, to produce a purified aluminum-deficient diet for rats. He concluded that if aluminum was required by the rat, this requirement could be met by as little as $1 \mu g$ daily.

Man's requirements are unknown, but are no doubt amply met in his diet. The aluminum content of food varies widely depending upon the contaminants introduced in commercial food processing and via cooking vessels used in the home. At present, aluminum pots and pans are used more commonly among the median- and lower-income groups. Kehoe *et al.*[8] found a mean aluminum content of 36.4 mg per day. Tipton *et al.*[9] reported daily dietary intakes of 18–22 mg. Zook and Lehmann[10] estimated daily dietary aluminum contents to range from 3.8 to 51.6 mg.

It is not unexpected that such levels of aluminum in the daily diet became a cause of concern and resulted in many investigative studies.

According to Bernheim and Bernheim[11] and others, the levels of aluminum present in the daily diet do not constitute much hazard to health. It was estimated than ten times the levels reported could be ingested with relative safety. Compounds of aluminum are absorbed poorly from the gastrointestinal tract. It should be borne in mind, however, that excessive intake, such as may occur through over medication with antacid products for example, can create problems. The occurrence of rickets due to aluminum interference with phosphate absorption has been reported.[12]

More recently, Rastogi and co-workers[13] showed that aluminum hydroxide given to control the hyperphosphatemia of chronic renal failure may be impairing the oral iron absorption in these patients. For example, the iron retention on the tenth day following its administration ranged between 31.7 and 81.6%, prior to instituting aluminum hydroxide [$Al(OH)_3$] therapy. While aluminum hydroxide was being administered, the amount of iron retained varied between 1 and 29%.

TOXICOLOGY OF ALUMINUM

High conentrations of aluminum may be toxic to the nervous system.

Kopeloff *et al.*[14] observed that aluminum cream applied to the frontal cortex of monkeys produced recurrent spontaneous clinical seizures. It was later shown that alumina, as well as cobalt and tungstic acid, will induce cortical foci in cats and rabbits but not in rats. Phenytoin and phenobarbital suppress aluminum seizures. Reasons for the epileptogenicity of aluminum hydroxide gels are not clear.

Crapper *et al.*[15] have shown that subarachnoid injections of aluminum salts in the cat resulted in a progressive encephalopathy characterized by development of neurofibrillary degeneration in the neurons. Similar, but not identical, cellular changes are observed in Alzheimer's disease in man, a disease occurring after the age of 40 which produces a progressive dementia. In the early stages in cats an apparently selective impairment in short-term memory and associative learning precedes the appearance of focal neurological signs. This decline in higher nervous functions resembles in part that noted in the human dementia of the Alzheimer type. In an earlier study with rabbits, Wisniewski *et al.*[16] observed forms of paresis, ataxia, and spasticity following suboccipital injections of aluminum phospate.

Aluminum concentrations approaching those used in Crapper's study[15] have been found in some regions of the brains of patients with Alzheimer disease. The grey matter of four normal cats in Crapper's study was found to contain 0.6–2.7 μg g^{-1} dry weight; various areas of normal human brain contained 0.43–2.4 μg g^{-1}. After injection of aluminum chloride the grey matter of the cat brain contained 3.2–32.2 μg g^{-1} dry weight. Cortical material obtained from a case of Alzheimer's disease contained from 5.4 to 11.5 μg g^{-1}.

Berlyne and Yagel[17] described an aluminum-induced porphyria in rats.

Occasionally patients, on long-term dialysis because of chronic renal failure, develop a psychosis referred to as dialysis dementia. The cause is unknown.

Since 1970, aluminum hydroxide used for preventing phosphorus retention has been implicated by investigations like those of Berlyne et al.[18,19] and Alfrey et al.[20] Dunea et al.[21] believe that to blame aluminum hydroxide would require an explanation of why the syndrome affects only some patients and not others. Most dialysis patients ingest large amounts of aluminum hydroxide. Occurrence of sporadic cases all over the world would suggest other factors. A slow viral infection, and drug intoxication are suggested as possible causes. These workers,[21] do however, support the view that aluminum may play a role in the cause of dialysis dementia.

In June of 1972 changes in the method of water purification in Chicago resulted in a higher aluminum content in the water used in dialysis. Prior to the change aluminum levels varied between 0 and 150 μg l^{-1}. After introducing the pure aluminum sulfate filtration method, peak water aluminum contents of 300 to 400 μg l^{-1} were attained; 100 μg l^{-1} of aluminum was the lowest level reported. The first cases of dialysis dementia appeared in September of 1972, three months after changing water treatment procedures and coinciding with a peak water aluminum level of 360 μg l^{-1}. Twenty patients in all became demented from September 1972 to January 1976; nineteen died. The periodic outbreaks of dialysis dementia coincided with the periodic rises in water aluminum levels. Since this was a retrospective investigation, there is no analytical data of autopsy tissues available for this population.

Chemical data presented in the literature are at present inconclusive. We have found poor correlation between serum aluminum or erythrocyte aluminum levels among dialysis dementia patients and others in chronic renal failure. Among the twenty-one renal dialysis patients apparently free from symptoms of dementia that were studied, serum aluminum levels ranged from 0.042–1.08 μg ml^{-1}. Erythrocyte aluminum varied between 0.043 and 0.304 μg ml^{-1}. The three chronic renal failure patients with dialysis dementia showed serum levels of 0.124–1.096 μg ml^{-1} and erythrocyte contents between 0.12 and 0.196 μg ml^{-1}. Twenty subjects with normal kidney function who were not being treated with aluminum hydroxide gel had aluminum levels of 9–67 ng ml^{-1} in the serum; erythrocyte aluminum varied between 42 and 88 ng ml^{-1}. Normal subjects using cookware other than aluminum showed serum levels below 20 ng ml^{-1}.

Aluminum contents in the above study were determined by flameless atomic absorption spectrometry following chelation of pH 6 with 1% aqueous cupferron and extraction into methylisobutyl ketone.

Clarkson et al.[22] employed neutron activation analysis in studying the effects of aluminum hydroxide on calcium, phosphorous, and aluminum balances in chronic renal failure and reported plasma aluminum levels in a similar range.

There must be some factor present among renal patients suffering from dialysis dementia which is not shared by other dialysis patients. For example, interrelationships of aluminum with essential metals such as copper, magnesium, zinc, cobalt, iron, etc. have not been investigated in depth.

Inhaling aluminum dusts can prove hazardous,[23] the degree depending upon the dose, i.e. concentration in inhaled air and length of exposure. Both local and systemic effects have been described.

Waldron-Edward and co-workers[24] observed that 30 of the 36 factory workers in an aluminum dust atmosphere exhibited significantly prolonged prothrombin times. Serum aluminum levels were increased two to three fold compared to those found in a control group. Furthermore, while inorganic phosphorus and total alkaline phosphatase levels were within normal ranges, serum intestinal alkaline phosphatase fractions, acid phosphatase, and adenosine triphosphate were reduced.

Numerous reports in the literature describe development of pulmonary fibrosis from exposure to dusts. Industrial applications of aluminum and its compound are wide. It is used alone or alloyed with other elements such as copper, zinc, silicon, magnesium, and manganese in manufacturing various construction materials, insulated cables and wiring, packaging materials, laboratory equipment, etc. Various aluminum oxides are used as abrasives and refractories in different industrial operations. This metal and its compounds find application in the paper, glass, and textile industries. Aluminum compounds are employed in water purification systems.

For the most part reports in the literature concerned with adverse effects developing from exposure to aluminum compounds in industry do not establish a clear dose-response relationship, i.e. between the degree of exposure and the magnitude of the resulting toxicity. The paper of McLaughlin[25] is an exception; it presents a graphic description of the clinical, radiological, pathological, and chemical features of a case of extensive aluminum fibrosis of the lungs in a 49-year-old man who had worked for $13\frac{1}{2}$ years in a ball-mill room of an aluminum powder factory. Symptoms were referable to the central nervous system as well. The subject died of bronchopneumonia following development of a rapidly progressive encephalopathy associated with epileptiform attacks.

Even at this point, 16 years later, the careful spectrographic analyses in McLaughlin's case are spectacular. As a point of interest, Table 2.1 below illustrates a comparison between levels reported in McLaughlin's case and normals of that day.

TABLE 2.1
Aluminum contents (µg g⁻¹ wet weight)

Tissue	McLaughlin's 49-year-old man[25]	Normals
Lung, upper lobe	430	16–24
Lung, lower lobe	340	not listed
Brain	5	0.23–0.29
Liver	90	0.61–0.76
Bone	30	not listed

Dust in the subject's work environment contained 60–71% of aluminum.

It has been reported that workers in cotton mills, employed in operations where there is long contact with aluminum in wet bobbin winding, develop acroanesthesia—a congestive numbness of the fingers.

Burns have been caused by the extremely reactive alkyl aluminum compounds being used as catalysts. Powdered aluminum, which ignites easily, has been responsible for many fires and explosions.

Aluminum dusts were once used in treating silicosis. This ill-advised practice has since been discontinued.

In August of 1977 we had occasion to analyze lung, kidney, and liver tissue from a 75-year-old man with possible black lung disease. He had been an industrial worker until 10 years previously. The nature of his employment was not clearly delineated in the history. Although only analyses for silicon contents were requested, determinations of bismuth, antimony, and aluminum were also done as a matter of curiosity. Findings are listed in Table 2.2.

By comparison, the aluminum content of tissues obtained from a subject diagnosed as a dialysis dementia more than a year before death were determined at about the same time. Aluminum levels are listed in Table 2.3.

TABLE 2.2
Trace metal contents—75-year-old man—black lung disease (µg g⁻¹ wet weight)

Tissue	Aluminum	Antimony	Bismuth	Silicon
Lung, left upper lobe	11.85	3.75	0.463	8.0
Lung, left lower lobe	6.57	0.24	0.17	2.2
Lung, right upper lobe	61.73	16.87	1.02	5.9
Lung, right middle lobe	2.03	11.04	0.30	8.0
Lung, right lower lobe	30.12	1.18	0.18	4.7
Liver	0.66	1.79	0.30	7.1
Kidney	22.22	8.33	0.328	1.5
Control lung	1.63	2.97	0.28	2.7

TABLE 2.3
Aluminum content dialysis
dementia (μg g⁻¹ wet weight)

Tissue	Aluminum
Brain	0.95
Liver	1.0
Kidney	0.9
Heart	0.98
Bone	1.0
Lung	not submitted

Had the subject in Table 2.2 been exposed in an aluminum industry? Perhaps as long ago he had been treated for silicosis with aluminum dusts!

METHODS OF ALUMINUM ANALYSIS

Underwood[26] states that published data for the aluminum content of animal tissue and fluids are extremely discordant. This is the case when comparing analytical data for metallic elements in general. Analytical technologies are ever changing. As a general rule values reported in the older literature are higher than those obtained fairly recently. Since the colorimetric procedures of the day lacked specificity, other metals in addition to aluminum were being measured in a reaction. However, certain investigators were able to obtain values considered realistic by current standards. For example, Myers and Morrison[27] in 1928 obtained the following values for the aluminum content in tissue of dogs fed aluminum:

Blood, 0.4–1.7 μg g^{-1} (wet weight)
Liver, 1.2–4.2 μg g^{-1} (wet weight)
Kidney, 0.3–1.3 μg g^{-1} (wet weight)
Brain, 0.4–1.8 μg g^{-1} (wet weight)

Fifty gram aliquots of tissue were first digested in sulfuric-perchloric acid, iron was removed, then any aluminum present was complexed with aurin tricarboxylic acids and measured with a Dubuoscq colorimeter.[28] The authors claim that with good technique one can detect 0.1–5 μg g^{-1} of aluminum with an error not above 10% by this method. The need for pure reagents free from both iron and aluminum was stressed.

Colorimetry

Colorimetric methods employing complexing agents, such as alizarin[29] and morin[30] had been described earlier. Cholak and co-workers[31] employed alizarin in their investigations in 1943. They also determined metals spectrographically.

Sandell[32] compared the relative sensitivities of various complexing agents used in aluminum analyses. All reactions were pH dependent and suffered from interferences from other metals coextracted from certain matrices. Reported detection limits for alizarin S were 3 ng ml^{-1} of aluminum as compared with limits for aurin tricarboxylic acid and eriochrome cyanine R of 2 ng and 6 ng l^{-1} of aluminum, respectively. Sensitivities ranging from 2.5 to 0.35 ng ml^{-1}, depending upon the wavelength at which the complex is measured are claimed for 8-hydroxyquinoline. The aluminum hydroxy quinolate complex in chloroform fluoresces strongly in the ultraviolet region. Pontachrome blue-black R forms a fluorescent compound with the metal also.

Quantitative extraction of aluminum (Al^{+3}) by 8-hydroxyquinoline in chloroform, benzene, carbon tetrachloride, xylene, toluene, etc. is obtained over the pH range 4.5–11. Copper, nickel, zinc, iron, cobalt, cadmium, zirconium are coextracted. Cyanide ion can mask the copper, cadmium, nickel and zinc. Iron can be masked by phenanthroline or by preliminary extraction with thiocyanate or cupferrate. The interference of fluoride ions with the aluminum 8-hydroxyquinoline reaction can be utilized for the indirect determination of fluoride.[33]

The colorimetric procedure of Gentry and Sherrington[34] for the determination of aluminum in blood is based on the formation of the aluminum 8-hydroxyquinoline complex.

About 90% extraction of aluminum by pure acetylacetone over the pH range 2–6 has been described.[35,36] Some investigators achieved complete recovery of aluminum by repeatedly extracting at pH 5–9 with 0.1M acetylacetone in various solvents.[37]

Eishelman and co-workers[38] investigating various complexing agents reported quantitative extraction at pH 5.5–6 by 0.1M thenoyltrifluoracetone in 4-methyl-2-pentanone. Aluminum was also completely extracted by a 0.05M solution of cupferron in chloroform.

Fluorimetry

Measuring the fluorescence of a metal complex is usually a more sensitive technique than colorimetry. For example, Good et al.[39] extended the sensitivity of the Gentry procedure by utilizing the fluorescence of the aluminum 8-hydroxyquinolate complex in chloroform. Rubino and Hagstrom[40] employed such fluorescence measurements in determining small amounts of aluminum in plant tissues. Poole and co-workers[41] used a modification of the Rubino-Hagstrom procedure for determining the aluminum content in blood. They found the method capable of detecting levels in the order of 1 μg ml^{-1} but not useful for measuring blood aluminum at normal levels. Nickel and iron interferences were removed by hydroxylamine and bathophenanthroline, respectively. The sample size required was fairly large—10 ml.

Numerous other compounds have been employed in the fluorescence analysis of aluminum. Representative applications include those of Will[42] and Dagnell *et al.*[43] Will determined the metal by complexing with morin. The latter group used salicylidene-*o*-aminophenol. Detection limits of 0.2 ng ml^{-1} and 0.3 ng ml^{-1} were claimed for the respective methods.

Chromatography

Chromatographic methods in conjunction with metal complexing techniques have been employed to some extent in trace metal analyses. To illustrate, Quereshi and co-workers[44] separated aluminum and titanium by paper chromatography. Zones, located by spraying with 0.1% alcoholic alizarin S and exposing to ammonia vapors, were eluted in weak acids. Individual metals were then reacted with suitable reagents and the resulting complexes were measured colorimetrically. For example, the aluminum eluate was complexed with trichrome cyanin R; titanium, with sulfosalicylic acid; and iron, with 1,10-phenanthroline. Mixtures of as many as seventeen metals, including molybdenum, tungsten, copper, chromium, cobalt, nickel, iron, aluminum, manganese and vanadium have been separated similarly.

Lee and Burrell[45] determined the aluminum content in marine and sea water by chelating the metal with trifluoroacetylacetone in toluene and measuring the complex by electron capture gas chromatography. Miyazaki and Kaneko[46] analyzed the aluminum content in a digest of liver tissue by a similar method.

Spectrography

Simultaneous detection of the numerous metallic elements present in a specific sample is the major advantage of the arc-spark spectrograph. Detection limits are about 1 μg g^{-1}. Rather extensive studies are found in the earlier literature. The work of Kehoe *et al.*[8] concerning the aluminum content in biological materials was cited previously. Wolf[47] applied a spectrographic technique in determining aluminum levels in blood. His values, 0.21–0.94 μg ml^{-1} (mean 0.54 μg ml^{-1}) are considerably higher than those obtained by Kehoe (mean, 0.13 μg ml^{-1}) or by Tipton and Cook[48] in a later study. The latter investigations agree fairly well with those of Kehoe. Perhaps Wolf investigated populations with greater exposure to aluminum.

As a rule, metal concentrations reported in the earlier literature tend to be considerably higher than values obtained more recently. Purity of reagents and analysts' awareness of, and ability to deal with, contaminants in the analytical environment no doubt contribute to this disparity.

Neutron activation analysis

The estimated detection limit for aluminum by neutron activation analysis is 0.02 μg. Included among the potential interferences are sodium, silicon, chromium, magnesium, nickel, iron, fluorine, lead and arsenic.

A few techniques involving destructive neutron activation analysis of aluminum appeared in the literature in 1971–72.[49-51] Subsequently, Blotcky et al.[52] described a method wherein the acid digest is subjected to cation exchange chromatography in order to remove the major interferences; namely, sodium, chloride, silicate and phosphate. Aluminum remaining in the resin is then irradiated and assayed. Aluminum content of apparently normal urine specimens determined by this technique varied from <0.05–0.13 μg ml^{-1}. The authors stress the need for a clean laboratory environment.

Garmestani compared neutron activation analysis with flameless atomic absorption spectrometry for determination of aluminum in blood and urine.[53] When the biological matrices were destroyed by wet ashing with nitric acid, results obtained with both techniques compared quite closely.

Atomic emission and absorption spectrometry

Detection limits of aluminum by flame atomic emission are about 0.01 μg ml^{-1} in a nitrous oxide–acetylene flame.[54] Approximate detection limits in the DC arc plasma jet are said to be 0.001–0.003 μg ml^{-1}.[55-57] Limits of 0.1 μg ml^{-1} and 1×10^{-1} μg ml^{-1} are claimed for flame[58] and flameless atomic absorption[59] respectively. The approximate detection limit for atomic fluorescence is 0.1 μg ml^{-1}.[60]

Perhaps because interferences due to matrix constituents were simpler to resolve, practical applications to the determinations of aluminum in biological materials have been made primarily with atomic absorption instrumentation. A flame atomic emission technique described requires a 25 ml blood sample, for example.[61]

Krishnan et al.[62] measured aluminum in tissue digests by flame atomic absorption quite successfully.

Fuchs et al.[63,64] describe the determination of aluminum in 25 μl of serum by direct analysis in a graphite furnace. It is assumed that matrix constituents were adequately destroyed in the charring cycle. A coefficient of variation of 5.3% is claimed for the analysis.

Dolinsek and co-workers[65] dissolved dental enamel in 6N nitric acid and measured the aluminum content by a carbon cup atomic absorption method.

McDermott and Whitehall[66] determined aluminum in brain tissue following wet digestion in perchloric and nitric acids. The matrix constituents remaining exerted only minimal effects upon the absorption signal.

LeGendre and Alfrey[67,68] preferred chelating aluminum in dried and defatted tissue with EDTA prior to analysis in the graphite furnace, and the results are said to be less erratic than those obtained following nitric acid digestion. The recovery of added aluminum seems acceptable. However, one does question the completeness and consistency with which the aluminum incorporated into the dried tissue matrix is removed by steeping in EDTA solution. Is that aluminum present in such a form that it can be chelated quantitatively in so simple a manner?

Valentin et al.[69] compared aluminum contents of urines and sera, from 110 factory workers exposed to dusts or metallic vapors, to those levels found among 40 control subjects and 33 dialysis patients receiving aluminum hydroxide. Serum levels among the exposed and control groups showed no significant difference. The aluminum content in the sera of the exposed workers varied from 6 to 164 μg l^{-1} as compared with levels of 4–34 μg l^{-1} in the control group. This finding is far different from that of Waldron-Edwards' study, for example.[24] Urinary aluminum excretions among the exposed workers were found to be 10.5–220 μg l^{-1} as compared with levels of 3.5–31.0 μg l^{-1} obtained in the control group. Dialysis patients had serum aluminum levels varying from 6 to 254 μg l^{-1}.

Levels reported by Valentin[69] are somewhat lower for all groups than those reported by other recent investigators, as shown in Table 2.4.

TABLE 2.4
Reported serum aluminum values in normal subjects

Al (μg l^{-1})	Number	Technique	Investigator	Year
150–530	21	AAS	Waldron-Edwards[24]	1971
65–79	10	NAA	Clarkson[22]	1972
10–90	29	AAS	Fuchs[64]	1974
4–34	40	AAS	Valentin[69]	1976
9–67	20	AAS	Berman	1977

Considering the differences in analytical techniques, these variations are not too disturbing. They will minimize with greater understanding of the nature of aluminum as well as of the limitations of the instrumentation applied to its analysis.

Laboratory ware, both glass and plastic, must be precleaned to eliminate possible contaminants in the analytical environment. The inclusion of blank determinations on reagents is a necessity.

Flameless sampling devices have extended the sensitivity of atomic absorption spectrometry many fold. Analyses can be performed on smaller samples. Many claims to the contrary notwithstanding, a degree of chemical pretreatment is required for most matrices prior to analysis in the graphite furnace. Deuterium background correction is partially effective, and even then only in eliminating interferences due to non-specific molecular absorption generated by matrix constituents.

Cerebrospinal fluid can be analyzed directly. However, because differences in viscosity between serum and aqueous standards will lead to sampling errors, sera must be diluted before introduction into the flameless atomizer.

Effects from constituents in the urine matrix can be eliminated by chelating the aluminum with cupferron and extracting into an organic solvent, such as methylisobutyl ketone. Some prefer to wet ash the urine. Background correction is used in addition.

Tissue samples can be ashed in nitric–sulfuric acids. Various investigators analyze the diluted digests directly. We prefer to chelate and extract the digests prior to analysis.

A review of flameless atomic absorption methods indicates that the 309.3 nm line is preferred. The graphite furnace is programmed according to the following parameters:

Dry, 90–100 °C for 40 s

Char, 1500–1600 °C for 60–180 s

Atomize, 2650 °C for 6–7 s

REFERENCES

1. Victor C. Myers and James W. Mull, *J. Biol. Chem.*, **78**, 605 (1928).
2. G. E. Hutchinson, *Soil Sci.*, **60**, 29 (1945).
3. C. Hackett, *Nature (London)*, **195**, 471 (1962).
4. B. L. Horecker, E. Stotz, and T. R. Hogness, *J. Biol. Chem.*, **128**, 251 (1939).
5. S. R. Stitch, *Biochem. J.*, **67**, 97 (1957).
6. W. H. Harrison, E. Codd, and R. M. Gray, *Lancet*, **2**, 277 (1972).
7. E. Hove, C. A. Elvehjem, and E. B. Hart, *Amer. J. Physiol.*, **123**, 640 (1938).
8. R. A. Kehoe, J. Cholak, and R. V. Story, *J. Nutr.*, **19**, 579 (1940).
9. I. H. Tipton, P. L. Stewart, and P. G. Martin, *Health Phys.*, **12**, 1683 (1966).
10. E. G. Zook and J. Lehmann, *J. Amer. Diet Assoc.*, **52**, 225 (1968).
11. F. Bernheim and M. L. P. Bernheim, *J. Biol. Chem.*, **127**, 353 (1939), and **128**, 79 (1939).
12. H. J. Deobald and C. A. Elvehjem, *Amer. J. Physiol.*, **111**, 118 (1958).
13. S. P. Rastogi, F. Padilla, and C. M. Boyd, *J. Arks. Med. Soc.*, **73**, 133 (1976).
14. L. M. Kopeloff, S. E. Barrera, and W. Kopeloff, *Amer. J. Psychiat.*, **98**, 881 (1942).
15. D. R. Crapper, S. S. Krishnan, and A. J. Dalton, *Science*, **180**, 511 (1973).
16. H. Wisniewski, O. Narkiewicz, and K. Wisniewska, *Acta Neuropathol.*, **9**, 127 (1967).
17. G. M. Berlyne and R. Yagel, *Lancet*, **2**, 1501 (1973).
18. G. M. Berlyne, J. Ben Ari, D. Pest, G. Weinberger, M. Stern, G. R. Gilmore and R. Levine, *Lancet*, **2**, 494 (1970).
19. G. M. Berlyne, R. Yagel, J. Ben Ari, G. Weinberger, E. Knopf and G. M. Danovitch, *Lancet*, **2**, 564, (1972).
20. A. C. Alfrey, G. R. LeGendre and W. D. Kaehny, *New Eng. J. Med.*, **294**, 184 (1976).
21. G. Dunea, S. D. Mahurkar, B. Mamdani and E. C. Smith, *Ann. Int. Med.*, **88**, 502 (1978).
22. E. M. Clarkson, V. A. Luck, H. V. Hynson, R. R. Bailey, J. B. Eastwood, J. S. Woodhead, V. R. Clements, J. L. H. O'Riordan and H. E. DeWardener, *Clin. Sci.*, **43**, 519 (1972).
23. A. Hamilton and H. L. Hardy, *Industrial Toxicology*, Publishing Sciences Group, Inc., Action, Massachusetts, 1974.
24. D. Waldron-Edwards, P. Chan and S. C. Skoryna, *Canad. Med. Assoc. J.*, **105**, 1297 (1971).

25. A. I. G. McLaughlin, G. Kazantis, E. King, D. Teare, R. J. Porter and R. Owen, *Brit. J. Ind. Med.*, **19**, 253 (1962).
26. E. J. Underwood, *Trace Elements in Human and Animal Nutrition*, Academic Press, New York and London, 1971.
27. V. C. Myers and D. B. Morrison, *J. Biol. Chem.*, **78**, 615 (1928).
28. V. C. Myers, J. W. Mull and D. B. Morrison, *J. Biol. Chem.*, **78**, 595 (1928).
29. F. W. Atack, *J. Soc. Chem. Ind.*, **35**, 936 (1915).
30. E. Schantl, *Mikrochem.*, **2**, 174 (1924).
31. J. Cholak, D. M. Hubbard and R. V. Story, *Ind. Eng. Chem. (Anal. Ed.)*, **15**, 57 (1943).
32. E. B. Sandell, *Colorimetric Determination of Traces of Metals*, Interscience Publishers, Inc., New York, 1959.
33. H. H. Willard and C. H. A. Horton, *Anal. Chem.*, **24**, 862 (1952).
34. G. H. R. Gentry and L. C. Sherrington, *Analyst*, **71**, 432 (1946).
35. J. F. Steinbach and H. Freiser, *Anal. Chem.*, **26**, 375 (1954).
36. A. Krishen and H. Freiser, *Anal. Chem.*, **31**, 923 (1959).
37. E. Abrahamczik, *Mikrochem. Mikrochim. Acta*, **33**, 209 (1947).
38. H. C. Eishelman, J. A. Dean, O. Menis and T. S. Rains, *Anal. Chem.*, **31**, 183 (1959).
39. E. Good, J. E. Petley, W. H. McMullen, S. E. Wiberley, *Anal. Chem.*, **25**, 608 (1953).
40. E. J. Rubino and G. R. Hagstrom, *J. Agr. Food Chem.*, **7**, 722 (1959).
41. J. W. Poole, L. A. Zeigler and M. A. Dugan, *J. Pharm. Sci.*, **54**, 651 (1965).
42. F. Will, *Anal. Chem.*, **33**, 1360 (1961).
43. R. M. Dagnell, R. Smith and T. S. West, *Talanta*, **13**, 609 (1966).
44. M. Quereshi, J. P. Rawat and F. Khan, *J. Chromatog.*, **34**, 237 (1968).
45. M. L. Lee and D. C. Burrell, *Anal. Chim. Acta*, **66**, 245 (1973).
46. M. Miyazaki and H. Kaneko, *Chem. and Pharm. Bull. (Tokyo)*, **18**, 1933 (1970).
47. H. Wolf, *Biochem. Z.*, **319**, 1 (1948).
48. I. H. Tipton and M. J. Cook, *Health Phys.*, **9**, 103 (1963).
49. K. Fritze and R. Robertson, *J. Radioanal. Chem.*, **7**, 213 (1971).
50. G. C. Goode, C. M. Howard, A. R. Wilson and V. Parsons, *Anal. Chim. Acta*, **58**, 363 (1972).
51. H. Thurston, G. R. Gilmore and J. D. Swales, *Lancet*, **1**, 881 (1972).
52. A. J. Blotcky, D. Hobson, J. A. Leffler, E. P. Rack and R. R. Recker, *Anal. Chem.*, **48**, 1084 (1976).
53. K. Garmestani, A. J. Blotcky and E. P. Rack, *Anal. Chem.*, **50**, 144 (1978).
54. C. Veillon in *Trace Analysis;* (ed. J. D. Winefordner) Wiley-Interscience, New York, 1976.
55. G. H. Morrison and Y. Talmi, *Anal. Chem.*, **43**, 809 (1970).
56. P. W. J. M. Boumans and F. J. de Boer, *Spectrochim. Acta*, **27B**, 391 (1972).
57. R. H. Scott, V. A. Fassel, R. N. Knisley and D. E. Nixon, *Anal. Chem.* **46**, 75 (1974).
58. G. D. Christain and F. J. Feldman, *Appl. Spectros.*, **25**, 660 (1971).
59. B. V. L'vov, *Atomic Absorption Spectrochemical Analysis*, Adam Hilger, London, 1970.
60. J. D. Winefordner and R. C. Elser, *Anal. Chem.* **43**, 25A (1971).
61. M. Ihnat, *Anal. Biochem.*, **73**, 120 (1976).
62. S. S. Krishnam, K. A. Gillespie and D. R. Crapper, *Anal. Chem.*, **44**, 1469 (1972).
63. C. Fuchs, M. Brasche, U. Donath, H. V. Henning, D. Knoll, H. Nordbeck, K. Paschen, E. Quelhorst and F. Scheler, *Ver. d. Deutsch Ges. f. Inn. Med.*, **79**, 683 (1973).

64. C. Fuchs, M. Brasche, K. Paschen, H. Nordbeck and E. Quelhorst, *Clin. Chim. Acta*, **52**, 71 (1974).
65. F. Dolinsek, J. Stupar and M. Spenko, *Analyst*, **100**, 884 (1975).
66. J. R. McDermott and I. Whitehall, *Anal. Chim. Acta*, **85**, 195 (1976).
67. G. L. LeGendre and A. C. Alfrey, *Clin. Chem.*, **22**, 53 (1976).
68. A. C. Alfrey, G. L. LeGendre and W. D. Kaehney, *New Eng. J. Med.*, **294**, 184 (1976).
69. H. Valentin, P. Preusser and K. H. Schaller, *Int. Arch. Occup. Environ. Health*, **38**, 1 (1976).

CHAPTER 3

ANTIMONY

Antimony is a toxic metal known since approximately 4000 BC. It is a constituent of ancient bronze and other alloys. Since very early times antimony was used as a medicine and cosmetic. Both Dioscorides and Pliny wrote of its use. Paracelsus in the sixteenth century prescribed antimony compounds for many diseases.

The obvious toxic effects induced by antimony caused the Faculty of Physicians at the University of Paris to prohibit its medicinal use. However, the ban was lifted in 1657, one hundred years later, when Louis XIV was ostensibly cured of typhoid with antimony.

Early in the twentieth century antimony compounds were introduced for use as parasiticides. Preparations have been employed as expectorants, also.

TOXICOLOGY OF ANTIMONY

Antimony(III) forms thioantimonates with sulfhydryl groups of cellular constituents. Symptoms of acute and chronic antimony poisoning are similar to those induced by arsenic. Upon local application, compounds of antimony are more caustic than those of arsenic. Salts of antimony, especially the tartrate, are powerful emetics, primarily because of the irritant action on the gastric mucosa. Toxic doses are also emetic by virtue of their central action on the medulla.

The Romans provoked vomiting by drinking wine that had been allowed to stand in a goblet fashioned from a rich antimony alloy and named the 'calices vomitorum'.

Subemetic doses of antimony exert a nauseant and expectorant action. Salivary and bronchial glands are stimulated reflexly. Antimony compounds are now considered too toxic for use as an expectorant.

Absorption of antimony compounds from the gastrointestinal tract is slow. Comparatively little is known of the distribution of trivalent antimony in tissues of man. Abnormally high concentrations have been found in the liver and thyroid.[1] Trivalent antimony is excreted for the most part by the kidney. Renal excretion is slow, however, because plasma antimony levels are comparatively low. Trivalent antimony is bound primarily to the erythrocytes. Pentavalent antimony (V), since it is not bound by red cells, is excreted more rapidly.

INDUSTRIAL USES AND EXPOSURES TO ANTIMONY

Antimony has various applications in industry. The metal is alloyed with tin, lead, and copper. It is incorporated into flame retardant compounds, paints, lacquers, and enamels. Antimony has application in the glass, pottery and rubber industries. A lead–antimony alloy is used in storage battery grids and in printers' type.

Printers' type contains far more lead than antimony. Verifying a case of antimony poisoning among printers is considered quite difficult, if not impossible. Symptoms of lead and antimony poisonings are similar. Both metals induce a metallic taste, vomiting, colic, indigestion, appetite loss and sometimes diarrhea. Sores in the mouth and throat help to distinguish industrial antimony poisoning from that of lead. It should also be remembered that in practically all industrial antimony there is some arsenic.

Certain industrial exposures to antimony can result in irritation of the skin and mucous membranes of the mouth, nose and throat. Eczema and other forms of dermatitis occur. Older authorities, especially those among the German, have described about 200 cases of industrial dermatitis in men who used antimony salts as mordants in dyeing cloth or who handled the dyed cloth.[2]

Significant exposure also takes place in antimony mining and smelting.

Taylor[3] reported a case of accidental exposure to antimony trichloride fumes. Upper respiratory irritation developed, followed by a gastrointestinal disturbance characterized by abdominal pain and persistent anorexia.

Urinary antimony excretion was 1 mg l^{-1} for one to two days following exposure. Trace quantities only were excreted by the third day, however!

Exposure to antimony pentachloride has been reported to evoke a severe pulmonary edema.[4]

Cooper et al.[5] described a pneumoconiosis developing among workers in an antimony industry.

Discharge residues from firearms contain significant quantities of antimony. As a consequence, personnel on police firing ranges, for example, are chronically exposed to dusts containing antimony, as well as lead. We have monitored firearms instructors of the Chicago Police Department for both lead and antimony exposure. Blood antimony levels among the group varied between 0 and 130 μg l^{-1}. Levels of blood antimony in an apparently unexposed population were found to be 10 μg l^{-1} or less. There seemed to be little correlation between the lead and antimony content in blood and urine among the exposed population. Both metals were rarely elevated concurrently. Representative blood antimony and lead levels obtained are listed in Table 3.1 below.

TABLE 3.1
Comparison of blood antimony and lead levels in Chicago Police Department firing range instructors

Subject	Antimony (μg l^{-1})	Lead (μg l^{-1})
1	20	690
2	0	720
3	10	480
4	0	660
5	130	650
6	10	560
7	30	550
8	40	630
9	80	410
10	0	560
11	20	610
12	40	430
13	70	570
14	0	450
15	80	650
16	10	420

ANALYSIS OF ANTIMONY

The quantitative estimation of small amounts of antimony was first reported by Schidrowitz and Goldsborough in 1911.[6] The method was based upon the extraction of antimony by boiling with copper and concentrated hydrochloric acid; the subsequent solution of antimony; and finally its conversion to antimony sulfide, a highly colored compound. The quantity of antimony in an unknown was estimated by comparing the colors of unknowns and standards. They were able to detect 100 μg quantities. Recoveries of 70–80% were claimed.

Among the many colorimetric methods for antimony that have been described subsequently are the Rhodamine B method, the phosphomolybdic acid procedure and the pyridine-iodide technique. Rhodamine B, because of its specificity and greater sensitivity, is usually preferred for the determination of trace quantities of antimony in biological materials. The other agents mentioned lack the sensitivity desired.

Gellhorn et al.[7] evaluated the Rhodamine B method. They applied both it and polarography in studying the tissue distribution of antimony compounds in hamsters.

Maren[8] has described a procedure wherein the organic matter is destroyed and the antimony converted to the pentavalent form by sulfuric-nitric-perchloric acids. Interferences from the ferric iron present can be avoided by extracting the acidic solution with isopropyl ether. Iron remains in the aqueous layer while antimony is removed into the solvent layer. Rhodamine B is then added to the isopropyl ether layer and forms a color complex with antimony. Incidentally, trivalent antimony does not react with Rhodamine.

The antimony–Rhodamine B complex can also be measured by fluorimetry (molecular luminescence spectrometry). A detection limit of 0.1 μg ml^{-1} is claimed.[9] Luminol (o-aminophthalhydrazide) forms a fluorescing compound with antimony. The sensitivity of the reaction is 0.05 μg ml^{-1}.[10]

Various antimony chelate complexes can be quantitatively extracted; for example, the thiooxinate, xanthate, diethyldithiophosphate, dithiocarbamate. Complexes with certain other metals are coextracted.

Malissa and Gomescek[11] describe the chelation of antimony, at pH around 1, by 0.2% ammonium pyrollidine dithiocarbamate and its subsequent extraction into chloroform. The diethyl dithiocarbamate of antimony is quantitatively extracted by carbon tetrachloride at pH 4–9.5. In the presence of EDTA and cyanide as masking agents, only bismuth, tellurium, and thallium are coextracted.[12]

Chromatography applications

Miketukova[13] separated gold, bismuth, cadmium, iron, mercury, molybdenum, antimony, thallium, vanadium, and tungsten by paper chromatography. Metals were then localized by spraying with Rhodamine B in hydrochloric acid, extracted by suitable solvents, and subsequently determined. Rhodamine B in hydrochloric acid reacts with metals to yield red or violet compounds.

Neutron activation analysis

Wester[14] determined 23 trace elements in human and bovine hearts. He found antimony levels in human heart tissue to vary from 1 to 4 ng g^{-1}; and in bovine heart, from 0.7 to 5 ng g^{-1}.

Bowen[15] found neutron activation analysis to be the most sensitive method (circa 1967) for the determination of antimony, cadmium, cerium, indium, and silver in biological material. The detection sensitivity of antimony by neutron activation was 10^{-10} g compared to 3×10^{-8} g for colorimetry, or 4×10^{-7} g for X-ray fluorescence, etc.

Polarography

Goodwin and Page[16] described a polarographic technique that could determine 2–10 μg of antimony, as well as larger quantities, with an error of only 2%. When applying this technique in investigating the metabolism of antimony, they observed that there was a conversion of ingested pentavalent antimony to trivalent antimony in the liver of rabbits and mice.

Van Dyck and Verbeek[17] discussed the mechanistics of determinations of traces of antimony in copper by anodic stripping voltammetry (ASV). Applications of ASV to analyses of antimony in biological determinations have not yet been reported.

Spectroscopic methods

Pyroantimonate complexes in the rat kidney have been determined by X-ray micro-analysis (analytical electron microscopy).[18]

Thompson[19] describes the determination of antimony, arsenic, selenium, and tellurium by combining hydride generation techniques with atomic fluorescence. The procedure is said to be 5–30 times more sensitive than atomic absorption spectrometric analysis. Possible interferences were not discussed. Previously Thompson and co-worker Thomerson[20] had utilized hydride generation with atomic absorption in determining arsenic, antimony, bismuth, germanium, lead, selenium, tellurium, and tin hydrides, generated by adding acidified samples to dilute sodium borohydride solutions. Liberated hydrides passed directly into a 17 cm long silica tube in an air–acetylene flame. The use of background correction was not considered necessary.

A. E. Smith[21] studied interferences in the determination. Using an atomic absorption spectrometer equipped with a hydrogen–argon flame, he found that gold, germanium, nickel, platinum and palladium could seriously interfere in arsenic and antimony analysis. Approximately a 50% suppression of the arsenic and antimony signals was observed. Silver, bismuth, cobalt, copper, selenium and tin caused appreciable but lesser suppression of the signal. Antimony suppresses the signal from arsenic and vice versa. Antimony moderately suppresses the signals generated by selenium, tellurium and bismuth as well.

Flameless atomic absorption spectrometry can be applied to antimony analysis following chelation by sodium diethyldithiocarbamate at pH 6–6.5 and extraction into methylisobutyl ketone. Sample preparation is minimal. Blood samples are hemolyzed in an acetate buffer before proceeding with the chelation–extraction steps. After pH adjustment, antimony in urine is chelated and extracted directly. Tissue samples are ashed in sulfuric-nitric acids. After dilution with water, an aliquot of the digestant is adjusted to proper pH then chelated and extracted. Reagent blanks and standards (0–1 μg ml^{-1}) are carried through the entire procedure concurrently with the unknown specimens.

Twenty microliters of the organic phase are introduced into the graphite furnace programmed as follows:
Dry at 100 °C for 20 s;
Char at 300 °C for 20 s;

Atomize at 2400 °C for 6 s.

Analyses are performed at the 207 nm line. Background correction is necessary.

REFERENCES

1. A. Gellhorn, N. A. Tupikova and H. B. Van Dyke, *J. Pharmacol. Expertl. Therap.*, **87**, 169 (1946).
2. A. Hamilton and H. L. Hardy, *Industrial Toxicology*, Publishing Sciences Group, Inc., Action, Mass., 1974.
3. P. J. Taylor, *Brit. J. Ind. Med.*, **23**, 318 (1966).
4. E. M. Cordasco and F. D. Stone, *Chest*, **64**, 182 (1973).
5. D. A. Cooper, E. P. Pendergrass, A. J. Vorwald, R. J. Maycock and H. Brieger, *Amer. J. Roentgen*, **103**, 496 (1968).
6. P. Schidrowitz and H. A. Goldsborough, *Analyst*, **36**, 101 (1911).
7. A. Gellhorn, M. E. Krahl and J. W. Fertig, *J. Pharmacol. Expertl. Therap.*, **87**, 159 (1946).
8. T. H. Maren, *Anal. Chem.*, **19**, 487 (1947).
9. A. I. Ivankova and D. P. Scherbov, *Issled. Razrab. Fotometrich. Metod. Opred Mikrokolichestv. Elem. Miner. Syr'l*, **138** (1967) cited in *Trace Analysis* (ed. J. D. Winefordner) John Wiley and Sons, 1976.
10. O. I. Komley and V. K. Zinchak, *Visn. l'viv. Univ. Ser. Khim*, **9**, 50 (1967) cited in *Trace Analysis*, (ed. J. D. Winefordner) John Wiley and Sons, 1976.
11. H. Malissa and S. Gomescek, *Z. Analyt. Chem.*, **169**, 402 (1959).
12. H. Bode, *Z. Analyt. Chem.*, **143**, 182 (1954).
13. V. Miketukova, *J. Chromatog.* **24**, 302 (1966).
14. P. O. Wester, *Acta Med. Scand. Suppl.*, **439**, 7 (1965).
15. H. J. M. Bowen, *Analyst*, **92**, 118 (1967).
16. L. G. Goodwin and J. E. Page, *Biochem. J.*, **37**, 198 (1943).
17. G. Van Dyck and F. Verbeek, *Anal. Chim. Acta*, **66**, 241 (1973).
18. C. Tisher, *Amer. J. Pathol.*, **69**, 255 (1972).
19. K. C. Thompson, *Analyst*, **100**, 307 (1977).
20. K. C. Thompson and D. R. Thomerson, *Analyst*, **99**, 595 (1974).
21. A. E. Smith, *Analyst*, **100**, 300 (1975).

CHAPTER 4

ARSENIC

Arsenic, a metalloid, is fairly widespread in nature, being present in soils, rocks, and all living things, plant and animal. Copper sulfur-arsenate ores used in making copper artifacts of approximately 5000 years ago contained as much as 12% arsenic.[1]

Through the ages arsenic and its compounds have found many uses. Ancient peoples, Greek, Roman, Arabic, Peruvian, to name but a few, used the compounds therapeutically and as poisons. Arsenic trioxide obtained during smelting copper is said to have been first prepared circa 2000 BC. It was the favorite agent of Medieval and Renaissance poisoners.

Arsenic in its elemental form is said to have been obtained first by Albertus Magnus in 1250 AD. Writings of Paracelsus contain directions for its preparation. Elemental arsenic is not toxic. Compounds of arsenic are another matter, however.

TOXICITY OF ARSENIC

Arsenic is a general protoplasmic poison. It is cumulative. All bodily systems are affected. Trivalent arsenicals react with sulfhydryl groups in cells and so inhibit sulfhydryl containing enzyme systems essential to cellular metabolism. The pyruvate oxidase system, for example, is susceptible. Arsenic uncouples phosphorylation. Arsine (arsenic hydride) combines with hemogloblin and is oxidized to a hemolytic compound that does not appear to act by sulfhydryl inhibition.

Soluble arsenicals are absorbed from all mucous membranes and parenteral sites of administration. Arsenic contained in ointments or lipid soluble vesicants can be absorbed, appreciably, from the skin. The more water soluble compounds are the least toxic, locally.

Non-allergic contact dermatitis and conjunctivitis are frequent among workers exposed to arsenic-containing dusts. Continued inhalation of arsenic dusts can cause perforation of the nasal septum. Other metallic dusts, e.g. chromium, also have this faculty.

Systemically, small doses of inorganic arsenic induce a mild vasodilation; large doses evoke rather pronounced effects upon the circulatory system. The injury, occurring in the splanchnic area and all capillary beds, results in transudation of plasma and a subsequent decrease in blood volume.[2] Blood pressure falls to shock levels.

Electrocardiographic abnormalities, indicating myocardial damage, may persist for months after recovery from acute arsenic intoxication. Changes are possibly due to action of arsenic on pyruvate–oxidase systems.[3]

Chronic ingestion of inorganic arsenic has been reported to cause a peripheral arteriosclerosis, commonly called 'black foot disease'.[4] The 'milk and roses' complexion credited to chronic ingestion of arsenic is due to cutaneous vasodilation. Prolonged ingestion of these compounds will produce a hyperkeratosis and hyperpigmentation.

Trivalent arsenicals, introduced systemically, interfere with the inflammatory response in skin and favor the occurrence of pyoderma. There also is interference with wound healing of the skin and other tissues.[5]

The nausea, vomiting, and diarrhea associated with acute or chronic ingestion of large doses of arsenic results from both local and systemic effects. Capillary transudation of plasma forms vesicles under the gastrointestinal mucosa which rupture and result in sloughing of the epithelium.

Renal damage of varying degrees is caused by the action of absorbed arsenic on capillaries of glomeruli and tubules. Mitochrondria of the proximal tubules were found to be highly sensitive to arsenic toxicity.[6]

Chronic exposure to inorganic arsenic may induce a peripheral neuritis resulting in motor and sensory paralysis of the extremities. Unlike the palsy of lead, legs are usually more severely affected than arms. Peripheral neuritis was a common finding among the several thousand in England who were poisoned in the 'arsenic in the beer' episode of 1900.[7] Heyman[8] reported on 41 cases of peripheral neuropathy due to arsenical intoxication. Hindmarsh et al.[9] discuss occurrence of electromyographic abnormalities among people drinking well water with arsenic concentrations above 0.05 ppm (50 μg 1^{-1}).

Arsenic is toxic to the liver, producing fatty infiltration and causing central necrosis and cirrhosis. Noncirrhotic portal hypertension due to chronic arsenic intake has been reported.[10,11]

Datta[12] measured arsenic contents in the tissues of a patient with noncirrhotic portal hypertension whose liver contained about 25 times the arsenic of control liver; kidney, 26 times; and the brain, almost 7 times that of corresponding control tissues.

Both acute and chronic arsenic poisoning are said to be reflected by changes in some protein fractions.[13]

Bone marrow is affected, and cellular elements of blood are altered. Erythrocytes are decreased by moderate doses. Larger amounts induce morphological changes resulting in the appearance of microcytes and megalocytes. A marked reticulocytosis upwards of 15% occurs. Myeloid elements are depressed.

Interference by arsenic with porphyrin metabolism is manifested by a marked increase in urinary coproporphyrin and uroporphyrin excretion.

An early toxic reaction to chronic inorganic arsenic intoxication is an apparent weight gain arising from an occult edema due to the capillary damage. This toxic sign was often misinterpreted as 'beneficial' in a former era when arsenicals were incorporated into tonic preparations.

Diagnosis of arsenical poisoning

The gastrointestinal symptoms of acute arsenic poisoning (intense pain, projectile vomiting, diarrhea) can be confused with those due to 'food poisoning' of a bacterial nature and vice versa. Arsenic poisoning also resembles thallium poisoning. Abdominal symptoms of arsenic poisoning may occur within an hour of ingesting a dose or may be delayed as much as twelve hours, depending upon the quantity of food in the stomach. Symptoms of shock appear as fluid loss progresses. Hypoxic convulsions and coma may occur terminally.

A form of acute intoxication without any gastrointestinal symptoms but with manifestations of headache, vertigo, restlessness, irritability, loss of memory, followed by stupor and coma may occur. Potential *sequelae* include muscle contractions and atrophy, uncoordination, and blurred and flickering vision. Acute arsenical poisoning is fatal, usually within 24 hours. However, deaths have been reported to occur as rapidly as one hour, or as long as four days after poisoning.

Mortality due to acute arsenical poisoning has declined.[14] Dimercaprol, British anti-Lewisite (BAL), is an effective antidote. Restoration of fluid and electrolyte balance is contributory. Previously patients would die from the effects of fluid loss after the onset of gastrointestinal symptoms.

Chronic arsenic poisoning

By virtue of the various effects exerted upon the bodily systems, chronic arsenic poisoning mimics many diseases. However, a history of weakness, languor, anorexia, and occasionally nausea, vomiting, diarrhea or constipation, associated with a patient exhibiting a melanosis of the lower eyelids and clavicular areas should arouse suspicions. Vitiligo may be observed in heavily pigmented individuals. Alopecia, a striking feature of thallotoxicosis, occurs occasionally. A garlic-like breath odor, excessive salivation, coryza, sore throat, numbness and tingling or burning of the extremities are contributing symptoms.

It should be mentioned that while skin changes like hyperpigmentation (melanosis) and hyperkeratosis are manifestations of systemic arsenic poisoning, these may be due to prolonged, direct contact of arsenic with the skin.[15]

INCIDENCE OF ARSENIC POISONING

Accurate statements cannot be made concerning the incidence of arsenic poisoning, acute or chronic. Only a portion of the cases occurring around the world are diagnosed; less are reported. Nevertheless, there are some rather dramatic accounts in the literature of the last two decades. For example, in western Japan in 1955, 9718 infants drinking powdered milk prepared by a certain plant, became poisoned with arsenic. Sixty-two supposedly died. Hamamoto[16] discusses, in depth, the symptoms, laboratory data, and treatment of infants admitted to the Okayama University Hospital. Nagai et al.[17] report their observations on 134 infants seen at the Kyoto University Medical School. A follow-up study, six months later, is included. Statistics compiled subsequently indicate that the Morinaga arsenic in milk incident involved a total of 12 083 children, 128 of whom did not survive.

Arsenic toxicosis observed in the Dermatology Department of Niigata University Medical School in 1959 led to the discovery of well water contaminated by industrial wastes as the cause.[18] The plants manufactured arsenic trisulfide, an important component of paints in Japan. About eight individuals were involved.

Hideo et al.[19] reported further studies among this population. It is interesting that the plant in question had a hundred-year history. However, inhabitants around the area drank river water. About three years before the detection of the first case, inhabitants began drinking well water.

An endemic peripheral vascular disease (black foot disease) has been described in regions of Taiwan where the arsenic content in the drinking water is high.[20] Rosenberg[21] describes systemic arterial disease and chronic arsenicism in infants in Autofugasta, Chile, a region with a high arsenic content in its drinking water.

Various arsenical compounds are incorporated into pesticide, rodenticide and herbicide preparations commonly used in the home environment.

Kjeldsberg and Ward[22] report the development of aplastic anemia in a sixty-seven year old male after many months' use of arsenic-containing pesticides.

We recall seven children in a family who exhibited mild to severe signs of neurological involvement including uncoordination, peripheral neuritis, vertigo etc., plus hyperpigmentation of the eyelids and clavicular region. During the summer they were accustomed to playing in a field of tomatoes being grown for a cannery. The field had been liberally sprayed with Paris green (cupric acetoarsenite).

The dangers inherent in both home and farm use of arsenic-containing insecticides and herbicides are not fully appreciated. The following history of a forty-five year old farm woman is a case in point.

The patient complaining of intense headache, vertigo, general malaise, some abdominal discomfort, was referred for neurological consultation and was hospitalized. Routine laboratory data was noncontributory. However, urinary coproporphyrin and uroporphyrin excretions were markedly elevated, and the patient was investigated for possible exposure to lead and arsenic. Blood and urine lead contents were within normal limits. Urinary arsenic excretions ranged between 400 and 500 μg l^{-1}. Arsenic was not detected in the blood. Hair contained 0.3 μg g^{-1}. Analyses of samples obtained from her husband revealed a urine arsenic content of 500 μg l^{-1}; hair arsenic, 0.1 μg g^{-1}; and an undetectable blood arsenic. The source of exposure was traced eventually to an ant killer, G-11, containing 10% sodium arsenite.

Certain herbicides marketed in the US contain varying amounts of arsenicals. For example, Pax is compounded with 3.5% lead arsenate and 47% arsenic trioxide. 'New Pax' contains 8.25% lead arsenite and 25% arsenic trioxide.

Kuruvilla et al.[23] report the occurrence of poisonings following a family picnic. A herbicide reported to contain 47% arsenic trioxide was accidentally spilled over a barbecue grill and the surrounding gravel below. The spill was cleaned up, more or less, and the picnic held. Severe gastrointestinal symptoms were subsequently experienced by five children. The authors postulate that possibly arsenic compound in the herbicide remaining on the grill sublimated when heated and then settled on the food to be consumed.

Dusts in Hawaian homes treated for pest control and termites have been reported to contain arsenic levels of 3–64 μg g^{-1}.[24] Arsenic content in dusts from homes of employees in pest control or wood treatment firms varied from 512 to 1,080 μg g^{-1}.

Back in 1933, Li and Yang[25] reported on an outbreak of chronic arsenicalism in Changsha province following the use of a mosquito incense containing orpiment (arsenic sulfide).

Various herbal preparations and health food supplements have been found to contain hazardous quantities of arsenic.

Tay and Seah[26] observed symptoms of arsenic poisoning among 74 patients in Singapore who gave a history of using anti-asthmatic herbal preparations. Forty percent of the group had taken these proprietaries for less than six months. The arsenic content of the 29 Chinese herbal preparations examined varied between 29 and 105 000 ppm.

Walkin and Douglas[27] report the appearance of symptoms of chronic arsenicalism in a food faddist consuming numerous kelp tablets daily. The preparation in question contained 30 μg per tablet. Symptoms abated when consumption of the kelp preparation ceased.

Occasionally, quack remedies may be the source of overt exposure. Robinson[28] discusses a polyneuropathy due to the application of a caustic arsenical paste as a cancer remedy!

Despite the apparent awareness of potential hazards in pigments containing arsenic compounds, one still upon occasion encounters arsenicalism traceable to their application. For example, recently a patient initially diagnosed as a porphyria was found to be excreting abnormal amounts of arsenic (500 μg l^{-1}). The subject, who painted in oils as a hobby, was accustomed to pointing her brushes with her mouth.

A rather bizarre incident which occurred in the Chicago area a few years ago is worthy of mention. A three year old girl had been diagnosed as a possible chronic arsenic poisoning. The home environment seemed free of possible sources of exposure. However, during an interview, the child's mother did mention that upon occasion she had come upon the child with crayons in her mouth.

Crayons submitted for analysis consisted of two brands, one widely sold interstate, and the other unknown to the analyst and of apparent local manufacture. Most of the latter were marred by multiple and deep teeth marks and scrapings. Crayons of the former group did not bear such marks. Analyses of representative samples of both crayon brands revealed that while the known brand was apparently free from arsenic, the strange brand, obviously favored by the child, contained 5–15 μg g^{-1} of arsenic.

No discussion concerning arsenic poisoning in a non-industrial population can be considered complete without presentation of at least one instance where exposure to arsenic was not accidental. We venture to state that many metal poisons are initially misdiagnosed. Most clinicians, at least in the US, are not oriented to consider arsenic as a possible cause of the signs and symptoms being observed. The following case study will serve to illustrate this point.

A few years ago, a thirty-five year old white male complaining of excessive fatigue, general malaise, painful limbs, and difficulty in walking short distances was admitted to a hospital. Prior to the onset of some symptoms a few months previously, the subject, a former college athlete, was accustomed to playing 36 holes of golf in a morning and experiencing little fatigue. The patient also reported that he had experienced great abdominal discomfort, often accompanied by vomiting, after meals.

A resident physician did see the marked meleinosis of the lower eyelids and clavicular areas, the hyperkatosis of the palmar surfaces, but, like Watson, he did not observe.

Hematologic studies revealed a mild anemia. Hemoglobin was 12 g%. Reticulocytes were 25%. Leukocyte count was 4000 with a relative lymphocytosis.

An electrocardiogram indicated some myocardial involvement.

After ten days of 'diagnostic workup', the possibility of collagen disease was considered. A blood lead determination was requested as an afterthought. Upon being informed of the clinical picture, the toxicologist requested that urine and hair samples also be submitted. Blood and urinary lead levels were found to be 100 μg l^{-1} and 20 μg l^{-1}, respectively, well within normal limits. On the other hand, the arsenic content of the blood was 300 μg l^{-1}.

Urinary arsenic excretion was 5 mg l^{-1}. The hair contained 10 μg g^{-1} of arsenic.

The blood arsenic content was of immediate concern. Arsenic clears the blood rapidly. Within four days after a fair dose, it cannot be detected with a method of the sensitivity of the Gutzeit procedure. Obviously, arsenic was in some way being administered in the hospital. Hair and urinary contents indicated exposure over a fairly long interval.

All edibles, toothpaste, soaps, mouthwash, etc. belonging to the patient were sent for analysis. Arsenic was found in chocolates and cookies (sweet biscuits) sent by his wife—it was about a week before Christmas. Arsenic was also present in his mouthwash.

Later history revealed that the patient, a railroad engineer, was accustomed to drinking a two-quart thermos of coffee prepared by his wife over the course of a 16-hour trip. Months later, a well-scrubbed thermos belonging to the patient was submitted for analysis and found to contain significant quantities of arsenic. After discharge from the hospital, the patient removed himself from his home environment. Recovery was subsequently fairly complete.

MEDICAL USES OF ARSENIC COMPOUNDS

Arsenic compounds have enjoyed a variety of therapeutic applications; as a tonic to promote weight gain, and also in treating chlorosis, a secondary anemia. Fowler introduced his 'ague drops', a 1% solution of arsenic trioxide, in 1786. Subsequently, his solution was employed until fairly recent times in treating leukemia and psoriasis. Cacodylate, the first organic arsenical known, appeared in 1842.

Erlich, in 1907, after many years of search for an efficient trypanocidal and spirochetocidal agent that was not too toxic for clinical use, discovered arsphenamine (606), and the age of chemotherapy began. Since the introduction of penicillin, however, the use of organic arsenicals in treating syphillis has declined. The compounds are however still employed in amebiasis.

Arsenicals have some application in veterinary medicine in treating non-parasitic skin diseases in horses, cattle, dogs, sheep and swine. Various compounds were used fairly extensively in treating chronic coughs, anemia, general debility, blood diseases, petechial fever of horses etc.

INDUSTRIAL USES OF ARSENIC

Arsenic and its various compounds are involved in numerous industrial processes. Elemental arsenic is used in the manufacture of glass, particularly infrared transmitting glass, and in the manufacture of electrical semiconductors and photoconductors. It is used in the manufacture of linoleum and oil cloth. Arsenic trisulfide is used as a depilatory for hides.

Arsenic pentasulfide, a brownish yellow, glassy, highly refractive material, is used in pigments and in thin sheets as a light filter.

The hemiselenide of arsenic is applied in glass manufacture.

Arsenic pentoxide is found in adhesives for metals, in wood preservatives, and as a herbicide and fungicide. It is used in the manufacture of colored glass.

Arsenic trioxide is a byproduct of many smelting operations. Other arsenic compounds are made from it. The use of arsenic trioxide in rodenticides, insecticides, and herbicides was mentioned previously. It is also used as a mordant in the textile industry and in the manufacture of glass and enamel.

Arsine (arsenic trihydride), a byproduct of many industrial processes involving arsenic, is probably the most poisonous of all arsenicals. It is said to be injurious at dilutions of $1:20\,000$.[29] Chronic arsine poisonings occur in the smelting and refining of metals, in galvanizing, soldering and etching, and in lead plating.[30] The compound can be liberated by the action of some fungi in sewage plants. G. Melin in 1839 first noted the evolution of the gas from molds.[31]

Teitelbaum and Kier[32] describe accidental poisoning from arsine gas in a petroleum industry. The massive hemolysis induced by arsine leads to renal failure. Some victims responded to hemodialysis and exchange transfusions.

Wilkinson et al.[33] describe, quite graphically, the symptomatology elicited among eight sailors exposed to arsine gas during an accident aboard a freighter.

ARSENIC AS A CARCINOGEN

Arsenic has long been credited with being a carcinogen, a view not necessarily accepted by all investigators. The dispute probably began around the time J. A. Paris, a physician practising in Cornwall (1813–17) stated that the arsenic fumes developing in smelting works in Cornwall and Wales occasionally caused a cancer of the scrotum similar to that which affected chimney sweeps. Sir Jonathan Hutchinson[34] in 1887 was the first observer to postulate the role of arsenical drugs as an etiological factor in cancer of the skin.

Blejer and Wagner[35] reviewed various epidemiological studies concerning cancer incidence in arsenic-exposed workers. A significant excess mortality from respiratory cancer was noted among the workers. Two studies seemed to indicate a dose-response relationship between exposure and lung cancer.

Workers inhaling inorganic arsenic compounds during copper smelting, pesticide manufacture, or in gold mine operations are at risk. Vintners and sheep-dip workers are also at great risk. There are three fairly recent and well-documented investigations concerning the incidence of respiratory cancer among workers in copper refineries.[36-38] Ott et al.[39] reviewed the incidence of respiratory cancer in arsenic pesticide manufacture.

ARSENIC CONTENT IN TISSUES

Arsenic is present in trace quantities in 'normal' tissue. The amounts reported are dependent upon the analytical method employed. Neutron activation analysis figures are lower than those obtained by other methods. The region from which tissues are obtained may also be a factor in the variability of the results.

Kingsley and Schaeffert[40] found between 3 and 3.9 μg arsenic in 100 g of fresh liver. Boylen and Hardy,[41] also in the US, reported a range of 35-102 μg per 100 g. Johnson et al.[42] found arsenic contents varying between 2 and 7 μg per 100 g in liver tissue from a New Zealand population.

Arsenic contents per 100 g of fresh tissues, as reported by Kingsley and Schaeffert,[40] are as follows: kidney, 2.6-3.7 μg; heart, 2.4-3.7 μg; brain, 2.4-3.7 μg; lung, 1.8-2.9 μg; thigh muscle, 3.1-5.8 μg. Wester[43] in his trace element analyses of heart tissue found levels of 0.097-1.2 μg and 0.17-0.71 μg per 100 g of human and bovine heart, respectively.

As a means of comparison, the levels found in tissues of subjects with known chronic arsenic exposure (the patients with non-cirrhotic portal hypertension of Datta[12] and Rosenberg[21] are listed in Table 4.1.

According to Kingsley and Schaeffert[40] the normal blood arsenic levels vary between 3.5 and 7.5 μg%. In our experience, blood levels among an unexposed population in the midwest US are less than 1 μg%.

Althausen and Gunther[44] in 1929 reported arsenic levels from 3.1 to 32 μg per 100 g in the hair of supposedly normal subjects. Their range of values seems a bit wide by current standards. Normal hair arsenic contents are usually found to be less than 1 μg per 100 g.

TABLE 4.1
Arsenic content in non-cirrhotic portal hypertension (μg% wet weight)

Tissue	Datta[12]	Rosenberg[21]
Liver	247	740
Kidney	157	710
Brain	40	330
Heart	246	not detected
Intestine	42	not detected
Aorta	114	570
Portal vein	36	not detected

METHODS OF ARSENIC ANALYSIS

Arsenic, probably because of its position in history as 'the poison', is one of the first elements for which methods capable of detecting minute quantities were developed.

The Marsh-Berzelius test introduced in 1836 still finds limited application. Arsine generated from an acidic solution is trapped on a mirrored surface. A black deposit is indicative of, though not specific for, arsenic. Metals which form hydrides leave deposits on mirrored surfaces also.

The Reinsch test, developed later, which is somewhat more sensitive than the Marsh-Berzelius, is based on the deposition of arsenic from solution as a copper arsenide. In an emergency situation, the test can be applied directly without prior destruction of organic matter. For example, a gastric sample or emesis, urine or any suspect solution is acidified and a coil of burnished copper wire is placed in the solution. The solution is simmered for about $\frac{1}{2}$–1 h. Then the copper coil is removed and examined. If the foil remains bright, arsenic is not present in more than trace quantities. A black or brown deposit indicates metal, such as arsenic, mercury, antimony, silver, bismuth or lead. Arsenic and mercury sublime with heat. The former sublimes to form tetrahedral crystals; the latter, globules. Antimony, silver, bismuth or lead will not sublime following application of heat. A reagent blank is carried through the procedure.

The Gutzeit method is based on the liberation of arsine, under carefully controlled conditions, from an arsenic solution.[45] Arsine reduces mercuric bromide impregnated into a filter paper strip. The yellow to brownish stain produced is proportional to the quantity of arsenic present.

A simple arsine generator (the Gutzeit apparatus) is familiar to all chemistry students and can be constructed from a small, wide-mouth bottle and some pieces of tubing. A sketch of the apparatus appears in Fig. 4.1.

A wide-mouth bottle, about 60 ml in size, is connected by means of a rubber stopper to a glass tube, 1 cm wide, about 6 cm long, and constricted

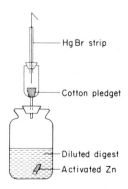

Fig. 4.1. Arsine generator

at one end. This tube contains a pledget of dental cotton (or clean sand) that has been saturated with 10% lead acetate to trap any hydrogen sulfide which might be generated along with the arsine gas (hydrogen sulfide can reduce mercuric bromide). This tube, in turn, is connected via a rubber stopper to a narrow glass tube about 2–3 mm in diameter and 10 cm long. A strip of paper, previously impregnated with 5% mercuric bromide in methanol, is placed in the tube. Activated zinc is added to the digest to generate arsine. After generation of gas has ceased (approximately 2 h), the strip is removed and the stain measured. The length of the stain is proportional to the concentration of arsenic (1 mm $\simeq$ 1 μg). Reagent blanks and standards are carried through the entire procedure.

Many modifications of the Gutzeit procedure have been published. Carefully controlled, this simple procedure is a fairly reliable method, even in this age of instrumentation.

Arsenic has been determined by different titrimetric procedures of varying degrees of sensitivity. The bromate method, for example, is applicable when the arsenic to be estimated is present in the order of 0.35 mg.[46] Arsenic is distilled as arsenious chloride from an acid digest. The distillate is titrated with standard bromate solution using methyl orange as an indicator.

A chlorometric procedure involving quantitative estimation of arsenic by titration with a standard hypochlorite solution has been described by Goldstone and Jacobs.[47] They also applied the method to the quantitative estimation of antimony.

The iodometric method of Cassel and Wichmann[48] is applicable for determining microgram quantities of arsenic. Arsine evolved from the acid digest is trapped in a mercuric chloride solution. The mercuric arsenide formed is oxidized by excess mercuric chloride with the formation of mercurous chloride and arsenious acid. The arsenious acid is then oxidized with a weak iodine solution.

One of the first colorimetric methods developed for arsenic analysis was the molybdenum blue method. Arsenic reacts with ammonium molybdate to form a complex molybdo-arsenate, a blue compound. In essence, arsenic in solution is evolved as arsine, trapped and oxidized by bromine water or sodium hypobromate solution. Ammonium molybdate and hydrazine sulfate are added and a blue color develops. Application of the method and its mechanistics have been discussed by many investigators.[49-52]

Phosphate is a major interference in the molybdenum blue method of arsenic analysis.

Various chelating agents which form solvent-extractable colored complexes have been applied in arsenic determinations. All suffer from interferences, more or less, when these complexes are measured spectrophotometrically.

At a pH of approximately 1, arsenic in the presence of ammonium pyrollidine dithiocarbamate is quantitatively extracted by chloroform.[53] Iron, cobalt, copper, nickel, vanadium, antimony, and tin are extracted concurrently.

Arsenic(III) can be completely extracted at pH 5–6 with carbon tetrachloride in the presence of 0.01 M sodium diethyl dithiocarbamate.[54] Lead, antimony, cadmium, thallium, copper, or iron, to name but a few are extracted concurrently. The chelate is quantitatively extractable into methylisobutyl ketone, also.

Arsenic(III), but not arsenic(V), is extractable from mineral acids by a solution of diethyl ammonium diethyldithiocarbamate in chloroform. Consequently, many interfering metals can be removed by preliminary extractions with the reagent, provided arsenic is in the pentavalent form.[55,56]

Arsenic(III), as a xanthate, can be quantitatively extracted by carbon tetrachloride from 0.1 M sulfuric acid. Xanthate extraction has been used in the isolation of arsenic from silicates, natural waters, food, and biochemical materials.[57,58]

The Analytical Standards Committee in the UK carried out a study comparing the Gutzeit, molybdenum blue, and silver diethyldithiocarbamate methods.[59] Somewhat higher recovery by the last-named method was indicated. Also, the procedure was more rapid and required less expertise than did the other two techniques.

A procedure utilizing dry ashing of material with magnesium oxide and magnesium nitrate prior to colorimetric estimation with silver diethyldithiocarbamate is described by George et al.[60] Good recoveries were obtained, apparently when the method was applied in a collaborative study on the arsenic analysis of animal tissues.

One of the challenges in arsenic analysis is that of achieving complete destruction of the organic matrix without much loss of arsenic. Results quoted in the literature indicate that losses are technique dependent. There is agreement that negligible losses occur during wet ash procedures, provided oxidizing conditions are maintained. Gorsuch[61] describes ashing procedures in depth. Recoveries are far less when dry ashing procedures are employed.[62] Pijck et al.[63] report a 23% recovery of arsenic added to blood when ashing was carried out at 400 °C. Other investigators report lesser losses, in general. Hamilton et al.,[64] who determined arsenic recoveries in tissue from rats previously injected with radionuclides, found losses to be matrix dependent. The chemical and physical form in which an element is present in a matrix influences that element's volatility.

Neutron activation analysis

Neutron activation analysis is probably the most sensitive technique available for estimating the arsenic content of small samples. For example, Hamilton[64] and Smith,[65] in 1959, described a procedure whereby a single hair weighing 0.5 mg could be analyzed. After nitric–sulfuric acid digestion of the activated sample, arsine was evolved in a Gutzeit separation and removed by passing it through a trap containing 1.6% mercuric chloride. The activity of the liquid sample was then estimated by a Geiger counter.

Neutron activation analysis was subsequently applied in analyzing Napoleon's hair. According to Smith and co-workers,[66] Napoleon was exposed to abnormally large amounts for a period of approximately four months. Distribution studies show the exposure was intermittent. Arsenic contents per cm length ranged between 2.79 and 11.0 ppm (2.79–11 $\mu g\ g^{-1}$).

As a matter of interest other critics, such as another Smith,[67] maintain that arsenic could have been added to the hair after death by spraying, dusting or dipping.

Lee and Murphy[68] discuss the determination of the arsenic content of American cigarettes by neutron activation analysis. Arsenic content per gram of tobacco varied between 5.5–12.8 μg. Arsenic in tobacco was implicated as a possible carcinogen among smokers and chewers of tobacco.

Chromatography

Chromatographic methods have been utilized in combination with other techniques in arsenic analysis.

Edwards et al.[69] describe a simple qualitative detection of arsenic in biological systems by thin layer chromatography and electrophoresis. Material applied to a nonfluorescent cellulose strip was subjected to electrophoresis in pH 7 acetate buffer, then dried and placed in a chromatography chamber containing ammonium sulfide. Arsenic was visualized by a spraying with Jüngnickel's[70] reagent.

Gas chromatography

Triphenyl arsine, which is produced quantitatively when arsenic diethyldithiocarbamate is reacted with magnesium diphenyl, can be measured by gas chromatography. Schwedt and Rüssel,[71] using a flame ionization detector and a column of 5% Carbowax treated with terephthalic acid, were able to detect about 0.4 ng of arsenic. Biological materials are dried and incinerated prior to analysis.

Talmi and Bostick[72,73] discuss the analysis of alkylarsenic acids in pesticides and environmental samples with a gas chromatography—microwave emission spectrometric detection system. The procedure involves gas chromatographic separation and elution into a low-power microwave argon plasma and the measurement of the resultant emission intensity of arsenic at 228.8 nm.

Braman and Foreback[74] reported that methyl arsenicals are excreted by some human subjects and also appear in various environmental samples. Daughtrey et al.[75] discuss the quantitative measurements of inorganic and methyl arsenicals in biological materials by gas chromatography employing electron capture detection. Both the inorganic and methyl arsenic complexes of diethylammonium diethyldithiocarbamate are chromatographed on a 5% OV-17 column. The authors stress the need for silanizing the chromatographic columns heavily in order to prevent organic arsenicals from sticking to the column walls.

Other applications of gas chromatography to the trace detection of arsenic have been reported,[76-78] some employing mass spectrometric detection.

Vycudelik[79] describes an interesting gas chromatographic procedure wherein arsine is produced by the reduction of organic arsenicals with sodium borohydride. The sample is introduced by head space technique into a gas chromatograph equipped with a nitrogen/phosphorus detector. The method has a detection sensitivity of toxicological interest.

Andreae,[80] utilizing a borohydride generation technique with gas chromatography, was able to determine arsenate, arsenite, mono-, di-, and trimethyl arsine, monomethyl arsenic and dimethylarsinic acid, and trimethyl arsine oxide in natural waters. Detection limits attained were in the ng l^{-1} range.

Polarography

Electroanalytical procedures have found limited application in arsenic analyses. Greater success has been attained with pure solutions than with digests of biological materials.

Whitnack and Brophy[81] described a single sweep polarographic method in 1969. Kaplin and co-workers[82,83] applied anodic stripping voltammetry (ASV) to the analysis of arsenic in trace quantities.

Forsberg et al.[84] compared ASV and differential pulse anodic stripping voltammetry (DPASV) using a gold wire electrode. They report detection limits of 0.02 ng ml^{-1} (0.02 ppb) when analyzing pure solutions.

Analyses of arsenic by ASV, especially at low levels, are subject to interferences from the biological matrix. Iron and copper pose major problems. In blood and tissue concentrations of these elements can be overwhelming.

Davis et al.[85] describe the determination of total arsenic at nanogram levels by high-speed ASV. They employ a gold film electrode. The technique was designed primarily for processing environmental air samples. The authors present minimal analytical data on NBS reference material (bovine liver and orchard leaves); however, comparative results of blood, urine, and fresh tissue are needed to better evaluate the practicability of the technique as a real life field method.

X-Ray fluorescence spectrometry

Pinta[86] discusses in depth the theory and practice of X-ray fluorescence in trace metal analysis. The technique has been applied to a variety of matrices. Arsenic has been determined in waters following extraction with ammonium pyrollidine dithiocarbamate.[87] Kato and Murano[88] claim detection at concentrations of less than 0.5 μg ml^{-1} after separation as a volatile hydride. Boiteau and Robin[89] separated arsenic from coexisting metals by dithizone extraction.

Campbell et al.[90] wet ashed plant and biological materials before determining arsenic. The method was also employed to analyze wine. A detection limit of 0.1–1 ppm is claimed.

X-ray fluorescence has been used, particularly for the detection of toxic elements in tissue.

Lubozynski et al.[91] found the detection limit for arsenic in blood to be 2–10 ppm.

Mathies[92] determined arsenic in urine. After wet ashing in sulfuric, nitric and perchloric acids, arsenic was converted to arsenic trihydride and collected on filter paper impregnated with silver nitrate. The method has been applied to the analysis of hair, tissues, and gastric fluids.

Atomic absorption spectrometry (AAS)

Detection limits for arsenic by flame atomic absorption are said to be 0.2 μg ml^{-1}.[93] An arsenic content of 0.05 μg ml^{-1} in blood or urine is considered significant. Consequently, concentration techniques must necessarily be employed to make atomic absorption spectrometric application to analyses practical. Interferences from matrix constituents, primarily sodium, which remain in a digest or, to a lesser degree, in an extract, are almost overwhelming. Furthermore, at 193.7 nm, the most sensitive resonance line for arsenic, absorption by the usual air–acetylene flame itself is upwards of 50%. Less absorption is encountered with an argon flame.

To further eliminate interferences and increase sensitivity in analyses, investigators began employing techniques which utilized the generation of arsine gas directly into the flame. One of the earlier attempts is that of Holak.[94] Orheim and Bovee[95] reported the determination of nanogram amounts of arsenic by arsine generation and flame atomic absorption spectrometry. Hoover et al.,[96] when employing 10 g aliquots, were able to detect arsenic at the 0.1 μg g^{-1} level in animal feed.

Florino et al.[97] applied the method in the sequential determination of arsenic, selenium, antimony and tellurium in foods. There have been other applications.[96]

The borohydride technique of arsenic analysis is not interference free. According to Aggett and Aspell,[99] cobalt, copper, iron, nickel, antimony, and zinc interfere with borohydride generation.

Flameless AAS

Flameless sampling devices extend the detection sensitivity of pure arsenic solutions many fold. Matrix effects, however, still remain an obstacle. Certain ions suppress, others enhance, the absorption signal.[100] Various investigators performing water analyses utilized borohydride generation methods to overcome matrix interference during flameless sampling.[101,102] Shaikh and Tallman[101] were able to attain a detection limit of 10 ng, corresponding to a concentration of 0.2 ppb when a 50 ml sample was used.

Tam[103] extracted water samples with diethylammonium diethyldithiocarbamate in carbon tetrachloride and measured the arsenic in an atomic absorption spectrometer equipped with a carbon rod analyzer. Arsenate, arsenite and any organoarsenic compounds soluble in carbon tetrachloride could be determined by the method described. Matrix interferences were not observed. Limits of detection were stated as 1 μg l^{-1}.

As regards analyses of clinical and toxicological samples of blood, urine, and tissues for arsenic content, flameless atomic absorption spectrometry, in truth, offered little advantage over a simple Gutzeit procedure; detection limits by the two methods were similar.

With the introduction of the electrodeless discharge lamps, practical arsenic analyses began approaching reality.

Blood proteins can be removed by trichloroacetic acid precipitation. After adjusting the pH to 6, the arsenic is chelated by sodium diethyldithiocarbamate and extracted into methylisobutyl ketone. Urine or gastric specimens, after pH adjustment, can be extracted directly. Tissue samples are digested in sulfuric-nitric acids. Arsenic standards containing 0–20 μg% (0–0.2 μg ml^{-1}) are determined concurrently with each matrix. Standards used in blood analyses are extracted from trichloroacetic acid. Standards for tissue samples are carried through the wet digestion procedure.

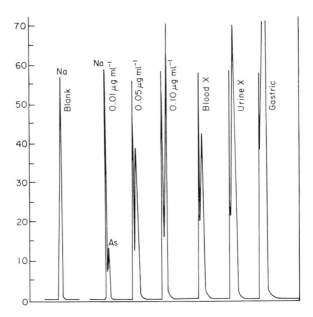

Fig. 4.2.   Arsenic DDC complex flameless atomic absorption

Arsenic contents of extracts are determined at the 193.7 nm line in a Perkin-Elmer 403 atomic absorption spectrometer equipped with an HGA 2000 graphite furnace and recorder. Background correction is employed. The graphite furnace is programmed as follows:

Dry at 100 °C for 40 s;

Char at 100 °C for 40 s;
Atomize at 2500 °C for 6 s.
Background correction does not completely eliminate the signal due to the non-specific molecular absorption of the sodium present in the extracts. As can be seen from the representative recorder tracing in Figure 4.2, the sodium signal emerges a few seconds before that of arsenic.

REFERENCES

1. H. A. Schroeder and J. J. Balassa, *J. Chronic Dis.* **19**, 85 (1966).
2. L. S. Goodman and A. Gelman (eds.), *The Pharmacological Basis of Therapeutics*, Macmillan, New York, 1975.
3. F. S. Glazener, J. G. Ellis and P. K. Johnson, *Calif. Med.*, **109**, 158 (1968).
4. K. H. Butzengeiger, *Klin. Wochenschr.*, **19**, 523 (1940).
5. D. J. Stone, *Texas Med.*, **65**, 40 (1969).
6. M. N. Brown, B. C. Rhyne, R. A. Goyer and B. A. Fowler, *J. Toxicol. Environ. Health*, **1**, 505 (1976).
7. The Final Report of the Royal Commission on Arsenical Poisoning, *Lancet*, December 12, 1674 (1903); December 19, 1746 (1903).
8. A. Heyman, J. B. Pfeiffer, R. W. Willett and H. M. Taylor, *New Eng. J. Med.*, **254**, 401 (1956).
9. J. T. Hindmarsh, O. R. McLetchie, L. P. M. Heffernan, O. A. Hayne, H. A. Ellenberger, R. F. McCurdy and H. J. Thiebaux, *J. Anal. Toxicol.*, **1**, 270 (1977).
10. J. S. Morris, M. Schmid, S. Newman, P. J. Scheuer and J. Sherlock, *Gastroenterology*, **66**, 86 (1974).
11. A. Viallet, F. Guillaume, J. Cote, A. Legarte and P. Lavoie, *Gastroenterology*, **62**, 177 (1972).
12. D. V. Datta, *Lancet*, **1**, 433 (1976).
13. S. Byczkowski, J. Gadomska and J. Byczkowski, *Acta Poloniae Pharm.*, **29**, 499 (1972).
14. J. B. Jenkins, *Brain*, **89**, 479 (1966).
15. H. Michael and D. Utidjian, *J. Occup. Med.*, **16**, 264 (1974).
16. E. Hamamoto, *Jap. Med. J. (Nihon iji shimpo)*, **1649**, 3 (1955).
17. H. Nagai, R. Okuda, H. Nagami, A. Yagi, C. Mori and H. Wada, *Ann. Ped. (Shonika Kiyo)* **2**, 124 (1956).
18. T. Yoshikawa, J. Utsumi, T. Okada, M. Moriuchi, K. Ozawa and Y. Koneko, *Therapy (Chiryo)* **42**, 1739 (1960).
19. T. Hideo, K. Kazuo, S. Tsutomu, S. Hideaki, S. Heiichiro, F. Katsuro, S. Chukichi, Y. Yoshiro, H. Shigeru, W. Giichi, H. Kazuo, O. Tatsuo and S. Chukichi, *Jap. J. Clin. Med. (Nihon rinsho)* **18**, 2394 (1960).
20. W. P. Tseng, *J. Formosan Med. Assoc.*, **69**, 1 (1970).
21. H. G. Rosenberg, *Archs. Path.*, **97**, 370 (1974).
22. C. R. Kjeldsberg and H. P. Ward, *Ann. Ant. Med.*, **77**, 935 (1972).
23. A. Kuruvilla, P. S. Bergeson and A. K. Done, *Clin. Toxicol.* **8**, 535 (1975).
24. H. W. Klemmer, E. Leitis and K. Pfenninger, *Bull. Environ. Contam. Toxicol.*, **14**, 449 (1975).
25. P. L. Li and C. S. Yang, *Chin. Med., J.*, **47**, 929 (1933).
26. C. H. Tay and C. S. Seah, *Med. J. Austral.* **2**, 424 (1975).
27. O. Walkin and D. E. Douglas, *Clin. Toxicol.*, **8**, 325 (1975).
28. T. J. Robinson, *Brit. Med. J.*, **3**, 139 (1975).
29. P. G. Stecher (ed.) in *The Merck Index*, Merck & Co., 1968.

30. B. A. Fowler and J. B. Weissberg, *New Eng. J. Med.*, **291**, 1171 (1974).
31. D. V. Frost, *Fed. Proc.*, **26**, 194 (1967).
32. D. T. Teitelbaum and L. C. Kier, *Arch. Environ. Health*, **19**, 133 (1969).
33. S. P. Wilkinson, P. McHugh, S. Horsely, H. Tubbs, M. Lewis, A. Thould, M. Winterton, V. Parsons and R. Williams, *Brit. Med. J.*, **3**, 559 (1975).
34. O. Neubauer, *Brit. J. Cancer*, **1**, 192 (1947).
35. H. P. Blejer and W. Wagner, *Ann. N.Y. Acad. Sci.*, **271**, 179 (1976).
36. A. M. Lee and J. F. Fraumeni, Jr., *J. Natl. Cancer Inst.*, **42**, 1045 (1969).
37. S. Milham, Jr., in *Health Effects of Occupational Lead and Arsenic Exposure*, (ed. B. Carnow), Department HEW, US Government Printing Office, 1976.
38. S. Tokudome and M. Kuratsune, *Int. J. Cancer*, **17**, 310 (1976).
39. M. G. Ott, B. B. Holder and H. L. Gordon, *Arch. Environ. Health*, **29**, 250 (1974).
40. G. R. Kingsley and R. R. Schaeffert, *Anal. Chem.*, **23**, 914 (1951).
41. G. W. Boylen, Jr. and H. L. Hardy, *Amer. Indust. Hyg. Assoc. J.*, **28**, 148 (1967).
42. C. A. Johnson and J. P. Lewin, *Anal. Chim. Acta*, **82**, 79 (1976).
43. P. O. Wester, *Acta Med., Scand. Suppl.*, **439**, 7 (1965).
44. T. L. Althausen and L. Gunther, *J. Amer. Med. Assoc.*, **92**, 2002 (1929).
45. M. B. Jacobs, *The Analytical Chemistry of Industrial Poisons, Hazards, and Solvents*, Interscience Publishers, New York, 1949.
46. *Methods Assoc. Off. Agr. Chemists*, 6th Edn., Association of Official Agricultural Chemists, Washington, DC, 1945.
47. N. J. Goldstone and M. B. Jacobs, *Ind. Eng. Chem. (Anal. Ed.)*, **16**, 206 (1944).
48. C. C. Cassel and H. J. Wichmann, *J. Assoc. Offic. Agr. Chemists*, **22**, 436 (1939).
49. G. Deniges, *Compt. Rend.*, **171**, 802 (1920).
50. W. R. G. Atkins and E. G. Wilson, *Biochem. J.*, **20**, 1225 (1926).
51. E. H. Maechling and F. B. Flinn, *J. Lab. Clin. Med.*, **15**, 779 (1930).
52. M. B. Jacobs and J. Nagler, *Ind. Eng. Chem. (Anal. Ed.)*, **14**, 442 (1942).
53. H. Malissa and A. Gomiscek, *Z. Analyt. Chem.*, **169**, 402 (1959).
54. H. Bode, *Z. Analyt. Chem.*, **143**, 182 (1954).
55. P. F. Wyatt, *Analyst*, **78**, 656 (1953).
56. P. F. Wyatt, *Analyst*, **80**, 368 (1955).
57. A. K. Klein and F. A. Vorhes, *J. Assoc. Offic. Agr. Chemists*, **22**, 121 (1939).
58. K. Sugawara, M. Tanaka and S. Kanamori, *Bull. Chem. Soc. Japan*, **29**, 620 (1956).
59. Analytical Standards Committee, *Analyst*, **100**, 54 (1975).
60. G. M. George, L. J. Frahm and J. P. McDonnell, *J. Assoc. Offic. Anal. Chemists*, **56**, 793 (1973).
61. T. T. Gorsuch, *The Destruction of Organic Matter*, Pergamon Press, Oxford, England, 1970.
62. D. J. Lisk, *J. Agr. Food Chem.*, **8**, 121 (1960).
63. J. Pijck, J. Hoste and J. Gilles, International Symposium on Microchemistry, Birmingham, 1958.
64. E. J. Hamilton, M. J. Minski and J. J. Cleary, *Analyst*, **92**, 257 (1967).
65. H. Smith, *Anal. Chem.*, **31**, 1361 (1959).
66. H. Smith, S. Forshufond and A. Wassen, *Nature (London)*, **194**, 725 (1962).
67. H. Smith, *J. Forensic Med.*, **8**, 165 (1961).
68. B. K. Lee and G. Murphy, *Cancer*, **23**, 1315 (1969).
69. T. Edwards, H. L. Merilees and B. C. McBride, *J. Chromatog.*, **106**, 210 (1975).
70. F. Jüngnickel, *J. Chromatog.*, **31**, 617 (1967).

71. G. Schwedt and H. A. Rüssel, *Chromatographia*, **5**, 242 (1972).
72. Y. Talmi and D. T. Bostick, *Anal. Chem.*, **47**, 1510 (1975).
73. Y. Talmi and D. T. Bostick, *Anal. Chem.*, **47**, 2145 (1975).
74. R. S. Braman and C. C. Foreback, *Science*, **182**, 1247 (1973).
75. E. H. Daughtrey, Jr., A. W. Fetchett and P. Mushak, *Anal. Chim. Acta*, **79**, 199 (1975).
76. M. Covello, G. Crampa and E. Ciamellio, *Farmaco. Ed., Prat.*, **22**, 218 (1967).
77. H. Brandenberger and B. Lusse-Schlatter, *Z. Klin. Chem. Klin. Biochem.*, **12**, 224 (1974).
78. H. Brandenberger, D. Frangi-Schnyder, B. Lusse-Schlatter and H. von Rechenberg, *Forensic Sci.*, **5**, 109 (1975).
79. W. Vycudelik, *Arch. Toxicol.*, **36**, 177 (1976).
80. M. O. Andreae, *Anal. Chem.* **49**, 820 (1977).
81. G. C. Whitnack and R. G. Brophy, *Anal. Chim. Acta*, **48**, 123 (1969).
82. L. F. Trushina and A. A. Kaplin, *Zh. Anal. Khim.*, **25**, 1616 (1970).
83. A. A. Kaplin, N. A. Veits and A. G. Stromberg, *Zh. Anal. Khim.*, **28**, 2192 (1973).
84. G. Forsberg, J. W. O'Laughlin, R. G. Megargle and S. R. Koirtyokann, *Anal. Chem.*, **47**, 1586 (1975).
85. P. H. Davis, G. R. Dulude, R. M. Griffin, W. R. Matson and E. W. Zink, *Anal. Chem.*, **50**, 137 (1978).
86. M. Pinta, *Modern Methods for Trace Element Analysis*, Ann Abor Science, Ann Arbor, Mich., 1978.
87. F. J. Marcie, *Norelco Rep.* **15**, 3 (1968) cited in M. Pinta, *Modern Methods for Trace Element Analysis*, Ann Arbor Science, Ann Arbor, Mich., 1978.
88. K. Kato and M. Murano, *Japan Analyst*, **22**, 1312 (1973).
89. H. L. Boiteau and M. Robin, *Coll. Rayons X Matiere*, Siemens, Monaco, May 1973, cited in M. Pinta, *Modern Methods for Trace Element Analysis*, Ann Arbor Science, Ann Arbor, Mich., 1978.
90. J. L. Campbell, B. H. Orr, A. W. Herman, L. A. McNelles, J. A. Thomson and W. B. Coor, *Anal. Chem.*, **47**, 1542 (1975).
91. M. F. Lubozynski, R. J. Baglan, G. R. Dyer and A. B. Brill, *Int. J. Appl. Radiat. Isotopes*, **23**, 487 (1972).
92. J. C. Mathies, *Appl. Spectros.*, **28**, 165 (1974).
93. R. Mavrodineaunu (ed.), *Analytical Flame Spectroscopy*, Springer-Verlag, New York, 1970.
94. W. Holak, *Anal. Chem.* **41**, 1712 (1969).
95. R. M. Orheim and H. H. Bovee, *Anal. Chem.*, **46**, 921 (1974).
96. W. L. Hoover, J. R. Melton, P. A. Howard and J. W. Bassett, Jr., *J. Assoc. Offic. Anal. Chemists*, **57**, 18 (1977).
97. J. A. Florino, J. W. Jones and S. G. Capar, *Anal. Chem.*, **48**, 120 (1976).
98. M. Ihnat and H. J. Miller, *J. Assoc. Offic. Anal. Chemists*, **60**, 813 (1977).
99. J. Aggett and A. C. Aspell, *Analyst*, **101**, 341 (1976).
100. P. R. Walsh, J. L. Fasching and R. A. Duce, *Anal. Chem.*, **48**, 7014 (1976).
101. A. U. Shaikh and D. E. Tallman, *Anal. Chem.*, **49**, 1093, (1977).
102. M. Fishman and R. Spencer, *Anal. Chem.*, **49**, 1599 (1977).
103. K. C. Tam, *Environ. Sci. Technol.*, **8**, 734 (1974).

CHAPTER 5

BARIUM

Barium is one of the rare earth metals, rather ubiquitous in nature, but with no known metabolic function. Barium is found in the earth's crust and in all living things. Marine animals concentrate barium from sea water. Barium is found in insects and fish. Douglas fir, oak, green ash and walnut trees accumulate barium.[1] Brazil nuts contain 3000–4000 ppm of the metal.

Barium, as well as strontium, is concentrated in the pigmented parts of the eye of mammals.[2] Concentrations up to 206 μg g^{-1} in the iris of cows and up to 1100 μg g^{-1} in the choroid have been reported.[3] Lesser amounts were detected in the eye of man and rabbit.

Barium was found in all tissues examined spectrographically by Tipton.[4] It does not seem to increase with age. Bone contains about 91% of the barium found in the human body.[1]

Barium enjoys various industrial applications. About the only medical application of barium compound is the use of the sulfate as a contrast medium in X-ray examinations.

Soluble barium salts are toxic. Barium acetate, used in lubricating oil and grease and as a mordant for printing fabrics, is poisonous. The bromate, used as a corrosion inhibitor for low carbon steel, is toxic, as is the benzene-sulfonate salt which is used as an additive in lubricating oils.

Barium carbonate is used in ceramics, paints, enamels, in the manufacture of marble substitutes, rubber, paper, electrodes and optical glasses. It is also a most effective rat poison.

The chlorate and chromate salts are used as pigments and in the manufacture of fireworks.

Other industrial uses of barium compounds include electroplating, glass manufacture, sugar refining, and use in photocells, television tubes and explosives.

Barium chloride was once used medically as a cardiac muscle stimulant in treating complete heart block.

Barium sulfide, used in some commercial depilatory preparations, is as toxic as barium carbonate when absorbed.

TOXICITY OF BARIUM

The systemic action of barium is characterized by very intense stimulation of muscles of all types, regardless of innervation. Vomiting, severe colic, diarrhea and hemorrhage result from its action on the gastrointestinal musculature. Muscle tremors are frequent, and paralysis ensues. Motor disorders including immobility of limbs and trunk are observed. There may be paralysis of the tongue and pharynx. A hypertension due to spasms of the musculature of the arterioles is seen early. Respiratory muscles may be paralyzed. Cardiac arrest and death occur frequently.

Poisonings have resulted from accidental or deliberate ingestion of soluble barium compounds as well as from exposure to the compounds during various industrial processes. There has also been described a granulomatous disease of the lung, barytosis[5], which occurs in workers engaged in the mining, sorting, grinding, and bagging of the ore barytes, i.e. barium sulfate.

The use of depilatories containing barium sulfide is hazardous. Barium can be absorbed from minute scratches on the skin.

Gould *et al.*[6] report a suicide attempt by a twenty-six year old black male who consumed a can of 'Magic Shave', a facial depilatory containing barium sulfide. The subject had a flaccid paralysis of the head, neck, arm, trunk, and thoracic muscle. He also had a spastic paralysis of muscles of mastication, other facial muscles, and muscles of the tongue and pharynx. The patient eventually recovered after treatment.

Two episodes of acute poisoning by the use of common table salt containing barium chloride were reported in 1943.[7,8] The condition, referred to as 'Pa Ping' or 'soft disease', had its onset following a meal and was characterized by flaccid paralysis of the legs and arms. There were no sensory disturbances. Deep tendon reflexes were absent. Magnesium sulfate was given to precipitate the barium. Recovery was complete within a few days.

Morton[9] describes two attacks of barium poisoning which affected 85 British soldiers in Persia in 1945. It seems that at the supply depot a sack of barium carbonate intended for use as a rat poison was placed in a flour tin which was subsequently filled with flour used for making pastry. The marmalade tart consumed by the affected soldiers was later found to contain 15 g of barium per portion of tart.

Sausage contaminated with barium carbonate caused severe food poisoning among 100 persons in Israel.[10]

Mechanism of toxic action

The mechanism by which barium exerts its action upon muscles is not quite clear. However, Huang[11] in 1943, reporting another episode of barium chloride poisoning, mentioned the dramatic relief of symptoms in patients treated with intravenous potassium citrate. His rationale for the use of potassium was the apparent similarities between barium intoxication and familial periodic paralysis. Hypokalemia in barium poisoning was first described by Diengott et al.[12] Two patients so poisoned were found to have potassium levels of 2 and 2.4 μg l[-1]. Both responded to potassium administered intravenously.

Roza and Berman[13] are of the opinion that the hypokalemia is not due to urinary or gastrointestinal losses but rather to the transfer of potassium ion from extracellular to intracellular compartments.

Barium and potassium were each found to be powerful antagonists of the cardiac toxicity induced by the other.

According to Berning,[14] the severe and rapid decrease of serum potassium seems to explain systolic cardiac arrest and early death within a few hours after the ingestion of barium.

BARIUM ANALYSIS

Barium can be isolated by precipitation as the sulfate or chromate. Strontium will be coprecipitated as the sulfate. Lead will be coprecipitated as the sulfate or chromate. Barium can be separated from strontium and calcium chromatographically, using aluminum oxide as an adsorbent.[15,16] These metals can be separated on cellulose powder as well.

The chromate method described for barium, wherein the precipitate is dissolved in acid and the color compared with that of an acidified barium chromate solution is not too specific. Lead and strontium yield similar colors.

The o-cresolphthalein method provides one of the few direct colorimetric methods for barium.[18] Barium must first be separated from all other elements, however. According to the authors, the determination of 2 μg of barium can be made to within a 10% error.

Barium can be extracted by thenoyltrifluoracetone in benzene.[19]

Applications of instrumental methods to the analysis of biological materials have been few. Tipton's investigations with the spectrograph have already been mentioned. Henke et al.[20] determined the barium, lanthanum, rubidium, and zinc contents in human brain by neutron activation analysis.

Tanaka et al.[21] determined the barium content of bone and tissues by atomic absorption spectrometry. Samples were ashed at 500–550 °C, then dissolved in nitric acid. Potassium was added as the ionization buffer.

Renshaw[22] applied flameless atomic absorption spectrometry to the analysis of barium in hair and in hand swabs. Hair was analyzed directly in the tube. Barium on swabs was solubilized and extracted with thenoyltrifluoracetone before analysis.

The sensitivity limit by flame atomic absorption spectrometry at the 553.5

nm resonance line is said to be 0.4 μg ml^{-1}. A fuel rich reducing flame is used.

Atomic absorption analysis of barium suffers from ionization interference, which can be controlled by the addition of alkali salts (1 to 2 mg ml^{-1}) to both samples and standards.

The air–acetylene flame absorbs somewhat at the analytical line for barium. Chemical interferences do occur. The sensitivity is not as good as that attained with a nitrous oxide–acetylene flame. Analyses in the latter flame seem free from chemical suppressions.

Detection capabilities are extended with flameless atomization. Background correction is advised.

REFERENCES

1. H. A. Schroeder, I. H. Tipton and A. P. Mason, *J. Chron. Dis.*, **25**, 491 (1972).
2. E. Snowden and A. Pirie, *Biochem. J.*, **70**, 716 (1958).
3. K. Kostial-Simonovic and A. Pirie, *Nature, (London)*, **199**, 1007 (1963).
4. I. H. Tipton, in *Metal Binding in Medicine*, M. J. Seven (ed.), Lippincott, Philadelphia, Pa., 1960.
5. C. K. Blum, *Scot. Med. J.*, **7**, 478 (1962).
6. D. B. Gould, M. R. Sorrell and A. D. Lupariello, *Arch. Int. Med.*, **32**, 891 (1973).
7. A. S. Allen, *Chin. Med. J.*, **61**, 296 (1943).
8. D. Y. Ku, C. K. Yen and C. C. Li, *Chin. Med. J.*, **61**, 302 (1943).
9. W. Morton, *Lancet*, **2**: 738 (1945).
10. Z. Lewi and Y. Bar-Khayim, *Lancet*, **2**, 342 (1964).
11. K. Huang, *Chin. Med. J.*, **61**, 305 (1943).
12. D. Diengott, O. Roza, N. Levy and S. Maummar, *Lancet*, **2**, 343 (1964).
13. O. Roza and L. B. Berman, *J. Pharmacol. Exptl. Therap.*, **177**, 433 (1971).
14. J. Berning, *Lancet*, **1**, 110 (1975).
15. H. Ballczo and H. Muthenthaller, *Mikrochem. Ver. Mikrochim. Acta*, **39**, 152 (1952).
16. H. Ballczo and W. Schenk, *Mikrochim. Acta*, 163 (1953).
17. J. Fourage, *Anal. Chim. Acta*, **12**, 342 (1955).
18. F. H. Pollard and J. V. Martin, *Analyst*, **81**, 348 (1956).
19. A. M. Poskanzer and B. M. Foreman, *J. Inorg. Nuclear Chem.*, **16**, 323 (1961).
20. G. Henke, H. Mollmann and H. Alfer, *Z. Neurol.*, **199**, 283 (1971).
21. G. Tanaka, H. Tawamura and Y. Ohyage, *Inter. Cong. Anal. Chem.* IUPAC, paper No. C-1605, Kyoto, 1972.
22. G. D. Renshaw, *Symposium Perkin-Elmer and Royal Institution*, paper No. 652, London, 1972, cited in M. Pinta, *Modern Methods of Trace Metal Analysis*, Ann Arbor Science, Ann Arbor, Michigan, 1978.

CHAPTER 6

BERYLLIUM

Beryllium is not a normal constituent of living matter. The element occurs in the earth's crust in certain ores. Representative among them are beryl, a beryllium aluminum silicate that is widely distributed, beryllium oxide or bromellate found in Sweden, and helvite, an ore containing manganese beryllium sulfur and silicate, found in New Mexico, Norway, Russia, Canada, Australia and Brazil.

During the first 88 years following its discovery in 1797 by Vanquelin, beryllium was regarded as biologically inert and of little consequence medically. Blake[1] in 1882 wrote that beryllium had the general physiological effects of aluminium and iron. Today it is appreciated that the element and all its compounds are highly toxic.

Weber and Englehardt[2] in 1933 presented the first significant evidence in the medical literature concerning occupational beryllium poisoning in workers. They described the occurrence of a clinical disease including such findings as pneumonitis with cyanosis, bronchitis, bronchiolitis, and subsequent roentgenologic changes in the chest among a group of employees in a German plant engaged in the extraction of beryllium from beryl ore. Dermatitis and conjunctivitis were also observed in some cases.

Similar incidents in beryllium industries in Italy and Russia were reported.[3,4]

However, until 1940 or so, despite the reports from European observers, beryllium was generally accepted as an innocuous substance. Specific problems began appearing in the US when large scale development and production of fluorescent lamps began in Ohio and Massachusetts. Van Ordstrand,[5,6] Hardy[7] and others discussed chemical pneumonia and aspects of beryllium poisoning in general among workers extracting beryllium oxide. Subsequently, a registry of beryllium cases was established in 1953.

The occurrence of beryllium disease, especially the acute cases, diminished dramatically at the end of the 1940s. However, a significant number of new chronic cases continued to appear in different areas around the world. By 1972, 832 incidents in the US had been registered with the beryllium registry at Massachusetts General Hospital.[8] A lesser number of cases was published in Europe and elsewhere. According to Hasan and Kazemi,[9] chronic berylliosis is a continuing epidemiologic hazard.

Beryllium and its compounds have many different industrial uses. Beryllium–copper, considered the master alloy, is used in electrical equipment. Other metals, including nickel, magnesium, zinc, and aluminum, to name a few, are also alloyed with it. Beryllium fluoride is used in manufacturing optical glass as well as alloys. Beryllium nitrate is utilized in stiffening mantles for gas and acetylene lamps. The oxide of beryllium finds use in glass and ceramic manufacture and as a catalyst in organic reactions.

Nuclear reactors employ the fluoride and oxide and also beryllium itself. The metal serves as a neutron reflector and moderator. Beryllium is a source of neutrons when bombarded with alpha particles.

Potassium beryllium sulfate is employed in chromium and silver plating.

Exposure to beryllium and its compounds induces dermal and respiratory manifestations. Skin effects consist of a contact dermatitis, ulcers, and subcutaneous granulomata. The pneumonitis may be of a fulminating or insidious nature.

Berylliosis, as recognized in Germany in 1933, Russia in 1936, and the US in the 1940s, was an acute disease caused by exposure to high concentrations, and it was considered self limited if the exposure ceased.[10] Dermatitis, conjunctivitis, or respiratory tract inflammation ranging from rhinitis to fatal pneumonitis were also observed.

Chronic disease, which often develops long after exposure has ceased, is characterized by granulomatous interstitial pneumonitis, granulomatous lesions of the skin, liver, lymph nodes, and muscle. Beryllium can often be detected in these lesions.

We had occasion to determine the beryllium content of granulomatous skin lesions obtained from a twenty-six year old black male with a history of employment in a beryllium industry in the south. Beryllium contents between 6 and 11.7 ng g^{-1} of tissue (wet weight) were found. Although the subject had not been exposed for more than a year previously, beryllium levels of 195 and 237 ng l^{-1} were detected in blood and urine, respectively. Beryllium contents were determined by flameless atomic absorption spectrometry after chelation with cupferron and extraction into methyl isobutyl ketone. These values are in the range of those reported in the literature.

Hasan and Kazemi,[8,9] when evaluating subjects many years after exposure, found beryllium contents of 0.014–0.3 μg g^{-1} in dried peripheral lymph node (2.8–60 ng g^{-1} wet); 0.25–8.5 μg g^{-1} mediastinal lymph node (dry) (50–1700 ng g^{-1} wet) and 0–0.49 μg g^{-1} dried lung tissue (0–98 ng g^{-1} wet weight). They consider a beryllium content of 0.02 μg g^{-1} in dried lung or lymph node tissue to be significant.

Stoeckle and co-workers[10] found beryllium excretion in urine to be 0.01–1 μg l^{-1} (10–1000 ng l^{-1}) several years after exposure. In fact, trace amounts of beryllium were present as long as 20 years after exposure. It is interesting, furthermore, that renal calculi of exposed individuals were found to contain beryllium.

Brokeshoulder et al.[11] demonstrated beryllium in tissue sections by micro-emission spectrography (laser microprobe). As little as 1.48×10^{-12} g of beryllium in a section 50 microns thick could be detected.

TOXICOLOGY OF BERYLLIUM

The pathogenesis of beryllium disease is understood poorly. However, most investigators today agree that immunologic mechanisms may play an important role.

The concept that the interaction of beryllium and living tissue involves an immunological component was first suggested by Dr Nardi and co-workers as early as 1949[12] when they reported the existence of allergy-like features on exposed and unexposed parts of the body of individuals with a beryllium dermatitis. Furthermore, they considered this allergy or hypersensitivity to be related to the pathogenesis of beryllium pneumonitis.

Subsequently, Sterner and Eisenbud[13] said beryllium acts like an antigen and stimulates the production of a specific antibody, thus sensitizing the host. As a consequence, a type of inflammatory reaction will result when beryllium comes in contact with target tissue.

Curtis[14] developed a patch test as a diagnostic aid in assessing beryllium intoxication.

Belman[15,16] described the existence of a beryllium bovine albumin complex at an acid pH. He also demonstrated that beryllium can combine with alkaline phosphatase and with nucleic acids of the epidermis at physiological pH. It was postulated that these combinations may well include the beryllium combinations that are involved in the hypersensitivity reaction.

Alekseeva,[17] investigating the effects of beryllium compounds upon guinea pigs, observed that the oxide and chloride, when injected intradermally, caused hypersensitivity reactions of the skin of the delayed type, a response being illicited in 24 hours or more. This hypersensitivity could be transfered passively via lymphoid cells from sensitized animals. However, it could not be transfered via blood sera obtained from patients with berylliosis. She observed, furthermore, that a delayed hypersensitivity reaction to beryllium salts may result when non-specific changes in the immunological status of an organism occur through vaccination or the injection of heterogeneous sera.

Alekseeva[17] also reported that beryllium chloride has a 'primary toxic reaction' upon the skin.

Immunoglobulin concentrations in berylliosis were investigated by Resnick and co-workers.[18] Subjects with resolved acute berylliosis, chronic berylliosis, and beryllium dermatitis were included in the study. Most individuals who had either the chronic pulmonary or the cutaneous forms of the disease manifested a hypergammaglobulinemia and an increased concentration of immunoglobulin (IgG). The abnormality was seen less frequently in patients with resolved acute berylliosis. Significant increase in IgG was found in several industrial workers who experienced long exposure to either the metal or its compounds, but who have never exhibited any evidence of beryllium toxicity.

Krivanek and Reeves,[19] from their investigations concerning the effect of chemical forms of beryllium on the production of the immunologic response, concluded that the intensity of skin reactions was not related to the concentration of ionic beryllium but to the form in which it was more active as an antigen. Among the compounds tested, beryllium albuminate produced the largest reaction, followed in order by the sulfate, citrate, and aurin tricarboxylate of beryllium.

ANALYSIS OF BERYLLIUM

Colorimetric methods of beryllium analysis are not sufficiently sensitive for practical application in analyses of biological material. Sensitivities claimed for fluorometric methods are more reasonable, but there also exists the matter of interferences from other trace metals as well as from miscellaneous biological constituents. For the most part these can be eliminated chemically.

Sill and Willis[20] describe an elegant, lengthy procedure for the determination of beryllium using morin (2,4,3,5,7-pentahydroxy flavone) to form a fluorescent compound after the metal had been isolated by repeated extractions with acetyl acetone. A detection limit of 0.04 ng of beryllium is claimed for the analysis. Recoveries of beryllium added to bone and urine matrices were 100%.

Beryllium can be separated by chelation—extraction methods and determined by other spectroscopic techniques. Gas chromatography has also been utilized.

Acetyl acetone, either in the pure state or dissolved in different organic solvents, forms a chelate with beryllium at pH 1.5–8, which is quantitatively extractable.[21-23]

T. Stiefel et al.[24] described a method for determining beryllium in the nanogram range. A 1 ml urine aliquot was first decomposed with nitric acid. Beryllium was separated by extraction with acetylacetone in benzene and determined by flameless atomic absorption spectroscopy.

We have employed 5% acetyl acetone in methyl isobutyl ketone for the direct extraction of beryllium from urine samples.

The cupferrate of beryllium can be quantitatively extracted by chloroform at a pH greater than 3.[25] In our experience pH 6–6.5 is optimum for the quantitative extraction of this chelate by methyl isobutyl ketone (MIBK). After pH adjustment, urine samples are chelated and extracted directly. Blood samples are wet ashed or deproteinized with 5% trichloroacetic acid prior to chelation–extraction. Tissue samples are wet digested in sulfuric–nitric acids before analysis. Beryllium is determined at 234.8 nm in an atomic absorption spectrometer equipped with a graphite furnace. The detection sensitivity achieved is dependent upon the sample size employed and the degree of concentration. For example, by extracting a 10 ml aliquot into 1 ml of MIBK, we were able to detect 2 ng ml^{-1} of beryllium in solution.

Quantitative extraction of beryllium by thenoyltrifluoroacetonate in benzene takes place from neutral solution.[26]

Foreman et al.[27] attained a detection limit of 1 ng ml^{-1} when determining beryllium in human and rat urine by electron capture gas chromatography following extraction of the metal as a trifluoroacetylacetonate. Samples were analyzed directly, as well as after wet digestion.

Wolf et al.[28] were able to detect picogram quantities of chromium and beryllium by gas chromatography/mass spectrometry following chelation by trifluoroacetylacetone in benzene.

Black and Sievers[29] applied the trifluoroacetylacetonate method to the analysis of beryllium in environmental samples.

Frame et al.[30] determined beryllium oxide in biological samples by using electron capture gas chromatography and trifluoroacetylacetone chelation.

REFERENCES

1. J. Blake, Ber. Deut. Chem. Ges., 14, 394 (1882).
2. H. H. Weber and W. E. Englehardt, Zentr. Gewerbehyg. Unfalverhüt, 10, 41 (1933), cited in Beryllium: Its Industrial Hygiene Aspects, (ed. H. E. Stokinger), Academic Press, New York, 1966.
3. E. M. Zamakhovskara, B. I. Martsenkovskii and E. E. Syroechkovskii, Gigiena Truda i Tekh. Bezopasnosti, 2, 23 (1934), cited in Beryllium: Its Industrial Hygiene Aspects, (ed. H. E. Stokinger) Academic Press, New York, 1966.
4. S. Marrack-Fabroni, Med. Lavoro, 26, 297 (1935) cited in Beryllium: Its Industrial Hygiene Aspects, (ed. H. E. Stokinger) Academic Press, New York, 1966.
5. H. S. Van Ordstrand, R. Hughes and M. G. Carmody, Cleveland Clinic Quarterly, 10, 10 (1943).
6. H. S. Van Ordstrand, R. Hughes, J. M. De Nardi and M. G. Carmody, J. Amer. Med. Assoc., 129, 1084 (1945).
7. H. L. Hardy and I. R. Tabershaw, J. Ind. Hyg. Toxicol., 28, 197 (1946).
8. F. M. Hasan and H. Kazemi, Amer. Rev. Resp. Dis., 108, 1252 (1973).
9. F. M. Hasan and H. Kazemi, Chest, 65, 289 (1974).
10. J. D. Stoeckle, H. L. Hardy and A. L. Weber, Amer. J. Med., 46, 545 (1969).
11. S. F. Brokeshoulder, F. R. Robinson, A. A. Thomas and J. Cholak, Amer. Ind. Hyg. Assoc. J., 27, 496 (1966).
12. J. M. De Nardi, H. S. Van Ordstrand and M. G. Carmody, Ohio State Med. J., 45, 467 (1949).
13. J. H. Sterner and M. Eisenbud, Arch. Ind. Hyg. Occup. Med., 4, 123 (1951).

14. G. H. Curtis, *Arch. Dermatol. Syphil.*, **64**, 470 (1951).
15. S. Belman, *J. Amer. Chem. Soc.*, **85**, 2154 (1963).
16. S. Belman, *J. Occup. Med.*, **11**, 175 (1969).
17. O. G. Alekseeva, *Fed. Proc. (trans. suppl.)*, **25**, 843 (1966).
18. H. Resnick, M. Roche and W. K. C. Morgan, *Amer. Rev. Resp. Dis.*, **101**, 504 (1970).
19. N. Krivanek and A. L. Reeves, *Amer. Ind. Hyg. Assoc. J.*, **33**, 45 (1972).
20. C. W. Sill and C. P. Willis, *Anal. Chem.*, **31**, 598 (1959).
21. J. F. Steinbach and H. Freiser, *Anal. Chem.*, **25**, 881 (1953).
22. J. D. Buchanan, *J. Inorg. Nuclear Chem.*, **7**, 140 (1958).
23. J. Stary and E. Hladky, *Anal. Chim. Acta*, **28**, 227 (1963).
24. T. Stiefel, K. Schulze, G. Tolg and H. Zorn, *Anal. Chim. Acta*, **87**, 67 (1976).
25. J. Stary and J. Smizanska, *Anal. Chim. Acta*, **29**, 546 (1963).
26. R. A. Bolomey and L. Wish, *J. Amer. Chem. Soc.*, **72**, 4483 (1950).
27. J. K. Foreman, T. A. Gough and E. A. Walker, *Analyst*, **95**, 797 (1970).
28. W. R. Wolf, M. L. Taylor, B. M. Hughes, T. O. Tiernan and R. E. Sievers, *Anal. Chem.*, **44**, 616 (1972).
29. M. S. Black and R. E. Sievers, *Anal. Chem.* **45**, 1773 (1973).
30. G. M. Frame, R. E. Ford, W. G. Scribner and T. Ctvrtnicek, *Anal. Chem.*, **46**, 534 (1974).

CHAPTER 7

BISMUTH

Before the age of penicillin, bismuth was an important therapeutic agent in the treatment of syphilis. Although it was first tried in human syphilis in 1889, years of laboratory investigation into animal trypanosomiasis, spirillosis and syphilis were required to establish its value.[1-3] Fournier and Guenot[4] then introduced bismuth into the therapeutics of human syphilis in 1922. Bismuth was found to be more effective than mercury but less so than the arsenicals. It usually was administered concurrently with the arsenicals or in between courses of the latter.

Like the other toxic metals mentioned, bismuth exerts its action by binding with the sulfhydryl groups of enzymes.

With the exception of glycobiarsol, a pentavalent arsenical containing bismuth, systemic use of bismuth preparations is now quite obsolete. A few compounds, such as the subcarbonate, subnitrate and subgallate, are still used today for their local action as a gastrointestinal astringent, protective and adsorbent. They also serve as skin protectives. Bismuth aluminate has been used as an antacid.

In veterinary practice, these bismuth preparations have been used for their local action only, e.g. to protect the gastrointestinal mucosa, or as a dusting powder in eczema, ulcers and wounds.

The nephrotoxicity of absorbable bismuth compounds, varying from asymptomatic proteinuria to acute or chronic renal failure, has been reported since the beginning of the last century.[5] Other manifestations of bismuth toxicity, including stomatitis, erythema and local pigmentation have been described.[6-9] Some fatalities have occurred.[10]

Until fairly recently, the poorly absorbed bismuth compounds, such as the subnitrate, gallate, salicylate, and subcarbonate, were considered fairly innocuous. However, in 1974 the occurrence of encephalopathy due to chronic ingestion of bismuth subnitrate was reported. More than 100 cases were reported subsequently from France, Belgium and Australia in what was considered almost epidemic proportions.

Monseu et al.[11] in 1976 described a 'bismuth encephalopathy' in a twenty-seven year old pregnant female, characterized by acute confusion, drowsiness alternating with visual hallucinations, and non-rhythmic myoclonic jerks in all limbs that increased during attempts at movements following visual and auditory stimuli. Coarse intention tremors and occasional choreic movements were also observed. Her language was severely dysarthritic. She experienced generalized epileptic seizures.

This patient, a physician, had taken 20 g of bismuth daily for eight months. Then for two months prior to admission, she increased the dose to 40 g a day.

Agitation, hallucinations, and myoclonic jerks decreased slowly and disappeared by the twelfth day. The severe ataxia with retropulsion persisted for five weeks. There was full recovery in about two months.

Admission blood bismuth levels were not reported, but levels of 487 $\mu g\%$ and 218 $\mu g\%$ were found on the eighth and fourteenth days, respectively.

Bismuth levels as high as 3 mg% have been reported in other cases of bismuth encephalopathy. Allain[12] found bismuth levels of 0.1 $\mu g\%$ in subjects not receiving bismuth.

The persistence of encephalopathy induced by the relatively insoluble bismuth salts is due to their retention in the kidney, bones, and gastric wall. Similar encephalopathies have never been described from the parenteral use of soluble bismuth salts (chloride and iodide) used in treating syphilis because these are quickly eliminated in the urine and saliva.

British Anti Lewisite (BAL) has not proved beneficial in treating bismuth encephalopathy.

Burns et al.[13] described five patients in Australia with a reversible encephalopathy probably induced by overuse of bismuth subgallate.

Other reports of bismuth encephalopathy have appeared subsequently.[14–17]

A percentage of the essentially normal tissue samples analyzed by Tipton[18] contained minimal amounts of bismuth. Since there is no known metabolic role for bismuth, this probably represents environmental contamination.

NON-MEDICAL USES OF BISMUTH

Bismuth and its compounds do have certain industrial uses. The metal itself is used in fusible alloys, electric fuses, and low-melting solders. The bromide oxide and iodide oxide are used in manufacturing dry cell cathodes. Bismuth chloride oxide is employed in face powders, as a pigment, and in manufacturing artificial pearls.

Bismuth trihydride (bismuthine), a highly toxic compound, finds use in the manufacture of semi-conductors.

Bismuth nitrate is a constituent in some luminous paints. The subnitrate and subcarbonate are employed in forming glazes for ceramics and in enamels.

Glass manufacture utilizes bismuth oxide and phosphate. The former is also used in rubber vulcanization, and in the fireproofing of papers and polymers.

Bismuth phosphate finds application in the separation of plutonium from fission products.

ANALYSIS OF BISMUTH

Colorimetric

Bismuth forms colored complexes with numerous chemical agents, these can be measured spectrophotometrically.

Diphenylthiocarbazone (dithizone) is probably the most familiar. Bismuth (III), as a primary dithizonate, can be quantitatively extracted by carbon tetrachloride[19] or chloroform over the pH range 3–10. While this analysis may be sensitive, it is not specific for bismuth. Dithizone forms colored complexes with approximately seventeen other heavy metals including lead, thallium, cadmium, mercury, and gold. True, these interferences can be eliminated chemically for the most part, but some bismuth is lost in the process.

The iodide method of bismuth analysis has also been employed in analyzing biological materials. As with the dithizone procedure, samples were wet ashed. After digestion, ascorbic acid, potassium iodide and sodium sulfite were added and the reaction mixture was set aside to permit the liberation of iodine by the bismuth present. The liberated iodine was extracted with amyl alcohol–ethyl acetate.[20] The absorbance of the extract is measured at 460 nm. Thallium, cadmium, and some lead are coextracted.

Bismuth (III) is quantitatively extracted at pH 4–11 with carbon tetrachloride containing sodium diethyldithiocarbamate. At pH 11, in the presence of EDTA and cyanide as masking agents, the extraction of bismuth is very selective.[21] Optimal pH for lead, thallium, and cadmium is 5.5–6.5. The complex is extractable by methyl isobutyl ketone.

Oxinates can be utilized for bismuth analysis. Quantitative extraction of the 8-hydroxyquinolate can be achieved over the pH range 2.5–11.[22] Bismuth also forms a precipitate with methyl oxinate which can be extracted at approximately pH 10 by chloroform.[23]

Salicylaldoximate forms a yellow complex with bismuth which can be extracted with chloroform.[24] At a pH greater than 2.5, bismuth can be quantitatively extracted by 0.25 M thenoyltrifluoroacetonate in benzene.[25]

As has been shown with other heavy metals, the bismuth–chelate complexes can be employed in bismuth analysis by other instrumental means.

Various instrumental techniques

Bismuth has been determined by neutron activation analysis,[26] spectrography,[18] and polarography,[27] among others.

Polarography is considered a low resolution technique as far as biological materials are concerned.[28] Levels of other elements in normal blood, especially copper, cause difficulty in the determination of bismuth. This is the situation with lead analysis also. Polarography can detect 0.01 μg ml^{-1} of bismuth. Lesser concentrations are detectable by anodic stripping voltammetry.

X-ray fluorescence analysis applications have proved quite sensitive. Bismuth is usually measured following chelation and extraction. Detection limits of 5 ng ml^{-1} and 10 ng ml^{-1} are claimed by different methods which have been reported.[29,30]

Judging from the numerous papers in the literature, atomic absorption spectrometry is the instrumental technique utilized most commonly in bismuth analysis.

Willis[31] determined bismuth in urine by flame atomic absorption spectrometry. Koirtyohann and Feldman[32], using a modified long path burner determined both bismuth and tellurium in biological materials. Sensitivity limits, according to these investigators, are based on the use of organic extraction for separations of the desired elements from sample matrices.

Kinser[33] determined bismuth and tellurium contents in tissue digests by flame atomic absorption spectrometry. Chemical separations were not used, and apparently concentrations of bismuth as low as 1.5 μg ml^{-1} could be detected. However, he did find that tissue constituents like sodium, calcium, potassium, magnesium, iron and phosphate did exert an effect upon absorbance. Analyses were performed at the 223.1 nm line.

Devoto[34] employing chelation with ammonium pyrollidine dithiocarbamate (APDC), at pH greater than 4, and extraction was able to determine bismuth in urine at a concentration of 0.02 ppm (20 μg l^{-1}).

Hall and Farber[35] determined bismuth in tissues after extraction of the APDC chelate from digestants buffered to pH 3. An air–hydrogen flame was employed.

Rooney[36] determined bismuth in blood and urine following generation of bismuth hydride from wet oxidized sample. Detection limits of less than 0.01 μg ml^{-1} were attained. Good recoveries are reported.

Allain[12] measured the bismuth content in blood in the graphite furnace. 0.5 ml aliquots of blood were chelated by APDC and extracted into hexane.

We are able to attain a detection sensitivity of less than 1 ng ml^{-1} in the graphite furnace. Sodium diethyl dithiocarbamate is the chelate preferred.

REFERENCES

1. A. E. Robert and B. Sauton, *Ann. Inst. Pasteur*, **30**, 261 (1916).
2. W. Kolle and H. Ritz, *Deutsche Med. Wchnschr.*, **45**, 581 (1919).
3. R. Sazerac and C. Levaditi, *Compt. Rend. Acad. Sci.*, **173**, 338 (1921).
4. L. Fournier and L. Guenot, *Ann. Inst. Pasteur*, **36**, 14 (1922).
5. A. W. Czerwinski and H. E. Ginn, *Amer. J. Med.*, **37**, 969 (1964).

6. L. Mayer and G. Baehr, *Surg. Gynec. Obstet.*, **15**, 309 (1912).
7. E. E. Peters, *Amer. J. Syph.*, **26**, 84 (1942).
8. J. J. Grund, *Arch. Derm. Syph.*, **41**, 1076 (1940).
9. J. D. Gryboski and S. P. Gotoff, *New Eng. J. Med.*, **265**, 1289 (1961).
10. T. L. Sterne, C. Whitaker and C. H. Webb, *J. Louisiana Med. Soc.*, **107**, 332 (1955).
11. G. Monseu, M. Struelens and M. Roland, *Acta Neurol. Belg.*, **76**, 301 (1976).
12. P. Allain, *Clin. Chim. Acta*, **64**, 281 (1975).
13. R. Burns, D. W. Thomas and V. J. Barion, *Brit. Med. J.*, **1**, 220 (1974).
14. A. Buge, G. Rancurel, M. Poisson and H. Dechy, *Nouv. Presse Med.*, **3**, 2315 (1974).
15. J. Cambier, P. Le Bigot, I. D. Thoyer-Rozat and M. Levardon, *Nouv. Presse Med.*, **4**, 2275 (1975).
16. P. Loiseau, P. Henry, P. Jollon and U. Legroux, *J. Neurol. Sci.*, **27**, 133 (1976).
17. H. L. Boiteau, J. M. Cler, J. F. Mathe, R. Delobel, J. R. Fevre and C. Boussicault, *Eur. J. Toxicol. Environ. Hyg.*, **9**, 233 (1976).
18. I. Tipton, in *Metal Binding in Medicine*, (ed. M. J. Seven) J. B. Lippincott Co., Philadelphia, Pa., 1960.
19. J. L. Kassner, S. F. Ting and E. L. Grove, *Talanta*, **7**, 269 (1961).
20. E. B. Sandell, *Colorimetric Determination of Traces of Metals*, Interscience Publishers, Inc., New York, 1959.
21. H. Bode, *Z. Anal. Chem.*, **143**, 182 (1954).
22. J. Stary, *Anal. Chim. Acta*, **28**, 132 (1963).
23. J. P. Riley and H. P. Williams, *Mikrochim. Acta*, 825 (1959).
24. G. Gorbach and F. Pohl, *Mikrochem.*, **38**, 258 (1951).
25. F. Hagemann, *J. Amer. Chem. Soc.*, **72**, 768 (1950).
26. M. Pinta, *Modern Methods for Trace Element Analysis*, Ann Arbor Science Publishers, Ann Arbor, Mich., 1978.
27. W. M. Plank, *J. Assoc. Offic. Anal. Chemists*, **55**, 155 (1972).
28. D. F. Boltz and M. G. Mellon, *Anal. Chem.*, **46**, 227R (1974).
29. J. D. Winefordner and R. C. Elser, *Anal. Chem.*, **43**, 25A (1971).
30. S. A. Clyburn, B. R. Bartschmid and C. Veillon, *Anal. Chem.*, **46**, 2201 (1974).
31. J. B. Willis, *Anal. Chem.*, **34**, 614 (1962).
32. S. R. Koirtyohann and C. Feldman, in *Developments in Applied Spectroscopy*, Vol. 3 (eds J. E. Forrette and E. Lanterman) Plenum Press, New York, 1964.
33. R. E. Kinser, *Amer. Indust. Hyg. Assoc. J.*, **27**, 260 (1966).
34. G. Devoto, *Boll. Della Societa Ital. di Biol. Sperment.*, **44**, 1253 (1968).
35. R. J. Hall and T. Farber, *J. Assoc. Offic. Anal. Chemists*. **55**, 639 (1972).
36. R. C. Rooney, *Analyst*, **101**, 749 (1976).

CHAPTER 8

BORON

Boron, the element, is not too toxic itself. However, its various compounds are considered moderately to highly toxic.

The element and its compounds are versatile. Boron is employed as a neutron absorber in nuclear chemistry. It is also used in alloys to harden other metals. Boron nitride is utilized in nuclear reactors, semiconductors, and in the manufacture of alloys.

The trichloride, trifluoride, and the trifluoride–etherate complex serve as catalysts in different organic reactions. Boron trifluoride protects magnesium and alloys from oxidation.

Boric anhydride (boron oxide) is utilized in metallurgy.

Sodium borate is employed in soldering metals, in aging wood artificially, and as a preservative. It is used in curing skins. Both the sodium salt and boric acid are used in manufacturing glazes and enamels, in weatherproofing wood and in fireproofing fabrics. Boric acid is used in printing, dyeing, and in photography.

Boroethane (diborane, a hexahydride), which is used as a rubber vulcanizer, in rocket propellants, and as a catalyst for olefin polymerization, can be hazardous. Exposure produces symptoms resembling metal fume fever. Irritation of lungs and pulmonary edema occur.

Decaborane (decaboron tetradecahydride), employed as a catalyst in olefin polymerization and in rocket propellants, is also a toxic substance. Symptoms following exposure include dizziness, nausea, vomiting and muscular tremors. It may be a liver toxicant.

TOXICOLOGY OF BORIC ACID AND BORAX

Boric acid and borax serve as mild topical astringents and antiseptics for both human and veterinary use. The earliest boric acid medication known, *sal sedativum*, was prepared by Wilhelm Holmberg in the early eighteenth century.[1] Fatal cases were recorded as early as 1881. Shoemaker[2] in 1905 quoted an M. Chevalier, who had collected 22 cases from the literature.

The first good description of the clinical features of boric acid poisoning was given by J. C. McWalter[3] in 1907. A mixture of borax and honey was applied to the mouth of an infant as a treatment for thrush. As therapy continued, the infant developed progressive wasting, and a marked erythematous eruption appeared on the palmar aspects of the hands and plantar surface of the feet. Distinct desquamation between toes and fingers appeared. Well-marked urticarial eruptions were present on arms and forearms; the abdomen was tender; there was a raw, pinky redness of the lips, tongue, palate, and throat; the child's face had a wizened look; joints, especially the knees, were tender, swollen, and somewhat stiff. Once borax and honey applications were stopped, the infant recovered.

According to Goldbloom and Goldbloom,[1] who correlated experimental, pathological, clinical, and laboratory findings in boric acid poisoning, the highest concentration of the substance is found in the brain. Congestion of the brain and meninges, and scattered pin vascular hemorrhages are observed upon autopsy. Boric acid is detected in the cerebrospinal fluid. Symptoms related to central nervous system involvement include headache, weakness, excitement or depression, delirium or coma, convulsions, collapse, and cyanosis.

Small amounts of boric acid are excreted by the gastrointestinal tract. Vomiting, diarrhea, and crampy abdominal pain are observed occasionally. Eighty to almost 100% of the compound absorbed is excreted in the urine. The glomerular and tubular change produced is indicated by the diminished urine output. Anuria is a rare occurrence. Boric acid causes congestion and fatty changes of the liver. The second-highest concentration of the substance is present in the liver.

Fisher and Freimuth[4] determined the average blood boron concentration in 34 patients exposed only to dietary boron to be 0.25 μg ml^{-1}. The highest level found in this group was 1.25 μg ml^{-1}. Blood boron concentrations among 37 patients exposed to boron-containing medications were not found to differ significantly. However, two children who accidentally drank boric acid solution developed blood boron levels of 7.89 mg and 7.44 mg%. Despite these levels, apparently neither of them exhibited signs of toxicity.

Three fatal cases of boric acid poisoning with pancreatic inclusions were reported subsequently.[5]

The minimum lethal dose of boric acid in man is not known.

Boron can be absorbed from the mucous membrane and injured skin surfaces. Skipworth et al.[6] reported the occurrence of boric acid intoxication from 'medicated talcum powder' for example.

Gordon et al.,[7] while investigating two infants with a seizure disorder found that both had been exposed to large amounts of a preparation containing honey and borax used in coating their pacifiers. One child had developed a profound anemia, as well. Urine boron concentrations as high as 12.3 mg% were found. Blood boron levels were 14.5 mg and 9.44 mg%. Since cessation of use of the preparation, containing about 10% borax, their seizures stopped.

A toxic alopecia following ingestion of boric acid was reported by Stein and co-workers[8] in a young female. The subject, diagnosed originally as an acute psychotic organic brain syndrome, was a chronic ingestor of mouthwash. Her blood boron level was 3.2 mg%. Once the patient stopped swallowing the mouthwash, her hair grew anew.

It is interesting that approximately 400 proprietary preparations sold in the US contain significant quantities of boric acid.

THE ESSENTIAL NATURE OF BORON

For more than fifty years it has been known that boron is essential for plants.[9-11] Cell maturation and differentiation in higher plants are boron dependent. Boron occurs in plant tissues in much higher concentrations than in animal tissues.[12] Legumes, containing 25–50 ppm, are generally richest in the element. Fruit and vegetables contain 5–20 ppm, while cereal grains and hay contain 1–5 ppm. Among the fruits, avocados are highest in boron (7–10 ppm).[13] By comparison, muscle and other soft tissues of the body contain 0.5–1 ppm boron or less (dry weight). Bones concentrate boron and so contain several times the tissue concentrations.[14,15]

Cows' milk normally contains 0.5–1 ppm boron, the level depending upon the boron content in the cows' food.[16]

Reported intakes of boron by humans are diet dependent. Diets rich in vegetables and fruits and rather low in animal protein would contain more boron than a meat-rich vegetable-poor diet. Schlettwein-Gsell and Mommsen-Straub[17] discuss the boron content of foods in depth, with tables listing the boron content in individual servings of meat, fruits, and vegetables.

The use of boric acid and borax as additives to food can cause toxicological problems.

To date no function for boron in the animal body has been established. Several attempts have been made to induce boron deficiencies in rats.[18-20] Animals fed highly purified diets containing 0.15–0.16 ppm of boron fared well. If this element is essential to the rat, it must be at a dietary level lower than that used in the experiments.

ANALYSIS OF BORON

Colorimetric procedures

Tumeric yellow or curcumin (diferuloylmethane) is the classical reagent for the detection of boron. A rapid qualitative procedure of use in an acute emergency situation involves placing a drop of the suspect material (a solution, gastric aspirate, water, serum, urine) on a filter paper strip impregnated with tumeric. Negative and positive controls are placed alongside. Then, after the spots dry, the paper is exposed to ammonia fumes. Boric acid becomes a flaming red. Mair and Day[21] adapted the curcumin method to a spectrophotometric determination of boron extracted from ashed animal tissues with 2-ethyl-1,3-hexanediol in chloroform. Boron in the organic phase is converted to a rosocyanin complex by the addition of curcumin and its absorbance is measured at 550 nm. According to the authors, the complex obeys Beer's law over the range 2 to 20 ng ml^{-1}.

Bassett and Matthews[22] complexed boron with salicylate and then complexed the borodisalicylate with ferroin, and the colored complex extracted into chloroform was measured at 516 nm. Copper, zinc, carbonate, and anionic detergents interfere, however.

Fluorimetry

Boric acid forms fluorescent complexes with several reagents, and the reactions have been utilized in boron analysis.

Liebich et al.[23] determined the boron content in waters and plant materials by forming a fluorescent complex with 2-hydroxy-4-methoxy-4-chlorobenzene in sulfuric acid. The procedure detects boron in the range of 5–50 ng ml^{-1}.

Vasileskaya[24] described formation of a Rhodamine 6G complex. A detection sensitivity of 10 ng ml^{-1} is claimed.

According to Monnier et al.[25] boron in concentrated sulfuric acid when complexed with hydroxy-2-methoxy-4-chloro-4-benzophenone is detectable at a concentration of 1 ng ml^{-1}.

Other instrumental methods

Boron has been determined by activation analysis, both neutron and proton. Bankert et al.[26] discussed the proton activation analysis of boron in water without chemical separation. Activation analysis methods for other biological determinations apparently have not yet been developed. Emission spectroscopy and atomic absorption spectrometry are more commonly available for this application.

Pierce and Brown[27] utilized atomic emission spectrometry in determining microgram quantities of boron in surface waters. Samples were fed directly into the instrument via an autoanalyzer manifold.

Holak[28,29] determined boron in foods after chelating the element, present in an acid digest, with 2-ethyl-1,3-hexanediol and extracting the complex into methyl isobutyl ketone prior to instrumental analysis. Spielholtz et al.[30] used chelate formation with the reagent for analysis of seawater. A nitrous oxide acetylene flame was used. Boron was determined at the 249.7 nm resonance line.

It is logical that the sensitivity of the analysis could be extended by applying flameless atomization of extracts.

Green et al.,[31] in an interesting study, discussed boron contamination, of samples for analysis, by borosilicate glass. Water solutions did not appear to extract boron after more than forty-five days of storage. However, 0.1M hydrochloric acid and 0.1M sodium hydroxide extracted 0.03 μg ml^{-1} and 0.08 μg ml^{-1}, respectively; 1M sodium hydroxide extracted 0.03 μg ml^{-1} of boron within five minutes.

Obviously, avoiding contact of samples with borosilicate glass is necessary.

REFERENCES

1. R. B. Goldbloom and A. Goldbloom, *J. Peds.* **43**, 631 (1953).
2. J. V. Shoemaker, *Med. Bull., Phila.*, **27**, 68 (1905).
3. J. C. McWalter, *Lancet*, **2**, 369 (1907).
4. R. S. Fisher and H. C. Freimuth, *J. Invest. Dermat.*, **30**, 85 (1958).
5. M. A. Valdes-Dapena and J. B. Arey, *J. Peds.*, **61**, 531 (1962).
6. G. C. Skipworth, N. Goldstein and W. P. McBride, *Arch. Dermatol.*, **95**, 83 (1967).
7. A. S. Gordon, J. S. Prichard and M. H. Freedman, *Canad. Med. Assoc. J.*, **108**, 719 (1973).
8. K. M. Stein, R. B. Odom, G. R. Justice and G. C. Martin, *Arch. Derm. Syph.*, **108**, 95 (1973).
9. A. L. Somner and C. B. Lipman, *Plant Physiol.*, **1**, 231 (1926).
10. K. Warrington, *Ann. Bot. (London)*, **37**, 629 (1923).
11. K. Warrington, *Ann. Bot. (London)*, **40**, 27 (1926).
12. E. J. Underwood, *Trace Elements in Human and Animal Nutrition*, Academic Press, New York, 1971.
13. E. G. Zook and J. Lehmann, *J. Amer. Diet. Assoc.*, **52**, 225 (1968).
14. G. V. Alexander, R. E. Nusbaum and N. S. McDonald, *J. Biol. Chem.*, **192**, 489 (1951).
15. R. M. Forbes, A. R. Cooper and H. H. Mitchell, *J. Biol. Chem.*, **209**, 857 (1954).
16. E. C. Owen, *J. Dairy Res.*, **13**, 243 (1944).
17. D. Schlettwein-Gsell and S. Mommsen-Straub, *Int. J. Vitamin Nutr. Res. (Internat. Z. Ern-Forschung)*, **43**, 93 (1973).
18. E. Hove, C. A. Elvehjem and E. B. Hart, *Amer. J. Physiol.*, **127**, 689 (1939).
19. J. D. Teresi, E. Hove, C. A. Elvehjem and E. B. Hart, *Amer. J. Physiol.*, **140**, 513 (1940).
20. E. Orent-Keiles, *Proc. Soc. Exptl. Biol. Med.*, **44**, 199 (1941).
21. J. W. Mair, Jr. and H. G. Day, *Anal. Chem.*, **44**, 2015 (1972).
22. J. Bassett and P. J. Matthews, *Analyst*, **99**, 1 (1974).
23. B. Liebich, D. Monnier and M. Marcantonatos, *Anal. Chim. Acta*, **52**, 305 (1970).
24. A. E. Vasilevskaya, *Nauch. T. Vses Inst. Miner. Resur.*, USSR, **5**, 22, 1971, cited in M. Pinta, *Modern Methods for Trace Element Analysis*. Ann Arbor Science, Ann Arbor, Mich., 1978.
25. D. Monnier, C. A. Menzinger and M. Marcantonatos, *Anal. Chim. Acta*, **60**, 233 (1972).

26. S. F. Bankert, S. D. Bloom and G. D. Sauter, *Anal. Chem.*, **45**, 692 (1973).
27. F. D. Pierce and H. R. Brown, *Anal. Chem.*, **48**, 670 (1976).
28. W. Holak, *J. Assoc. Offic. Anal. Chemists*, **54**, 1138 (1971).
29. W. Holak, *J. Assoc. Offic. Anal. Chemists*, **55**, 890 (1972).
30. G. I. Spielholtz, G. C. Toralbolla and J. J. Willsen, *Mikrochim. Acta*, 649 (1974).
31. G. H. Green, C. Blincoe and H. J. Weeth, *J. Agr. Food Chem.*, **24**, 1245 (1976).

CHAPTER 9

CADMIUM

The discovery of cadmium as a distinct element was made by Stromeyer of Germany in 1817.[1] As cadmium and zinc are similar in atomic structure and chemical behavior, they often occur together in nature. Zinc has been shown to be an essential element. Cadmium, on the other hand, is a highly toxic metal with no known function in animal metabolism.

In 1858, Sovet[2] described what are probably the first reported cases of cadmium poisoning. Three servants, polishing silverware with cadmium carbonate, apparently inhaled great quantities of the dust and developed respiratory and gastrointestinal problems. However, Marme's paper regarding the toxic effects of cadmium[3] seems to be the first important work on that subject. Subsequently, the possible toxic effects of cadmium in zinc smelters and other industries using the metal were discussed. Prodan's review[1] of the literature prior to 1932 is a classic.

INDUSTRIAL USES OF CADMIUM

Cadmium and its salts are used in numerous and varied industrial processes. Electroplating is the major use of the metal itself. Cadmium is a constituent of fusible alloys and aluminum solder. It is utilized in such different applications as process engraving, amalgam in dentistry, and nickel–cadmium storage batteries as well.

The acetate is employed for producing iridescent effects on pottery and porcelain. Cadmium bromide and iodide are used in photography, process engraving, and lithography. The oxide is utilized in glass manufacture, in ceramic glazes, as well as in the manufacture of silver alloys and in electroplating.

Cadmium selenide is found in photoconductors, photoelectric cells, rectifiers, and in phosphors.

Cadmium sulfide, also known as cadmium yellow or Juane brilliant, being a pigment fast to light, is used in coloring glass, soaps, textiles, paper, rubber, in printing inks, ceramic glazes and fireworks.

Some cadmium salts, the oxide and iodide, were used to a limited extent as anthelminthics in swine and poultry. During the early 1900s cadmium salts were used sporadically in treating human syphilis and tuberculosis.[4]

TOXICOLOGY OF CADMIUM

Cadmium is an inhibitor of sulfhydryl enzymes. It also has affinity for other ligands in cells, such as hydroxyl, carboxyl, phosphatyl, cysteinyl and histidyl side chains of proteins, purines and porphyrin. It can disrupt pathways of oxidative phosphorylation.

Cadmium interacts or competes with other metals. For example, in animal studies high dietary levels of this element have been shown to depress copper uptake and to change the distribution of tissue copper.[5,6] Reduced weight gains, induced in mice and rats by excess cadmium in the diet, are largely overcome by zinc and copper supplements.[7]

Axelsson and Piscator[8] observed that rabbits fed cadmium developed a hyperplastic bone marrow and a hypochromic microcytic anemia similar to that induced by iron deficiency. Development of anemia in other animal studies has been demonstrated also. The mechanism of this cadmium effect is not known. High intakes of cadmium, as well as zinc, copper and manganese, interfere with iron absorption possibly through competition for protein-binding sites in the intestinal mucosa. In the studies of Jacobs et al.[9] with Japanese quail, an increase in plasma transferrin was noted to occur concurrently with the severe anemia. Cadmium interference with the release of iron by transferrin has been suggested as a possible mechanism.

Cadmium has been shown to aggravate zinc deficiencies in various animals.[10,11] A partial replacement of zinc by cadmium in various tissues has been demonstrated.[12-14]

The investigations of Pulido et al.,[15] wherein it was demonstrated that the sum of zinc and cadmium bound to renal metallothionein (a low molecular weight protein first described by Margoshes and Vallee[16]) remains constant while the concentrations of the individual metals vary, suggest that these elements compete for protein-binding sites. Friberg et al.[17] postulated that metallothionein acts as the transport protein for cadmium.

TOXICITY OF CADMIUM

Industrial exposures to cadmium occur via the respiratory tract, though bad hygienic practices may result in some gastrointestinal absorption.

Inhalation of fumes or dusts containing cadmium and its compounds primarily affects the respiratory tract, but there are subsequent systemic effects as well. Some hours after exposure a dryness of the throat, a sense of constriction, and difficulty in breathing are experienced. There may be headache, vomiting, muscle cramps. Because of the delayed onset of symptoms following even massive exposures, workmen have been known to inhale fatal concentrations without experiencing much discomfort.

In fatal cases, a pulmonary edema, acute inflammatory changes in the kidney, and fatty degeneration of the liver have been noted.

The symptomatology of chronic cadmium poisoning was recognized only in recent years. Friberg[18] reported that the main findings in workmen employed at an accumulator factory for over 20 years were: emphysema of the lungs, mild liver damage, anemia, proteinuria, renal tubular damage, some dental changes, and impairment of the sense of smell (anosmia).

Smith et al.[19] in 1961 described the excretion of a low molecular weight albumin as an effect of exposure to cadmium dusts and fumes.

Ingestion of cadmium compounds produces symptoms suggestive of food poisoning of microbial origin. Nausea, salivation, vomiting, followed by diarrhea with abdominal discomfort and pain may appear almost suddenly or can be delayed a few hours. While some cases of poisoning due to ingestion of cadmium have occurred in industry because of improper hygienic practices, most instances result from common household items.

When cadmium-plated cooking utensils, ice cube trays, and coffee urns were in use, circa World War II, outbreaks of 'food poisoning' due to cadmium occurred with frequency. Acidic foods and beverages solubilized the cadmium plating.

Cadmium-based paints can be as hazardous in the environment of a child as their lead counterparts. Poisoning due to cadmium has been misdiagnosed as lead poisoning, primarily because of physicians' unawareness. The following is a case in point.

A few years ago a two-year-old white male, afebrile, with an encephalopathy, and unconsciousness was admitted to a sister hospital in the city. X-ray examination revealed the presence of radiopaque material in the abdomen. Blood and urine lead levels were not significant, being 50 $\mu g\%$ and 250 $\mu g\,l^{-1}$, respectively. The blood cadmium was 25 $\mu g\%$ and the urinary excretion, 1000 $\mu g\,l^{-1}$.

History revealed that the child's parents, quite cognizant of the dangers of leaded paint, had painted his crib with a red, lead-free paint containing 0.5% cadmium. The child had chewed off most of the paint. Furthermore, he liked to lick white shoes, amply dusted with cadmium sulfate. In addition, his sisters cooked little delicacies for him in their cadmium-plated toy pans and dishes.

The subject did not recover.

Cadmium is present only in minute amounts in tissues at birth, but concentrations increase as a function of age.[20] Apart from industrial exposures, the increase in the body's cadmium burden stems from the environment. All foods, seafood, meat, milk, grains, and water contain the element, and there is cadmium in cigarette smoke, etc.[21] Accumulation is continuous since cadmium has a long biological half life, about 38 years.[17] The element accumulates at the rate of 40 μg per day. An adult of 40–60 years probably has a body burden of 300 mg, 45% of which can be recovered from the liver and kidney.

The role cadmium may play in causing hypertension is in dispute. There is supportive evidence on both sides of the controversy.[22]

Itai-itai disease is a form of chronic cadmium poisoning manifest by renal dysfunction in combination with osteomalacia or severe osteoporosis. The disease, quite prevalent in some areas of Japan, was caused by consuming rice grown in fields close to smelters. Concentrations of cadmium in rice in polluted areas were found to range between 0.6 and 1 μg g^{-1}.[23]

Normal cadmium concentrations

Cadmium in blood reflects current exposure, but cadmium in urine may reflect body burden when the exposure is low.[24] Normal levels are less than 1 μg% in blood.

Chicago children in high risk areas for lead poisoning were shown to have blood cadmium levels between 1 and 5 μg%. Blood levels among cigarette smokers, both normotensive and hypertensive, ranged between 0.7 and 4.8 ng ml^{-1} (0.07–0.48 μg%).[25]

Glauser et al.[26] found blood cadmium levels to be 3.4 ± 0.5 ng ml^{-1} in normotensive subjects as compared to 11.1 ± 1.5 ng ml^{-1} in untreated hypertensives.

Hammer et al.[27] compared tissue cadmium contents of smokers and non-smokers. For example, it was estimated that the cadmium burden in adult male cigarette smokers was more than twice that of comparable nonsmokers. McKenzie[28] in her investigations also found that cadmium contents of tissues from subjects with hypertensive disease did not differ significantly from those dying from other causes. Tissues from smokers had greater cadmium contents than tissues from nonsmokers.

Tsuchiya et al.,[29] while investigating Tokyo inhabitants for cadmium exposure, found the average urinary excretion to be 1.7 μg l^{-1} in all ages. Cadmium excretion in a non-industrial Chicago population is less than 5 μg l^{-1}.

ANALYSIS OF CADMIUM

Colorimetric

Colorimetric procedures for cadmium analysis, while possessing fairly good sensitivity in many instances, suffer from interferences from other elements present in the matrix. Rather large-sized samples, by current standards, were required in order to attempt more adequate separation of cadmium from its coextractants. For example, the di-β-naphthylthiocarbazone procedure of Cholak and Hubbard[30] requires 50–100 ml of urine, or 5–20 g of blood, tissue, etc.

Initially, organic matter is destroyed by digestion with sulfuric-nitric acids. The digestant is diluted with an ammonium citrate solution, adjusted to a pH of about 8.3, and extracted with a chloroform solution of di-β-naphthylcarbazone. Zinc, lead, nickel, copper, and cobalt are some of the elements coextracted. Shaking the chloroform extract with 0.2N hydrochloric acid removes the cadmium along with the zinc, lead, bismuth, and nickel. Tartrate and sodium hydroxide are added to the acid solution. The now strongly basic solution is extracted with a chloroform solution of dithizone to isolate the cadmium from lead, zinc, and bismuth. Cadmium is reextracted into the aqueous phase by 0.2 N hydrochloric acid, and extracted by di-β-naphthylcarbazone in chloroform. Cadmium in the final extract is determined by comparing its absorbance with extracts of standard solutions at 540 nm.

Various procedures employing dithizone in cadmium analysis have appeared in the literature.[31-33]

Different chelating agents were investigated, also. To cite a few, cadmium can be extracted by thenoyltrifluoracetonate in chloroform.[34] Salicylaldoxime forms a yellow complex with the metal, which is extractable by organic solvents.[35]

Quantitative extraction of cadmium with carbon tetrachloride and other solvents takes place in the presence sodium diethyl dithiocarbamate over the pH range 5–11. EDTA and potassium cyanide do not interfere.[36] We found a pH of 6–6.5 to be optimal. Thallium and lead are coextracted.

Complete extraction of cadmium as the 8-hydroxyquinolate complex occurs at pH 5.5–9.5.[37]

Fluorimetry

Fluorimetry has had limited application in determining cadmium. Among the complexing agents used are morin and 8-hydroxyquinoline,[38] p-tosylaminoquinoline,[39] 8-hydroxyquinoline-5-sulfonic acid,[40] and picolinealdehyde-2-quinolylhydrazone.[41]

Polarography

Knockaert et al.[42] describe a polarographic method for the simultaneous determination of lead and cadmium. Samples were wet ashed in nitric and perchloric acids prior to polarographic analysis. Franke and de Zeeuw[43] determined cadmium in urine by differential pulse anodic stripping voltammetry, using both the hanging mercury drop electrode and the glassy carbon electrode plated with a mercury film. Polarographic methods can be applied to wet digests of blood, tissues, and other biological materials.

Neutron activation analysis

Chueca *et al.*[44] reported a method for the simultaneous determination of zinc and cadmium, which is suitable for the analysis of small samples of biological material. Following neutron activation and a simple ion exchange separation, zinc and cadmium are identified and assayed by gamma ray spectroscopy. The sensitivity of detection for cadmium is 3×10^{-8} g (0.03 ng ml^{-1}). The method has been applied to the determination of cadmium and zinc in subcellular fractions prepared from rat testes and kidney.

Lieberman and Kramer[45] described an elegant technique for cadmium determination in biological tissue with a detection limit of 50 ppb. Time required for analysis was more than eight days.

Spectroscopic analysis

Before 1940 spectroscopic applications to analyses of cadmium in biological materials were limited for the most part to flame and arc spectrographic techniques. Sheldon and Ramage[46] in 1931, using an oxyacetylene flame and a Hilger quartz spectrograph, were able to both detect and determine cadmium and other elements in small pieces of tissue and 100 μl aliquots of blood. Prior to spectrographic analyses, tissues were washed, dried to constant weight, and then ground into powder. Blood specimens were soaked into filter paper and dried to constant weight before burning in an oxyacetylene flame.

Tipton[47] employed the spectrograph to investigate the distribution of cadmium in various tissues.

X-ray emission spectroscopy, a highly useful tool for the direct non-destructive measurement of elements in biological materials, enjoyed but limited applications in the determination of cadmium and other elements, both anions and cations. Sample size requirements are small. Grebe and Esser[48] investigated cadmium content in tissues by this technique.

X-ray fluorescence spectrometry, in combination with different separation techniques, has been utilized in the determination of the cadmium content in simple matrices, like water and atmospheric dusts. Boiteau and Robin[49] extracted cadmium as a dithizonate from a liter of water at pH 11. Lochmueller *et al.*[50] bound cadmium by immersing a cationic exchange membrane in a liter water sample. Dzubay and Stevens[51] improved sensitivities of X-ray fluorescence in determining cadmium and other metals in atmospheric dust.

Classical flame emission spectrometry is not sufficiently sensitive for cadmium analysis in biological materials. The detection limit at the 228.8 nm resonance line is 4.2 μg ml^{-1}.[52] However, cadmium levels of 0.05 μg ml^{-1} in blood or urine are worthy of further investigation. Greater sensitivity can be attained with a plasma source, but the question of interferences remains.

The detection limit for the metal by flame atomic absorption spectrometry is stated to be 1 ng ml^{-1}. Probably the first application of the technique to the analysis of biological materials was reported by Pulido *et al.*[53] These investigators merely diluted unknown and control sera 1:10 with water and aspirated them into a long path flame atomizer.

Cadmium appears not to be subject to interferences from other metals; nevertheless, the constituents of biological materials enhance the absorption signal many fold.[54] For accurate analysis, cadmium must be isolated from its matrices. The sodium diethyldithiocarbamate (NDDC) chelate of the element is quantitatively extractable into methylisobutyl ketone (MIBK) at pH 5.5–6.5. Lead and thallium, which are co-extracted, do not interfere. Prior to extraction blood proteins are removed by trichloroacetic acid precipitation; tissues are wet ashed in nitric–sulfuric acids. Urine samples can be extracted directly.[55] Cadmium is determined in a lean oxidizing flame.

Delves and coworkers[56] determined eleven metals, including cadmium, by sequential solvent extraction of various chelates formed. All eleven analyses were performed with 1 ml of an oxidized blood sample.

Lieberman[57] extracted an ammonium pyrollidine dithiocarbamate–cadmium complex with methylisobutyl ketone. Utilizing the Delves' cup technique, he measured the element in 20 μl of extract. A sensitivity of 1 ng ml^{-1} and a coefficient of variation of 3% are claimed for the analysis.

The determination of cadmium in the graphite furnace following chelation with NDDC and extraction into MIBK is not feasible. There appears to be some sort of interaction of the complex with the graphite tube, for it has been observed that absorbance signals decrease stepwise with repeated injections of aliquots of the same extract. The phenomenon is illustrated in Fig. 9.1.

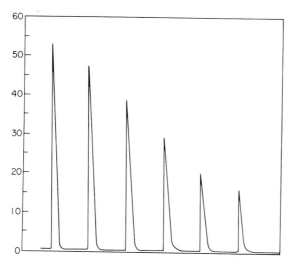

Fig. 9.1. Sequential atomizations of CdDDC (0.1 μg ml^{-1}) flameless atomic absorption

Cadmium in blood can be determined in the furnace, on the supernatant liquid from trichloroacetic acid, or nitric acid precipitation. Background correction is advisable. Plastic pipette tips used in the analysis must be precleaned with nitric acid. Plastic tips are contaminated, generously, with cadmium.

Ure and Mitchell[58] determined cadmium in plant material and soil extracts by the carbon rod technique after isolating the metal as a dithizonate in chloroform. Interference effects were found to be negligible.

Other modifications in addition to those discussed previously have been described.[59-62] The work of Lund et al.[63] is worthy of mention. These investigators determined cadmium in urine by a flameless technique in which the metal was electrolyzed onto a platinum wire and then atomized by electrical heating of the wire.

REFERENCES

1. L. Prodan, *J. Ind. Hyg.*, **14**, 132 (1932).
2. Sovet, *Presse Med. Belge*, **10**, 69 (1858) cited in L. Prodan, *J. Ind. Hyg.*, **14**, 132 (1932).
3. W. Marme, *Ztschr. f. rationel. Med.*, *S3*, **29**, 125 (1867) cited in L. Prodan, *J. Ind. Hyg.*, **14**, 132 (1932).
4. L. S. Goodman and A. Gelman, *The Pharmacological Basis of Therapeutics*, 2nd edn, Macmillan Co., New York, 1955.
5. C. H. Hill, G. Matrone, W. L. Payne and W. C. Barker, *J. Nutr.*, **80**, 227 (1963).
6. D. R. Van Campen, *J. Nutr.*, **88**, 125 (1966).
7. C. R. Bunn and G. Matrone, *J. Nutr.*, **90**, 395 (1966).
8. B. Axelsson and M. Piscator, *Arch. Environ. Health*, **12**, 374 (1966).
9. R. M. Jacobs, M. R. Spivey-Fox and M. H. Aldridge, *J. Nutr.*, **99**, 119 (1969).
10. G. W. Powell, W. J. Miller, J. D. Morton and C. M. Clifton, *J. Nutr.*, **84**, 205 (1964).
11. W. C. Supplee, *Science*, **139**, 119 (1963).
12. G. C. Cotzias and P. S. Papavasilou, *Amer. J. Physiol.*, **206**, 287 (1964).
13. D. H. Cox and D. L. Harris, *J. Nutr.*, **70**, 514 (1962).
14. S. A. Gunn, T. C. Gould and W. A. D. Anderson, *Arch. Pathol.*, **71**, 274 (1961).
15. P. Pulido, J. H. R. Kagi and B. L. Vallee, *Biochemistry*, **5**, 1768 (1966).
16. M. Margoshes and B. L. Vallee, *J. Amer. Chem. Soc.*, **79**, 4813 (1957).
17. L. Friberg, M. Piscator, G. F. Nordberg and T. Kjelstrom, *Cadmium in the Environment*, 2nd edn, CRC Press, Cleveland, 1974.
18. L. Friberg, *Acta Med. Scand.*, **138**, suppl., 240 (1950).
19. J. C. Smith, A. R. Wells and J. E. Kench, *Brit. J. Ind. Med.*, **18**, 70 (1961).
20. H. A. Schroeder, in *Metal Binding in Medicine*, (ed. M. J. Sevin) J. B. Lippincott Co., Philadelphia, Pa., 1960.
21. G. F. Lewis, W. J. Jusko and L. R. Coughlin, *J. Chronic Dis.*, **25**, 717 (1972).
22. G. Buell, *J. Occup. Med.*, **17**, 189 (1975).
23. T. Kjelstrom, in *Effects and Dose-Response Relationships of Toxic Metals*, (ed. G. F. Nordberg) Elsevier, Amsterdam, 1976.
24. R. R. Lauwerys, J. P. Buchet and H. Roels, *Int. Arch. Occup. Environ. Health*, **36**, 275 (1976).
25. D. G. Beevers, B. C. Campbell, A. Goldberg, M. R. Moore and V. M. Hawthorne, *Lancet*, **2**, 1222 (1976).

26. S. C. Glauser, C. T. Bells and E. M. Glauser, *Lancet*, **1**, 717 (1976).
27. D. T. Hammer, A. V. Calocci, V. Hasselbad, N. E. Williams and C. Pinkerton, *J. Occup. Med.*, **15**, 956 (1973).
28. J. M. McKenzie, *New Zealand Med. J.*, **79**, 1016 (1974).
29. K. Tsuchiya, Y. Seki and M. Sugita, *Keio J. Med.*, **25**, 83 (1976).
30. J. Cholak and D. M. Hubbard, *Ind. Eng. Chem.*, *(Anal. Ed.)*, **16**, 333 (1944).
31. H. Fischer and G. Leopoldi, *Mikrochim. Acta*, **1**, 30 (1937).
32. A. K. Klein and H. J. Wichmann, *J. Assoc. Offic. Agr. Chemists*, **28**, 257 (1945).
33. B. E. Saltzman, *Anal. Chem.*, **25**, 493 (1953).
34. G. K. Schweitzer and D. R. Randolph, *Anal. Chim. Acta*, **26**, 567 (1967).
35. G. Gorbach and F. Pohl, *Mikrochem.*, **38**, 258 (1951).
36. H. Bode, *Z. Analyt. Chem.*, **143**, 182 (1954).
37. F. Umland and W. Hoffmann, *Z. Analyt. Chem.*, **168**, 268 (1959).
38. C. E. White and R. J. Argauer, *Fluorescence Analysis—A Practical Approach*, Marcel Dekker, New York, 1970.
39. D. E. Ryan, A. E. Pitts and R. M. Cassidy, *Anal. Chim. Acta*, **34**, 491 (1966).
40. D. E. Ryan and B. K. Pal, *Anal. Chim. Acta*, **44**, 385 (1969).
41. E. R. Jensen and R. T. Pflaum, *Anal. Chem.*, **38**, 1268 (1966).
42. O. E. Knockaert, G. L. Maes and M. H. Fair, *Amer. Ind. Hyg. Assoc. J.*, **28**, 595 (1967).
43. J. P. Franke and R. A. de Zeeuw, *J. Anal. Toxicol.*, **1**, 291 (1977).
44. A. Chueca, M. Worwood and D. M. Taylor, *Int. J. Appl. Radiation Isotopes*, **20**, 335 (1969).
45. R. W. Lieberman and H. H. Kramer, *Anal. Chem.*, **42**, 266 (1970).
46. J. H. Sheldon and H. Ramage, *Biochem. J.*, **25**, 1608 (1931).
47. I. H. Tipton, in *Metal Binding in Medicine*, (ed. M. J. Seven) J. B. Lippincott Co., Philadelphia, Pa., 1960.
48. V. L. Grebe and F. Esser, *Fortschr. Gebiete Roentgenstrahlen*, **54**, 185 (1936).
49. H. L. Boiteau and M. Robin, *Coll. Rayons X Matiere*, Siemens, Monaco, May 1973, cited in M. Pinta, *Modern Methods for Trace Element Analysis*, Ann Arbor Science, Ann Arbor, Mich., 1978.
50. C. H. Lochmueller, J. W. Galbraith and R. L. Walter, *Anal. Chem.*, **46**, 440 (1974).
51. T. G. Dzubay and R. K. Stevens, *Environ. Sci. Tech.*, **9**, 663 (1975).
52. R. Mavrodineau (ed.), *Analytical Flame Spectroscopy*, Springer-Verlag, New York, 1970.
53. P. Pulido, K. Fuwa and B. L. Vallee, *Anal. Biochem.*, **14**, 393 (1966).
54. E. Berman, in *Applied Atomic Spectroscopy*, (ed. E. L. Grove) Plenum Press, New York, 1978.
55. E. Berman, *Atomic Abs. Newsletter*, **6**, 57 (1967).
56. H. T. Delves, G. Shepherd and P. Vinter, *Analyst*, **96**, 260 (1971).
57. K. W. Lieberman, *Clin. Chem. Acta*, **46**, 217 (1973).
58. A. M. Ure and M. C. Mitchell, *Anal. Chim. Acta*, **87**, 283 (1976).
59. R. T. Ross and J. G. Gonzalez, *Anal. Chim. Acta*, **70**, 443 (1974).
60. A. A. Cernik and M. H. P. Sayers, *Brit. J. Ind. Med.*, **32**, 155 (1975).
61. F. D. Pasma, J. Balke, R. F. M. Herber and E. J. Stuik, *Anal. Chem.*, **47**, 834 (1975).
62. H. T. Delves, *Analyst*, **102**, 403 (1977).
63. W. Lund, B. V. Larsen and N. Gunderson, *Anal. Chim. Acta*, **81**, 319 (1976).

CHAPTER 10

CHROMIUM

Until about twenty-five years ago, interest in chromium was confined to its toxic effects. However, the element is now considered essential. In 1954 Curran[1] demonstrated that the synthesis of cholesterol and fatty acids from acetate by rat liver was enchanced when chromium ions were present. In the rat experiments of Schroeder *et al.*[2-4] addition of chromium to a low-chromium diet suppressed serum cholesterol levels and inhibited the tendency of levels to increase with age.

In 1959 Schwarz and Mertz[5] observed impaired glucose tolerance in rats fed various diets. The cause was traced to a chromium deficiency. Chromium is now considered essential for maintaining normal glucose metabolism.[6] Mertz later hypothesized that Cr (III) acts as a cofactor with insulin at the cellular level through the formation of a ternary complex between membrane sites, insulin, and chromium.[7] Evidence has accumulated that chromium is involved in glucose tolerance in man.[8-11]

Rats fed diets deficient in protein and chromium exhibit impaired capacity to incorporate the amino acids, glycine, serine, methionine and α-aminobutyrate into the protein of their hearts.[12,13] The slight improvement in incorporation achieved with insulin alone was significantly enhanced by supplementation with trivalent chromium. Subsequently, Mertz and Roginski[14] demonstrated that raising rats in plastic cages on a low protein, low chromium diet resulted in moderate depression of growth that could be alleviated by chromium supplements.

Chromium occurs in a variety of foods in the form of a complex with nicotinic acid and possibly glycine, glutamic acid, and cysteine. Chromium in this complex, termed 'glucose tolerance factor', is absorbed better than in an inorganic form.[15] Animal protein is the best and most reliable source of chromium.

The chromium content in all tissues decreases with age. It has been postulated that this decline in chromium may be involved in the etiology of disease states with aging.[16]

The concentration of chromium in human blood is estimated to be 0.5–5 μg l^{-1} and in urine, 5–10 μg l^{-1}. The daily chromium intake by man ranges from 5–100 μg.[17]

Morgan[18] determined hepatic chromium contents in 92 subjects suffering from different diseases. The mean chromium content in livers of diabetics was 8.59 μg g^{-1} of ash as compared to 10.2 μg g^{-1} for hypertensives, 9.96 μg g^{-1} for arteriosclerotics, and 12.7 μg g^{-1} for controls.

Hambridge et al.[19] compared chromium levels in the hair of 33 normal children and of 19 juvenile diabetics. Normal children showed a mean chromium concentration of 0.85 μg g^{-1} (range of 0.36–1.87 μg g^{-1}). The mean chromium content in hair from diabetics was 0.56 μg g^{-1} (range 0.26–1.19 μg g^{-1}).

Martin et al.[20] demonstrated the development of corneal lesions in squirrel monkeys maintained on a low chromium diet. The role of chromium in the development of these lesions is unclear.

Jiejeebhoy and co-workers[21] described the development of weight loss and peripheral neuropathy in a forty year old female on parenteral nutrition for more than five years after a complete enterectomy. The condition, which developed after three and a half years of the regimen, was corrected by chromium supplementation. The subject's blood chromium level was 0.55 ng ml^{-1} as compared to 4.9–9.5 ng ml^{-1} in controls. The chromium content of her hair was 160 ± 15 ng g^{-1}. Normally, hair contains more than 500 ng g^{-1} of chromium. Incidentally, her hair zinc, copper, and manganese contents resembled those of the controls.

TOXICITY OF CHROMIUM

The oxidizing agent chromic acid and the dichromate salts are irritating to mucous membranes, skin, and conjunctiva. Allergies and dermatitis are induced through exposure to these compounds. Chrome holes, i.e. penetrating ulcers which occur around the fingernails, the surface of exposed fingerjoints, eyelids, and occasionally forearms, are not painful and do little if any permanent damage. Lesions on the nasal mucosa are more troublesome. In some instances the septum is perforated, and breathing through the nose makes a whistling sound. Some nasal lesions are painful, and nasal breathing is difficult.

A higher rate of bronchitis among chromate workers has been reported.[22] 'Chronic chromate lung' has also been described.[23]

Alkali bichromates can be absorbed through skin lesions in sufficient quantity to cause renal damage.[24]

For many years it was not realized that chromium compounds could be carcinogenic. However, certain epidemiological studies seem to have established a significant increase in pulmonary malignancy in workers exposed to chromium and all its compounds.[25] Hueper[26] and Payne[27] have induced carcinomas in animals with chromite.

Hypersensitivity to chromium is not restricted to industrial workers. Tazelaar[28] reported a case of chromium dermatitis in a man with a red and blue tatoo. Large amounts of chromium were detected spectroscopically in the blue areas. Also the patch test with potassium dichromate solution elicited a strongly positive reaction. It was necessary to excise the tatoo in order to relieve the chromium eczema of the hands. Chromium present in detergents has been implicated as the cause of dermatitis in housewives in Europe and Israel.[29]

USES OF CHROMIUM AND ITS COMPOUNDS

Chromium itself is used in the manufacture of stainless steel. The trioxide finds use in chromium plating, copper stripping, as a corrosion inhibitor, and in photography.

Chromium sulfate is used as a mordant in textile manufacture, in tanning leather, in manufacturing green varnishes, paints, inks, and glazes for porcelain.

Chromic potassium sulfate and oxalate are used in tanning leathers and in dyeing fabrics. Chromic oxide is used in coloring glass, among others.

Chromic acetate is used in dyeing and tanning.

ANALYSIS OF CHROMIUM

Colorimetric analysis

Colorimetric analyses of chromium as a chromate or complexed with diphenylcarbazide are fairly sensitive. The latter is said to be ten times as sensitive as the former. Detection limits of 12 ng ml^{-1} and 1.2 ng ml^{-1}, respectively have been claimed for the two methods.[30] Both procedures suffer from interferences of other metals. The chromate method is impractical for analyses of biological materials.

Since Moulin[31] first used diphenylcarbazide in 1904, many procedures applying this complexing agent in the analysis of biological materials for chromium have been described.[32-35] In general a solution of the reagent is added to a digest of the sample in sulfuric acid (about 0.2N). Diphenylcarbazide reacts with chromium (VI). The absorbance of the red-violet chromium complex is measured at 540 nm. Methods differ primarily in the means of sample preparation and the oxidation of chromium in solution. Iron is a major interference in the analysis of biological materials. Vanadium, while an interference in analysis of ores, steels, etc., is of minor if any consequence where biological materials are concerned.

Blair and Pantony[36] described a chromatographic–spectrophotometric procedure. Chromium was precipitated by adding excess 8-hydroxyquinaldine. The precipitate was collected, dried, and dissolved in chloroform–benzene, and the solution was then run through an activated alumina column. Chromium was eluted with the chloroform–benzene, but the other metals co-precipitated (iron, aluminum, nickel, copper, molybdenum, manganese, cobalt, zinc, vanadium, thallium, etc.) remained on the alumina. The chromium-8-hydroxyquinaldine complex in the solvent mixture is measured spectrophotometrically.

Chromium (III) is extractable as an oxinate,[37] a methyl oxinate,[38] and as a thenoyltrifluoroacetonate.[39]

Chromium (III) forms a quantitatively extractable chelate with acetyl acetone at pH 6, but only after boiling in the reagent.[40] At room temperature, chromium III is not extracted at any pH by acetyl acetonate or solutions of the chelate in other solvents; therefore, other metals which form chelates with acetyl acetone at room temperature can be separated from chromium in this way.

Miscellaneous instrumental methods

Polarography has had a few applications in the analysis of chromium in biological materials.[41,42] Crosmum and Mueller[43] utilized differential pulse polarography in attaining detection limits of 10 ng ml^{-1} of chromium (VI) in natural waters.

Gutierrez et al.[44] proposed the application of electron spin resonance (ESR) to the measurement of chromium in tissues.

Gas chromatography

Among the first methods described for the gas chromatographic analysis of chromium in biological materials is that of Savory et al.[45] Urine and sera were wet ashed in nitric, perchloric and sulfuric acids. Any chromium present was converted to a trifluoroacetate, extracted into benzene, and determined by electron-capture gas chromatography. In a subsequent modification of this method the chromium was co-precipitated with serum proteins in ethanol at 4 °C, isolated, dried and reacted with trifluoroacetyl acetone.[46] Recoveries of added chromium were 75%. Apparent normal serum values obtained by the procedure ranged between 2.7 and 24 μg l^{-1}.

Hansen et al.[47] did not consider preconcentration of samples to be necessary since the electron capture detector responded so well to picogram amounts of chromium. 50 μl aliquots were reacted with a hexane solution containing 1,1,1-trifluoro-2,3-pentanedione. Chromium was detected as a trifluoroacetyl acetone. Mean recoveries of 91% were attained.

Booth and Darby[48] detected 20 ng g^{-1} of the metal in liver after wet ashing and chelating with trifluoroacetyl acetone. They claim 95% recoveries of added chromium.

Activation analysis

Activation analysis has been utilized in the determination of chromium in various materials. In earlier applications, sea water, for example, was irradiated, evaporated, and the element determined.[49,50] Chromium in waters has also been determined directly.[51] Neutron activation analysis has been employed in analyses of serum and other biological materials[52] and in determining the chromium content of foods.[53]

Bankert et al.[54] used proton activation to determine chromium and other metals in waters directly. Delmas and co-workers[55] applied both neutron and proton activation techniques in non-destructive analyses of rocks and silica minerals.

Spectroscopic methods

Different types of spectroscopic instrumentation, in combination with varying degrees of chemistry, have been applied successfully to chromium analyses.

Beyermann et al.[56] was able to measure chromium in urine at levels above 5 ng by X-ray fluorescence spectroscopy. Prior to analysis, chromium was extracted as an oxinate by chloroform. Other metallic elements co-extracted were removed by re-extracting the organic solvent with 3 M hydrochloric acid.

Hambridge[57] attained an absolute sensitivity of 1 ng in analyses of serum, urine, and hair with a DC arc spectrograph in a static argon atmosphere. Organic material was removed by low temperature ashing. It was noted that sodium had a suppressive effect, but phosphate and sulfate were not detrimental to chromium sensitivity. Analyses were performed at the 425.4 nm resonance line.

By flame emission the metal can be detected at levels of 0.1 μg ml^{-1}, and by atomic absorption, at levels of 5 ng ml^{-1}.

Williams and co-workers[58] first described the determination of chromium by atomic absorption.

In 1966 Pierce and Cholak[59] described a procedure for the determination of chromium, lead and molybdenum in blood by flame atomic absorption spectrometry. Samples were ashed, then chelated with ammonium pyrollidine dithiocarbamate (APDC) and extracted into methylisobutyl ketone.

Cary and Allaway[60] were able to estimate 4 ng ml^{-1} of chromium by a method which combines wet digestion of biological materials (plants, etc.), and extraction of the chromium (III) chelate of 2,4-pentanedione prior to analysis. Wet ashing was preferred to dry ashing because chromium can volatilize under some dry ashing conditions. In addition, the fusion or binding of chromium to glassware was noted in some cases.

The authors investigated 8-hydroxyquinoline and APDC as chelating agents for chromium. Ammonium pyrollidine dithiocarbamate was found to be an inferior reagent. 8-hydroxyquinoline proved as effective as 2,4-pentanedione, but pH was more critical with the former.

Davis and Grosman[61] preferred 10% tributylphosphate in methylisobutyl ketone as a chelating agent.

With the advent of flameless atomization devices, numerous papers concerning the determination of chromium in biological materials resulted. For the most part, the direct analysis techniques described are of little merit, as little thought seems to have been given to matrix effects and interferences. Signals generated by constituents in plasma, urine or tissue digestants often occur close to, or are superimposed upon, the chromium signal. Deuterium background correction is only partially effective in eliminating this non-specific molecular absorption.

The work of Davidson and Secrest[62] represents a fortuitous combination of good chemistry and spectroscopy. They were able to eliminate many interferences by adapting a double atomization procedure, wherein after an initial drying and ashing cycles the sample was atomized at 1400 °C, then subjected to additional drying and ashing cycles and finally atomized at 2400 °C. The initial atomization temperature is not sufficient to volatilize chromium or its salts.

The authors, after comparing direct analysis with wet ashing procedures, concluded that plasma samples could not be directly analyzed reliably except by the method of standard additions. Also, because of severe matrix interferences generated by urine, the use of standard additions with that type of sample was precluded.

A greater precision error was noted with direct analysis, due to a larger pipetting error as well as to uneven drying of the sample and splattering in the graphite tube. The residue which builds up in the graphite tube after several firings also entraps some chromium.

The authors recommend wet ashing of the specimen followed by the double atomization procedure described. Many chemical interferences due to matrix composition are then eliminated; analysis time in the furnace is markedly decreased, and analyses become more precise and accurate.

REFERENCES

1. G. L. Curran, *J. Biol. Chem.*, **210**, 765 (1954).
2. H. A. Schroeder, W. H. Vinton, Jr. and J. J. Balassa, *Proc. Soc. Exptl. Biol. Med.*, **109**, 859 (1962).
3. H. A. Schroeder and J. J. Balassa, *Amer. J. Physiol.*, **209**, 433 (1965).
4. H. A. Schroeder, *J. Nutr.*, **97**, 237 (1969).
5. K. Schwarz and W. M. Mertz, *Arch. Biochem. Biophys.*, **72**, 515 (1957).
6. K. Schwarz and W. M. Mertz, *Arch. Biochem. Biophys.* **85**, 292 (1959).
7. W. M. Mertz, *Fed. Proc., Fed. Amer. Soc., Exptl. Biol.*, **26**, 186 (1967).

8. W. H. Glinsman and W. M. Mertz, *Metab. Clin. Exp.*, **15**, 510 (1966).
9. R. A. Levine, D. H. Stretten and R. J. Doisy, *Metab. Clin. Exp.*, **17**, 114 (1968).
10. H. A. Braig and J. C. Edozien, *Lancet*, **2**, 662 (1965).
11. L. L. Hopkins, Jr., O. Ransome-Kuti and A. S. Majaj, *Amer. J. Clin. Nutr.*, **21**, 203 (1968).
12. E. E. Roginski and W. M. Mertz, *Fed. Proc.*, *Fed. Amer. Soc.*, *Exptl. Biol.*, **26**, 301 (1967).
13. E. E. Roginski and W. M. Mertz, *J. Nutr.*, **97**, 525 (1969).
14. W. M. Mertz and E. E. Roginski, *J. Nutr.*, **97**, 531 (1969).
15. W. M. Mertz, E. W. Toepfer, E. E. Roginski and M. M. Polansky, *Fed. Proc.*, **33**, 2275 (1974).
16. O. A. Levander, *J. Amer. Diet. Assoc.*, **66**, 338 (1975).
17. W. M. Mertz, *Proc. Nutr. Soc.*, **33**, 307 (1974).
18. J. M. Morgan, *Metabolism*, **21**, 313 (1972).
19. K. M. Hambridge, D. O. Rodgerson and D. O'Brien, *Diabetes*, **17**, 517 (1968).
20. G. D. Martin, J. A. Stanley and I. W. F. Davidson, *Invest. Ophthalmol.*, **11**, 153 (1972).
21. K. N. Jiejeebhoy, R. C. Chu, E. B. Marliss, G. R. Greenberg and A. Bruce-Robertson, *Amer. J. Clin. Nutr.*, **30**, 531 (1977).
22. T. F. Mancuso, *Ind. Med. Surg.*, **20**, 358 (1951).
23. G. K. Sluis-Cremer and R. J. DuToit, *Brit. J. Ind. Med.*, **25**, 63 (1968).
24. R. H. Major, *Johns Hopkins Hosp. Bull.*, **33**, 56 (1922).
25. H. L. Hardy, in *Industrial Toxicology*, (eds. A. Hamilton and H. L. Hardy) Publishing Sciences Group, Inc., Acton. Pa., 1974.
26. W. C. Hueper, *Arch. Ind. Health*, **18**, 284 (1955).
27. W. W. Payne, *Arch. Environ. Health*, **1**, 20 (1960).
28. D. J. Tazelaar, *Dermatologica*, **141**, 282 (1970).
29. Editorial, *Food Cosmetic Toxicol.*, **11**, 506 (1973).
30. E. B. Sandell, *Colorimetric Determination of Traces of Metals*, Interscience Publishers, Inc., New York, 1959.
31. A. Moulin, *Bull. Soc. Chim. (France)*, **31**, 295 (1904).
32. P. F. Urone and H. K. Anders. *Anal. Chem.*, **22**, 1317 (1950).
33. H. J. Cahnmann and R. Bises, *Anal. Chem.*, **24**, 1341 (1952).
34. B. E. Saltzman, *Anal. Chem.*, **24**, 1016 (1952).
35. G. Grogan , H. J. Cahnmann and E. Lethco, *Anal. Chem.*, **27**, 983 (1955).
36. A. J. Blair and D. A. Pantony, *Anal. Chim. Acta*, **14**, 545 (1956).
37. T. Moeller, *Ind. Eng. Chem. (Anal. Ed.)*, **15**, 346 (1943).
38. K. Motojima and H. Hashitani, *Anal. Chem.*, **33**, 239 (1961).
39. A. K. Majumidar and A. K. De, *Anal. Chem.*, **32**, 1337 (1960).
40. J. P. McKaveney and H. Freiser, *Anal. Chem.*, **30**, 1965 (1958).
41. M. Brezina and R. Zuman, *Polarography in Medicine, Biochemistry and Pharmacy*, Interscience Publishers, New York, 1958.
42. W. C. Purdy, *Ann. N.Y. Acad. Sci.*, **137**, 390 (1966).
43. S. I. Crosmum and T. R. Mueller, *Anal. Chim. Acta*, **75**, 199 (1975).
44. P. L. Gutierrez, I. Sarna and H. M. Swartz, *Phys. Med. Biol.*, **21**, 949 (1976).
45. J. Savory, P. Mushak, F. W. Sunderman, Jr., R. H. Ester and N. O. Roszel, *Anal. Chem.*, **42**, 294 (1970).
46. J. Savory, M. I. Glenn and J. A. Ahlstrom, *J. Chromatog. Sci.*, **10**, 247 (1972).
47. L. C. Hansen, W. G. Scribner, T. W. Gilbert and R. E. Sievers, *Anal. Chem.*, **43**, 349 (1971).
48. G. H. Booth, Jr. and W. J. Darby, *Anal. Chem.*, **43**, 831 (1971).
49. D. E. Robertson, *Anal. Chem.*, **40**, 1067 (1968).

50. D. Z. Piper and G. G. Goles, *Anal. Chim. Acta*, **47**, 560 (1969).
51. G. F. Clemente and G. G. Mastinu, *J. Radionucl. Chem.*, **20**, 707 (1974).
52. R. Cornelius, A. Speecke and J. Hoste, *Anal. Chim. Acta*, **68**, 1 (1973).
53. R. R. Becker, A. Veglia and E. R. Schmid, *Radiochem. Radioanal. Letter*, **19**, 343 (1974).
54. S. F. Bankert, S. D. Bloom and G. D. Sauter, *Anal. Chem.*, **45**, 692 (1973).
55. R. Delmas, J. N. Barrandon, J. L. Debrun, *Analusis*, **4**, 339 (1976).
56. K. Beyermann, H. J. Rose and R. P. Christian, *Anal. Chim. Acta*, **45**, 51 (1969).
57. K. M. Hambridge, *Anal. Chem.*, **43**, 103 (1971).
58. C. H. Williams, D. J. David and O. Iismaa, *J. Agr. Sci.*, **59**, 381 (1962).
59. J. O. Pierce and J. Cholak, *Arch. Environ. Health*, **13**, 208 (1966).
60. E. E. Cary and W. H. Allaway, *J. Agr. Food Chem.*, **19**, 1159 (1971).
61. M. H. Davis and V. B. Grosman, *Anal. Biochem.*, **44**, 339 (1971).
62. I. W. F. Davidson and W. L. Secrest, *Anal. Chem.*, **44**, 1808 (1972).

CHAPTER 11

COBALT

Cobalt occurs in a variety of ores, among them an arsenic–cobalt mineral, quite plentiful in Saxony. Early German miners exposed to the corrosive action of its dusts thought the toxic effects exerted to be the work of goblins. Hence the German 'kobelt', a term for gnomes and goblins became associated with the mineral.

In *De Re Metallica*, Agricola[1] describes measures to prevent ill effects of exposure to cobaltum: 'Further there is a certain kind of *cadmia* which eats away the feet of the workmen when they become wet and similarly their hands and injures their lungs and eyes. Therefore for their digging they should make for themselves not only boots of rawhide, but gloves, long enough to reach to the elbow and should fasten loose veils over their faces; the dust will then neither be drawn through these into their windpipes and lungs, nor will it fly into their eyes. Not dissimilarly, among the Romans the makers of vermillion took precautions against breathing its fatal dust.'

Cobalt, the element discovered in 1735, is a gray, hard, magnetic, ductile, somewhat malleable metal. Salts of cobalt are highly colored and are used as pigments. The chloride is blue; cobaltous arsenite, pink to blood red; the bromide, bright green; cobalt carbonate, pale red; the chromate, brilliant green; cobalt hydroxide, blue green; the nitrate, red; the oxalate, pink; cobalt oxide, olive green to red; the phosphate, pink to lavender; the sulfate, pink to red; and the thiocyanate, yellow brown. Cobalt blue, or *zaffre*, was known and used prior to 1540.

USE OF COBALT AND ITS SALTS

The metal cobalt itself is used in the manufacture of alloys and in nuclear technology. Cobalt-60 is used in medicine as a source of radiation in treating malignancies. Salts of cobalt have applications other than their use in pigments. Radioactive cobalt is manufactured from the chloride salt. The acetate, carbonate, and chloride were used as foam stabilizers in the brewing of malt beverages. Cobalt carbonate and chloride are utilized in trace element supplement preparations for ruminants. The oxide is used in grinding wheels.

TOXICITY OF COBALT

Inhalation of cobalt-containing dusts may cause pulmonary symptoms. Skin contact with powders or effluents in various industrial processes can cause dermatitis.[2] Ingestion of soluble cobalt salts induces nausea and vomiting because of local irritation of the gastric mucosa.

Barborik and Dusik[3] report the occurrence of fatal cardiomyopathy following industrial exposure to cobalt. Cobalt contents found in different organs of the case in question and in tissues of control subjects (a uremic and a cor pulmonale) are listed in Table 11.1 below. Differences between the cardiomyopathy and the control are five- to ten-fold.

TABLE 11.1
Tissue cobalt (μg% wet weight)

Organ	Cobalt cardiomyopathy	Uremic	Cor pulmonale
Heart	37	5	2
Lung	17	7	3
Liver	34	3	3
Spleen	19	4	1
Kidney	29	3	1

Tissue levels reported in the literature, for example Tipton's[4] spectrographic studies, show higher values. However, it must be born in mind that Tipton's subjects were chosen at random. Nothing definitive was known concerning the occupational history, habits, etc. of subjects. Furthermore, analytical technology has changed markedly over the last two decades. Tipton utilized the spectrograph in her analyses whereas the other workers employed atomic absorption spectrometry and neutron activation analysis.

Non-industrial toxicities

From August 1965 to April 1966, upwards of fifty patients with a sudden onset of cardiomyopathy were seen in Quebec City, Canada. The cause was traced to some beers being consumed which contained ten times more cobalt than was usual.[5] Underwood[6] suggested that the myopathy occurred because of an accompanying dietary protein and thiamine deficiency.

Sullivan et al.[7] determined cobalt, zinc, magnesium, and manganese contents of the heart tissues of patients dying from 'beer drinkers' myocardiopathy, by both neutron activation analysis and atomic absorption spectroscopy. The mean cobalt content of heart muscle was found to be 0.69 μg g^{-1} wet weight as compared to 0.04 μg g^{-1} for control cardiac tissue. Manganese, magnesium, and zinc in the former were all decreased compared to levels found in controls. The relationship between these changes is as yet uncertain.

Mechanisms of myocardial toxicity induced by cobalt salts have been investigated in rats.[8] Mohiuddin et al.[9] induced lesions in the pericardium, myocardium, and endocardium of guinea pigs by feeding 20 mg of cobalt per kg body weight/day. The lesions observed were strikingly similar to those seen in Quebec beer drinkers' cardiomyopathy. Addition of ethanol to the cobalt salt regimen failed to modify the incidence or severity of disease in the animals.

Rona[10] after investigating some experimental aspects of cobalt cardiomyopathy concluded that the damage observed possibly reflects an enzymatic block of oxidative decarboxylation at the pyruvate and ketoglutarate levels.

The occurrence of cobalt intoxication in patients with uremic myocardiopathy has been described.[11] Curtis et al.,[12] using neutron activation analysis, studied blood and tissue cobalt concentrations in normal subjects and in patients with terminal renal failure. Patients on maintenance dialysis given cobalt chloride in an attempt to treat the existing anemia showed a rise in blood cobalt from 3.1 to 138 μg l^{-1}. However, the maximum rise observed in hemoglobin levels was only 1 μg%. The myocardial cobalt content of renal failure patients treated with the salt rose to 165 μg%. Cobalt levels in the myocardial tissue of renal failure patients (non-cobalt treated) and in the tissue of cobalt-treated subjects without renal failure were 5 and 6 μg% respectively. The authors question the use of cobalt chloride in the management of the anemia of patients with renal failure.

Incidentally, 'normal values' reported by Curtis and co-workers[12] are within the range of those reported earlier by Barborik and Dusik.[3]

Cobalt has been reported to cause a hyperlipemia in rabbits.[13] A similar effect in rats was observed by Shabaan et al.[14] while investigating the role of the salt in the induction of fibrosarcomas. The effect of cobalt chloride on blood lipids in man has not yet been investigated.

ESSENTIAL ROLE OF COBALT

The first evidence that cobalt was a dietary essential was obtained more than 50 years ago as the result of researches regarding the cause of two debilitating diseases of sheep and cattle known as 'coast disease' and 'wasting disease', or 'bush sickness'.[15,16] Underwood and Filmer[17] demonstrated that a deficiency of cobalt was the cause. Normal growth in sheep and cattle could be secured by 0.1 mg of cobalt per day. Later it was shown that cobalt deficiencies in lambs could be cured by vitamin B$_{12}$.[18]

In 1948, it was discovered that the antipernicious anemia factor in liver is a compound containing 4% of cobalt.[19,20] The compound is now known as vitamin B_{12}.

ANALYSIS OF COBALT

Colorimetry

The more sensitive colorimetric methods for cobalt analysis are based on reagents containing a nitroso group.[21] Sensitivities reported for nitroso R salt; 2-nitroso-1-napthol-4-sulfonic acid; 2-nitroso-1-napthol; and o-nitroso-resorcinol are in the range of 2 to 4 ng ml^{-1}. Iron and copper are major interferences as regards biological materials. Copper is removed from cobalt by extraction with dithizone from a fairly acidic medium. Extraction of cobalt dithizonate from a basic citrate solution separates the cobalt from iron.

Middleton and Stuckey[22] chromatographed digestants on alumina prior to reacting the cobalt with 1-nitroso-2-naphthol.

Cobalt forms other extractable chelates. Cobalt (III) is quantitatively extractable as an acetyl acetonate in the pH range 0.3–2.0.[23] Cobalt (III) is quantitatively extractable as a diethyldithiocarbamate at pH 4–11.[24] Cobalt (II) forms quantitatively extractable chelates with cupferron,[25] dithizone,[26] and furoyltrifluoroacetone.[27]

About one third of the methods proposed for the colorimetric analysis of cobalt in biological materials employ 1-nitroso-2-naphthol-3,6-disulphonic acid. Van Klooster[28] first suggested using the reagent for detecting small quantities of cobalt in 1921. The latest modification of his method appeared in 1971.[29]

Fluorimetry

Limited applications of fluorimetry in the determination of cobalt in solution have been reported. Benzamido(p-dimethyl benzilidene)acetic acid forms a luminescence compound with cobalt at an apparent sensitivity of 0.2 ng ml^{-1}.[30]

The determination of the concentration of cobalt (II) at the 0.4 ng ml^{-1} level by the chemiluminescent oxidation of gallic acid has been reported recently.[31] Practical methods for the analysis of biological materials by fluorimetry have not appeared.

Neutron activation analysis

Activation analysis is a good, sensitive technique for the analysis of cobalt. Procedures for use with various matrices have been described. For example, seawater has been irradiated, evaporated, and the cobalt determined by spectrometry.[32] The element has also been measured directly in water.[33] Cobalt has been determined in extracts of biological materials[34] and in ash[35] and serum[36] by non-destructive analysis.

X-Ray fluorescence spectrometry

Chelation extraction procedures have been combined with X-ray fluorescence spectrometry in the analysis of cobalt in various materials. Marcie[37] utilized chelation with ammonium pyrollidine dithiocarbamate and chloroform extraction. Armitage and Zeitlin[38] isolated cobalt as an oxinate (8-hydroxyquinolate) prior to analysis.

Emission and atomic absorption spectrometry

The detection limit of cobalt by spark spectrometry at the 345.4 nm line is 2 μg ml^{-1}, while limits by flame emission and plasma spectrometry are 30 and 3 ng ml^{-1}, respectively.[39] By flame atomic absorption the detection limit using the 240.7 nm line are 20 ng ml^{-1}.[40] Analyses of the cobalt content in biological materials will require a concentration step prior to instrumental analysis. Extraction of cobalt as a dithiocarbamate complex into methylisobutyl ketone is a common method.

Cobalt has been estimated in a wide variety of materials by flame atomic absorption. Included are soils,[41] sediments,[42] waters,[43,44] plants,[45,46] and human tissues.[7]

Jago et al.[47] determined cobalt in animal and plant tissue at the 0.05–1 μg ml^{-1} range by flame atomic absorption following extraction of the cobalt-1-nitroso-2-napthol complex into ethylmethyl ketone.

A few applications of flameless atomization techniques to cobalt analyses have been reported. The element has been estimated in blood directly on a tantalum strip,[48] or using the mini-Massmann carbon rod atomizer.[49] Cobalt in animal food was estimated in the graphite tube[50] following ashing and extraction as a 1-nitroso-2-naphthol complex.

REFERENCES

1. G. Agricola, *De Re Metallica*, trans. by H. C. Hoover and L. H. Hoover from the first Latin edition of 1556, Dover Publications Inc., New York, 1950.
2. S. Fregert, B. Grewberger and A. Heijer, *Acta Derm. Venerol. (Stockh.)*, **52**, 221 (1972).
3. M. Barborik and J. Dusik, *Brit. Heart J.*, **34**, 113 (1972).
4. I. H. Tipton, in *Metal Binding in Medicine*, (ed. M. J. Seven) J. B. Lippincott, Philadelphia, Pa., 1960.
5. Y. Morin and P. Daniel, *Canad. Med. Assoc. J.*, **97**, 976 (1967).
6. E. J. Underwood, *Nutri. Rev.*, **33**, 65 (1975).
7. J. Sullivan, M. Parker and S. B. Carson, *J. Lab. Clin. Med.*, **71**, 893 (1968).
8. H. C. Grice, T. Goodman, I. C. Munro, G. S. Wiberg and A. B. Morrison, *Ann. N.Y. Acad. Sci.*, **156**, 189 (1969).

9. S. M. Mohiuddin, P. K. Taskar, M. Rheault, P. E. Roy, J. Chenard and Y. Morin, *Amer. Heart J.*, **80**, 532 (1970).
10. G. Rona, *Brit. Heart J. Suppl.* **33**, 171 (1971).
11. I. E. Lins and K. Pehrsson, *Lancet*, **1**, 1191 (1976).
12. J. R. Curtis, G. C. Goode, J. Herrington and L. E. Urdaneta, *Clin. Nephrol.*, **5**, 61 (1976).
13. R. M. Caplan and W. D. Block, *J. Invert. Derm.*, **40**, 199 (1963).
14. A. A. Shabaan, V. Marks, M. C. Lancaster and G. N. Dufen, *Lab. Animals*, **11**, 43 (1977).
15. H. R. Marston, *J. Council Sci. Ind. Res. (Austral.)*, **8**, 111 (1935).
16. E. W. Lines, *J. Council Sci. Ind. Res. (Austral.)*, **8**, 117 (1935).
17. E. J. Underwood and J. F. Filmer, *Austral. Vet. J.*, **11**, 84 (1935).
18. S. E. Smith, B. A. Koch and K. L. Turk, *J. Nutr.*, **44**, 455 (1951).
19. E. L. Rickes, N. G. Brink, F. R. Koniusky, T. R. Wood and K. Folkers, *Science*, **108**, 134 (1948).
20. E. L. Smith, *Nature (London)*, **162**, 144 (1948).
21. E. B. Sandell, *Colorimetric Determination of Traces of Metals*, Interscience Publishers, Inc., New York, 1959.
22. G. Middleton and R. E. Stuckey, *Clin. Chim. Acta*, **1**, 135 (1956).
23. J. P. McKaveney and H. Freiser, *Anal. Chem.*, **29**, 290 (1957).
24. F. A. Pohl and H. Demmel, *Anal. Chim. Acta*, **10**, 554 (1954).
25. J. Stary and J. Smizanska, *Anal. Chim. Acta*, **29**, 546 (1963).
26. E. Shulik and J. Laszlovsky, *Mikrochim. Acta*, **41** (1961).
27. R. T. McIntyre, E. W. Berg and D. N. Campbell, *Anal. Chem.*, **28**, 1316 (1956).
28. H. S. Van Klooster, *J. Amer. Chem. Soc.*, **43**, 746 (1921).
29. D. W. Dewey and H. R. Marston, *Anal. Chim. Acta*, **57**, 45 (1971).
30. J. B. Allred and D. G. Guy, *Anal. Biochem.*, **29**, 293 (1969).
31. S. Stieg and T. A. Niemann, *Anal. Chem.*, **49**, 1322 (1977).
32. D. E. Robertson, *Anal. Chem.*, **40**, 1067 (1968).
33. G. F. Clemente and G. G. Mastinu, *J. Radioanal. Chem.*, **20**, 707 (1974).
34. T. E. Henzler, R. J. Korda, P. A. Helenke, M. R. Anderson, M. M. Jimenez and L. A. Haskin, *J. Radioanal. Chem.*, **20**, 649 (1974).
35. B. Maziere, J. Gros and D. Comar, *J. Radioanal. Chem.*, **24**, 279 (1975).
36. R. Cornelis, A. Speecke and J. Hoste, *Anal. Chim. Acta*, **68**, 1 (1973).
37. F. J. Marcie, *Norelco Rep. (USA)* **15**, 3 (1968).
38. B. Armitage and H. Zeitlin, *Anal. Chim. Acta*, **53**, 47 (1971).
39. M. Pinta, *Modern Methods for Trace Element Analysis*, Ann Arbor Science Publishers, Ann Arbor, Mich., 1978.
40. R. Mavrodineanu (ed.), *Analytical Flame Spectroscopy*, Springer-Verlag, New York, 1970.
41. A. M. Ure and R. L. Mitchell, *Spectro chim. Acta*, **23B**, 79 (1967).
42. D. C. Burrell, *Atom. Absorpt. Newsletter*, **4**, 328 (1965).
43. S. L. Sachder, J. W. Robinson and P. W. West, *Anal. Chim. Acta*, **38**, 499 (1967).
44. J. Nix and T. Goodman, *Atom. Absorpt. Newsletter*, **9**, 119 (1970).
45. A. L. Gelman, *J. Sci. Food Agr.*, **23**, 299 (1972).
46. W. J. Simmons, *Anal. Chem.*, **45**, 1947 (1973).
47. J. Jago, P. E. Wilson and B. N. Lee, *Analyst*, **96**, 349 (1971).
48. P. A. Ullucci and J. Y. Hwang, 19th Spectroscopy Symposium, Montreal, Canada, 1972.
49. F. J. M. J. Maessen, F. D. Posma and J. Balke, *Anal. Chem.*, **46**, 1445 (1974).
50. L. R. Hagerman, L. Torma and B. E. Gunther, 88th Annual Meeting of the Association of Official Analytical Chemists, Washington, D.C., 1974.

CHAPTER 12

COPPER

Copper, one of the earliest known metals, is lustrous, ductile, malleable, and a good conductor of heat. It may have been the first metal worked by ancient man 8000 years ago. Artifacts hammered from copper that date back to 6000 BC have been found in the Near and Middle East. Tubal-Cain, 'who forged implements of copper and iron',[1] was but seven generations removed from Adam and is the first metallurgist of record in western civilization. Melting and casting of copper became common in the Near East after 4000 BC. Bronze, an alloy of copper and tin, was invented about 2500 BC. Brass, an alloy containing copper and zinc primarily, was not developed until the Roman era.

Copper is an essential element. It is present in all organisms, land and marine. The early literature is rich with accounts of investigations concerning the presence of copper in different life forms. Some of the earliest observations were made upon marine organisms.[2,3] By 1847, copper was shown to be combined with the blood proteins of snails.[4]

The essential role of copper became established when McHargue[5,6] observed its value in diets for rats, and Hart et al.[7] demonstrated that copper, in addition to iron, was necessary for blood formation. Soon reports began to appear from various regions that certain disorders of grazing sheep and cattle were due to deficiencies of copper and had responded to copper supplements.[8-11]

Copper containing enzymes include tyrosinase, ascorbic acid oxidase, cytochrome oxidase, monamine oxidase, uricase, and δ-aminolevulinic acid dehydratase. Several of the manifestations of copper deficiency in animals appear to be related to decreased tissue concentrations of certain of these enzymes.

Various aspects of copper metabolism and transport in man and animals have been studied in depth in both health and in disease states. Copper occurs in all bodily tissues. The distribution varies with age, species and diet. Liver, heart, brain, kidney and hair contain high concentrations compared to other tissues. Glandular tissues, such as, pituitary, thyroid and prostrate, contain low levels. Tissues containing intermediate concentrations include bones, skin, muscles, pancreas and spleen.[12,13] Brain copper increases with age while levels in the liver, spleen and lung decrease.[14]

Abnormally high copper levels in liver are characteristic of certain disease states in man. Wilson's disease, or hepatolenticular degeneration comes to mind first of all. Other conditions which manifest increased copper content in liver include thalassemia (Mediterranean anemia), hemachromatosis, cirrhosis, yellow atrophy of liver, tuberculosis and carcinomas.[15]

Blood copper levels of healthy animals vary rather widely, but for the higher mammals are of a similar magnitude. According to Beck,[16] most values lie between 80 and 120 μg%. Levels ranging between 23 and 35 μg% have been reported for domestic fowl.[17] Other investigators report similar blood copper levels.[15,18,19]

The mean copper levels for normal human females is somewhat higher than for males; there may be some hormonal involvement. Increases in serum or plasma copper levels of women taking oral contraceptives containing an estrogen have been reported.[20,21] Copper levels ranging around 280 μg% have been reported in women in late pregnancy.[22]

Hambridge and Droegemueller[23] observed changes in plasma and hair zinc and copper concentration during pregnancy. By sixteen weeks, copper had increased to 162 μg% from a 107 μg% level; zinc decreased from 88 to 68 μg%. After thirty-eight weeks, plasma zinc fell to 56 μg% and copper rose to 192 μg%.

At delivery copper levels in maternal sera are four to five times those found in newborns.[24] Levels in the latter range between 45 and 70 μg%.

Elevated serum copper levels in humans are manifest in certain other disease processes also, for example, infections, leukemias, Hodgkin's disease, hemochromatosis, and hyperthyroidism. Increased serum copper levels have also been reported in artherosclerosis.[25] Serum zinc is significantly decreased and copper and iron levels markedly increased in thalassemia; copper levels ranging about 250 μg% have been reported.[26] Elevated copper levels were also found in epilepsy.[27]

A summary of serum and urine copper concentrations found in health and disease states appears in Table 12.1. Values listed were taken from the literature as well as from our own experience.

As can be seen from Table 12.1, decreased serum copper levels and increased urinary copper excretions are associated with Wilson's disease and nephrosis. Comparable hypocupremia also occurs in cystic fibrosis and kwashiorkor, diseases with a coexisting protein deficiency. Copper deficiencies have also been observed following long-term parenteral nutrition. A rather rare case, described by Karpel and Peden[28], of an infant on parenteral therapy for eight and a half months did respond to copper supplementation.

TABLE 12.1
Serum and urinary copper concentrations in health and disease

Status	Serum copper (μg per 100 ml)	Urine copper (μg per day)
Normal	83–125	0–50
Wilson's disease	45–70	> 500
Cirrhosis	130–210	25–300
Nephrosis	45–85	90–490
Epilepsy	130–155	
Females on contraceptives	200–240	
Females at delivery	235–325	
Normal newborn	50–70	

According to Graham[29] marked copper depletion in severely malnourished infants with chronic diarrhea is not as uncommon as was previously supposed. The usual diet of modified cow's milk preparations are poor in copper. Incidentally, the infant discussed[28] had a serum copper level of only 14 μg% at three months. We have found serum copper and zinc levels of 25 μg and 40 μg% respectively, in infants classed as 'failing to thrive'. By comparison normal serum copper levels in a newborn are 50–70 μg%.

The determination of serum and urinary copper contents is a reliable diagnostic tool. Where atomic absorption spectrometry is available, analytical techniques are fairly rapid, not too difficult and quite reliable. The analyses can be performed with microliter-sized samples when necessary.

Serum copper measurement is a useful auxiliary test in the management of patients with lymphomas[30] or leukemias. Levels increase with progression of the disease and decrease with improvement. The Thorlings[31] observed that copper levels in Hodgkin's disease were reduced to within normal limits at complete remission.

HEREDITARY DISTURBANCES IN COPPER METABOLISM

Wilson's Disease

Hepatolenticular degeneration, or Wilson's disease, is characterized by the progressive development of neurological disturbances as manifest by widespread tremor and rigidity and, at times, dementia. Liver cirrhosis is also present. Copper metabolism is abnormal, and there are indications that other metals may be involved. Aminoacid excretion is increased.

According to Cumings,[32] most cases of the disease become manifest between age 11 and 40. A percentage of subjects are less than ten and over forty-one years of age. Any race may be involved. Often in the past this treatable disease was misdiagnosed. The first case which could be categorized as hepatolenticular degeneration was described in 1858. Many cases were reported subsequently.

Wilson, in 1912, postulated that the cause might be a non-microbial toxin.[33] In 1913, Rumpel[34] noted an increased copper content in the liver of a subject with the disease.

A rather interesting anomaly seen in some cases is a zone or ring of pigmentation in the cornea, referred to as a Kayser-Fleischer ring.[35,36] Early investigators suggested that the greenish granules in the ring consisted of silver, melanin, malarial pigment, etc. The granules were identified to be copper containing by Gerlach,[37] Rohrschneider,[38] and Fleischer,[39] using both chemical and spectroscopic methods. Policard et al.[40] identified copper in the ring of pigmentation by histospectrography.

Since Rumpel[34] reported his findings concerning an increased copper content in the liver of a subject with hepatolenticular degeneration, numerous cases have been studied. Interestingly, investigators agree fairly well. Cumings has reviewed the literature prior to 1968 in depth.[32,41] The copper content in the brain ranged from 1.3 to 3.9 mg per 100 g of fresh tissue; liver, from 4.6 to 29.0 mg; spleen, 1.4 to 5.2 mg; and kidney, 1.7 mg.

TABLE 12.2
Comparison of tissue copper deposition in controls, Wilson's and Parkinson's diseases

Tissue	Copper content (mg per 100 g wet weight)		
	Control	Wilson's	Parkinson's
Caudate nucleus	0.345	4.91	1.11
Cerebellum	1.3	7.591	1.15
Cerebral cortex	0.616	4.236	1.55
Dentate nucleus	1.3	16.182	1.16
Globus pallidum	1.15	6.903	1.66
Putamen	0.35	6.859	4.463
Substantia nigra	0.98	10.312	1.4
Liver	0.25	24.807	0.744
Serum	0.09	0.026	

Table 12.2 is a comparison of copper deposition in tissues obtained in the Chicago area from control subjects and from those with Wilson's and Parkinson's diseases. Copper was determined by atomic absorption spectrometry.[42]

As a matter of interest, the calcium, magnesium, manganese, zinc, and iron contents, as well as copper, in the tissues from the Wilson's disease subject are listed in Table 12.3.

According to Rupp and Weser[43] copper metabolism in Wilson's disease and in newborn infants is strikingly similar. Both have large quantities

TABLE 12.3
Trace metal contents of tissues from Wilson's disease

Tissue	Calcium	Magnesium	Copper	Manganese	Zinc	Iron
			(mg per 100 g wet weight)			
Cerebellum	0.510	2.471	7.591	ND	2.240	2.394
Cerebral cortex	0.212	1.864	4.236	ND	0.830	3.186
Caudate nucleus	0.521	1.507	4.910	ND	2.014	2.648
Dentate nucleus	0.231	1.620	16.182	0.837	2.243	2.895
Cervical cord	0.634	2.997	4.565	ND	2.368	4.804
Globus pallidum	0.508	1.744	6.903	1.395	2.411	1.072
Putamen	0.541	2.561	6.859	ND	1.559	3.594
Substantia nigra	0.604	3.238	10.312	ND	5.074	10.820
Thalamus	0.389	2.343	8.000	ND	2.371	5.771
Kidney	0.852	1.196	5.882	ND	2.355	2.691
Liver	0.750	1.336	24.807	ND	4.327	6.860

of copper in the liver and a decreased copper level in the serum. In healthy human infants these convert to normal within the first months after birth. However, in homozygous carriers of the gene causing the disease, the metabolic state persists and abnormal depositions of copper in the brain, kidneys, and cornea occur. Further accumulation of copper gives rise to serious damage to the liver. The well-known clinical symptoms are the result.

Wilson's disease is treatable within limits. D-Penicillamine is the agent of choice. Copper stores from tissues are mobilized and urinary copper excretion increased. With adequate treatment there will be neurological improvement and arrest of the liver damage.

Penicillamine, incidentally, also induces a cupriuria in normal subjects and patients with active liver disease.[44] Urinary copper levels in controls were reported to increase from levels of 16 to 140 μg per day to levels of 832–1325 μg per day after a dose of penicillamine. Excretion levels in active liver disease patients increased to 1390 μg per day. Following penicillamine, patients with Wilson's disease excreted between 1800 and 7000 μg per day.

In 1962, Menkes et al.[45] first described the 'kinky hair syndrome', a sex-linked recessive disorder with retardation of growth, peculiar hair, and local cerebral degeneration. Danks et al.,[46] noting the similarity between this syndrome and copper deficiency in sheep, demonstrated that serum copper and copper oxidase (ceruloplasmin) were extremely low in all infants with kinky hair disease.

Garnica[47] subsequently reported that the brain in this condition is smaller than that in the normal newborn. While the grey matter is apparently of normal quantity, there is a significant decrease in the white matter. Both the substantia nigra and locus coerulus are depigmented. Copper contents in the brain were found to be half the concentration in similar tissue of a normal infant succumbing to a trauma. Liver, kidney and spleen copper contents in the control were approximately twice that found in the kinky hair syndrome. Copper levels in the duodenal tissue of the latter, however, were more than twice those of the controls.

Menkes,[48] utilizing radioactive copper, showed the syndrome to be a defect in copper absorption.

Garnica[47] is of the opinion that this generalized disorder of copper metabolism is already manifest in the developing fetus and can be demonstrated in amniotic fluid, liver and skin fibroblasts.

COPPER TOXICITY

Toxic signs of copper intake have been known for centuries. As early as 1785, for example, Thomas Percival[49] described the symptomatology and sequelae of copper ingestion in a young girl who had eaten pickles containing a large quantity of copper salts. Other instances of poisoning through food and beverage contaminated with the metal have been reported in the past.[50,51] A rather unusual incident is that reported in 1968 by Nicholas[52] wherein 30 workmen suffered acute copper poisoning after drinking morning tea made with water from an unserviced gas water heater. The copper concentration of the water in the brew pan was 3 mg per 100 ml.

Salts of copper have found limited application as germicides, fungicides, insecticides and astringents. At one time, these were also used as emetics. Salts are used as pigments in ceramics and in the textile industry.

During a limited interval a few years ago when it was the fashion for inner-city women to spray metallic streaks on their hair, a number of children were admitted as emergencies with a syndrome resembling metal fume fever. Symptoms included a pulmonary edema and transitory fever. These subjects had inhaled a spray known as 'Nestle's Copper Streak', a preparation consisting of fine particles of metallic copper, about 10–50 μ in size. Serum copper levels and urinary copper excretions were elevated for approximately three days. Children apparently recovered within a week. 'Copper Streak' and similar metallic sprays were eventually withdrawn from the cosmetic market.

A condition named 'vineyard sprayer's lung' has been reported in vineyard workers who sprayed Bordeaux mixture (1–2.5% copper sulfate solution neutralized with hydrated lime) on grape vines to prevent mildew.[53] Patients exhibited diffuse pulmonary pathology suggestive of a pulmonary infection. Presenting symptoms included weakness, malaise, loss of appetite, and weight loss, and occasionally cough. Vineyard workers are also reported to show a high incidence of lung carcinoma.

Recently, liver pathology with inclusion of copper was recognized in rural workers with 'vineyard sprayer's lung'. Morphologically, the hepatic lesions resemble those previously reported in workers exposed to inorganic arsenic and vinyl chloride. Abundant deposits of copper were demonstrated in hepatic and pulmonary lesions by histochemical techniques. Copper was also identified in the lymphatic system, spleen and kidney.[54]

A number of cases, some lethal, of copper poisoning following dialysis have been described. Copper was leached from the tubing in the dialysis bath.[55-57] Serum copper levels rose above 2 mg per 100 ml in some instances.

Copper sulfate is said to be a fairly common agent for suicide or homicide in some European countries. Certain sections in India had reported that one third of all acute hospital admissions were due to copper poisonings. Copper salts are more usually the weapon of women. Mutal[58] reviewed the literature on the subject prior to 1972.

Suicide attempts by swallowing copper sulfate are a rarity in the US. However, such a subject was admitted to Cook County Hospital a few years back. The patient, a fifteen year old female, had ingested approximately four ounces of 'Blue Stone Cleaner', a crude copper sulfate preparation used in whitewashing walls and other surfaces. Spontaneous emesis began minutes after ingestion. The patient was seen at her neighbourhood hospital before being transferred to the above-named institution. She appeared to be in no distress. On the following day, however, house officers noted that the subject's urine had become 'tea colored'. As the day progressed, the urine became port wine-like in color and oliguria developed. Red cells were not seen microscopically, but the specimen was benzidine positive.

Examination of blood sample revealed a hematocrit of 28. The serum was brown in color. Profound hemolysis continued. Within a few hours the patient's hematocrit fell to 15.

The patient became confused. A hepatomegaly had developed, sclerae were icteric, and her temperature rose to 103.8 °F. The toxicology section was informed of the problem.

Serum and erythrocyte copper levels were 180 μg and 120 μg%, respectively. Urine copper content was 1 mg l^{-1}. Serum iron was 1960 μg%, and the iron binding capacity only 128 μg%. Normal serum iron levels ranged between 90 and 150 μg%; iron binding capacities, between 250 and 450 μg%. Urine iron excretion was 14.7 mg l^{-1}. The serum and urinary pigments were found to be primarily myoglobin.

The patient improved on a regimen of penicillamine, forced diuresis, and blood transfusions as required. She was discharged approximately two weeks after admission.

Serum and urinary copper and iron contents determined during her stay are listed in Table 12.4.

Exposure to copper dusts and fumes generated during various industrial processes is manifested by respiratory and dermatologic complaints.[59] Salts have been known to induce a conjunctivitis and edema of the eyelids.

Copper ion has been shown[60] to be an inhibitor of δ-aminolevalinic acid dehydratase (ALAD). Zinc activates ALAD.

TABLE 12.4
Copper and iron concentrations in copper poisoning

Date	Serum (μg per 100 ml)		Urine (μg per day)	
	Copper	Iron	Copper	Iron
6/23	180	1960		
6/24	155	1530	1850	9850
6/25	200	1452	1210	24220
6/26			830	10250
6/27			420	2250
6/28	180	qns.[a]	880	300
7/4			102	100
7/7	150	10	31	50

[a] qns. = quantity not sufficient for analysis

Interactions between metals

Chronic copper toxicity in animals can be ameliorated by feeding ammonium molybdate. Harker[61] suggests the use of molybdenum for prevention of nutritional copper poisoning in housed sheep. On the other hand, copper deficiencies in animals seem aggravated by feeding molybdenum.

Both metals are components of enzymes that mediate the oxidation and reduction of iron and are therefore involved in its utilization.[62]

Manganese may increase retention of copper in the body.[63]

Serum copper is elevated and zinc decreased in leukemic states. As therapy becomes effective, levels of both elements approach normal.

Effects of copper and lead upon ALAD are additive, presumably. Both elements together effect greater inhibition than either singly.

ANALYSIS OF COPPER

Colorimetric procedures

Colorimetric methods for the determination of copper are quite sensitive. As examples, detection limits found for the reagents dithizone, diethyldithio-carbamate and salicylaldoxime are 2.2 ng ml^{-1}; 5 ng ml^{-1}, and 0.8 ng ml^{-1}, respectively.[64] Copper is also extractable as a cupferrate,[65] an oxinate[66] and furoyltrifluoroacetonate,[67] among others. However, all colorimetric analyses for copper measurement, the classical as well as the more recent,[68-74] suffer from interferences. Copper determination in biological materials are especially difficult to control.

Fluorimetry

Fluorimetric analyses of copper in biological materials prove more satisfactory. Ritchie and Harris[75] applied the reaction in excess acid between copper and 1,1,3-tricyano-2-amino-1-propene to the analysis of copper in biological materials. The reaction is simple, sensitive, and particularly specific since calcium, manganese, sodium, zinc, iron and magnesium do not interfere.

Anglin et al.[76] have utilized the fluorescence of the copper-cysteine complex in determining copper contents in skin extracts and other biological materials.

Neutron activation analysis

Applications of neutron activation analysis to determine copper contents of different biological materials tend to rival those of atomic absorption spectrometry in number. Measurements have been made of such diverse materials as hair,[77] tissues,[78-80] warts,[81] psoriatic skin,[82] plants[83] etc.

Non-destructive analysis[84] was employed in some instances. Others utilized wet ashing techniques,[85] or extraction methods.[86-88] Maziere et al.[89] separated copper from other elements by ion exchange resins prior to activation analysis.

Smeyers-Verbeke et al.[90] compared neutron activation analysis with atomic absorption spectrometry in the analysis of biological materials. Agreement between the two instrumental techniques was good.

Polarography

Few applications of polarography to the measurement of copper in biological materials have been described. Franke and de Zeeuw[91] determined copper in urine directly by differential pulse anodic stripping voltammetry. Penicillamine used in treating patients with Wilson's disease was found not to interfere with the analysis.

Jones et al.[92] determined copper, cadmium, lead and zinc in foods by voltammetry. Samples were dry ashed, dissolved in nitric acid, and buffered to pH 4.3 with acetate buffer prior to analysis. The estimated detection limit was 5 ng g^{-1}.

X-Ray fluorescence, atomic fluorescence

Limited applications of X-ray fluorescence and atomic fluorescence of copper in biological materials have been reported. The work of Campbell et al.[93] with the former technique is especially noteworthy.

Both flame[94] and flameless[95,96] atomization techniques have been applied to atomic fluorescence analysis of diluted sera. The method is sensitive.

Emission and atomic absorption spectroscopy

At the 324.8 nm resonance line, the detection limits for copper in aqueous solutions are 0.1 μg ml^{-1} by flame emission, and 5 ng ml^{-1} by flame atomic absorption. Limits are extended about tenfold by flameless atomization. Atomic absorption is the method of choice.

Herrmann and Lang's paper[97] concerning the determination of copper in serum by atomic absorption spectrometry was the first of many such to appear.[98-100]

There appear to be no significant interferences with copper analysis other than that of non-specific absorption due to constituents in the biological matrices.

For accurate, precise analyses of serum by flame atomization, it is preferred to isolate copper from the matrix by chelation–extraction. Proteins are first removed by ashing or by precipitation with trichloroacetic acid. A specific concentration of copper in methylisobutyl ketone, aspirated into a flame, will produce an absorbance signal approximately four times as great as the same aqueous concentration. With flameless atomization, however, the signals from either medium are equivalent.

For analysis in the graphite furnace only precipitation of proteins is required. Aqueous standards and sera are diluted 1 : 5 with trichloroacetic acid. Background correction is not necessary because the non-specific molecular absorption signal emerges a few seconds before that of the copper. The two signals emanating from an aliquot of a protein-free supernatant do not superimpose upon each other.

Because the sodium content in urine is about 10^5 times that of copper, the metal must be isolated by chelation–extraction before analysis in either the flame or the furnace. Background correction cannot eliminate the signal emanating from an untreated sample.

Tissue is wet ashed, then chelated and extracted.

The graphite furnace is programmed to dry at 100 °C for 40 s; char at 300 °C for 40 s; and atomize at 2500 °C for 6 s.

Delves[101] determined copper in plasma protein fractions by combining electrophoresis on cellulose acetate with electrothermal atomic absorption spectrometry. After protein fractions were isolated, the individual copper contents were determined.

Ninety-five percent of the copper in serum was found in the α_2 fraction. Only 2 μl of sample were required for the analysis.

REFERENCES

1. *Genesis*, Chapter 4: verse 22.
2. A. S. Margraf, *Chemische Schriften*, Weber Berlin, 1768, cited in H. A. Schroeder *et al.*, *J. Chronic Dis.*, **19**, 1007 (1966).
3. B. Bizio, *J. Chem. Med.*, **10**, 102 (1833) cited in H. A. Schroeder *et al.*, *J. Chronic Dis.*, **19**, 1007 (1966).
4. E. Harless, *Arch. Anat. Physiol.*, 148 (1847) cited in E. J. Underwood, *Trace Elements in Human and Animal Nutrition*, Academic Press, New York, 1971.
5. J. S. McHargue, *Amer. J. Physiol.*, **72**, 583 (1925).
6. J. S. McHargue, *Amer. J. Physiol.*, **77**, 245 (1926).
7. E. B. Hart, H. Steenbock, J. Waddell and C. A. Elvehjem, *J. Biol. Chem.*, **77**, 797 (1928).

8. H. W. Bennetts and F. E. Chapman, *Austral. Vet. J.*, **13**, 138 (1937).
9. W. M. Neal, R. B. Becker and A. L. Shealy, *Science*, **74**, 418 (1931).
10. B. Sjollema, *Biochem. Z.*, **267**, 151 (1933).
11. B. Sjollema, *Biochem. Z.*, **295**, 372 (1938).
12. I. J. Cunningham, *Biochem. J.*, **25**, 1267 (1931).
13. H. Smith, *J. Forensic Sci. Soc.*, **7**, 97 (1967).
14. H. A. Schroeder, A. P. Nason, I. H. Tipton and J. J. Ballassa, *J. Chronic Dis.*, **19**, 1007 (1966).
15. G. E. Cartwright, in *Symposium in Copper Metabolism*, (eds W. D. McElroy and B. Glass) Johns Hopkins Press, Baltimore, 1950.
16. A. B. Beck, *Austral. J. Agr. Res.*, **12**, 743 (1961).
17. A. B. Beck, *Austral. J. Zool.*, **4**, 1 (1956).
18. A. L. Nielsen, *Acta Med. Scand.*, **118**, 87 (1944).
19. J. G. Schenker, E. Jungreis and W. Z. Polishuk, *Amer. J. Obstet. & Gynec.*, **105**, 933 (1969).
20. A. Clemetson, *Austral. Vet. J.*, **42**, 34 (1966).
21. J. A. Halsted, B. Hackley and J. C. Smith, *Lancet*, **2**, 278 (1968).
22. J. Fay, G. E. Cartwright and M. N. Wintrobe, *J. Clin. Invest.*, **28**, 487 (1949).
23. K. M. Hambridge and W. Droegemueller, *Obstet. Gynecol.*, **44**, 666 (1974).
24. J. G. Schenker, E. Jungiers and W. Z. Polishuk, *Biol. Neonate*, **20**, 189 (1972).
25. J. B. Bustamante, M. Mateo, J. Fernandez, B. de Queros and O. Ortiz-Machado, *Biomedicine*, **25**, 244 (1976).
26. A. Arcasoy, *Acta Haematol. (Scand.)*, **53**, 341 (1975).
27. C. H. Brunia, *Epilepsia*, **13**, 621 (1972).
28. J. T. Karpel and V. H. Peden, *J. Peds.*, **80**, 32 (1972).
29. G. G. Graham, *New Eng. J. Med.*, **285**, 857 (1971).
30. S. H. Mortazani, A. Bani-Hashemi, M. Mozafari and A. Raffi, *Cancer*, **29**, 1193 (1972).
31. E. B. Thorling and K. Thorling, *Cancer*, **38**, 225 (1976).
32. J. N. Cumings, *Heavy Metals and the Brain*, Charles C. Thomas, Springfield, 1959.
33. S. A. K. Wilson, *Brain*, **34**, 295 (1911/12).
34. A. Rumpel, *Dtsch. Z. Nervenheilk*, **49**, 54 (1913).
35. B. Kayser, *Clin. Mol. Augenheilk*, **40** (2), 22 (1902).
36. B. Fleischer, *Münch. Med. Wschr.*, **56**, 1120 (1909).
37. W. Gerlach and W. Rohrschneider, *Klin. Wscher.*, **13**, 48 (1934).
38. W. Rohrschneider, *Arch. Augenheilk*, **108**, 381 (1934).
39. B. Fleischer and W. Gerlach, *Klin. Wschr.*, **13**, 255 (1934).
40. A. Policard, P. Bonnet and G. Bonamour, *Compt. Rend. Soc. Biol. (Paris)*, **122**, 1120 (1936).
41. J. N. Cumings, *J. Clin. Pathol.*, **21**, 1 (1968).
42. E. Berman (unpublished data).
43. H. Rupp and U. Weser, *Biochem. Biophys. Res. Comm.*, **72**, 223 (1976).
44. R. E. Lynch, G. R. Lee and G. E. Cartwright, *Proc. Soc. Exptl. Biol. Med.*, **142**, 128 (1973).
45. J. H. Menkes, M. Alter, G. K. Steigleder, D. R. Weakley and J. H. Sung, *Pediatrics*, **29**, 764 (1962).
46. D. M. Danks, P. E. Campbell, B. J. Stevens, V. Mayne and E. Cartwright, *Pediatrics*, **50**, 188 (1972).
47. A. D. Garnica, *Clin. Genet.*, **11**, 154 (1977).
48. J. H. Menkes, *Pediatrics*, **50**, 181 (1972).
49. T. Percival, *Med. Trans. Coll. Phys.*, **3**, 80 (1785) cited in J. N. Cumings, *Heavy Metals and the Brain*, Charles C. Thomas, Springfield, 1959.

50. A. Devergie, *Ann. Hyg. Publ., Paris*, **24**, 136 (1840) cited in J. N. Cumings, *Heavy Metals and the Brain*, Charles C. Thomas, Springfield, 1959.
51. I. von Hattingberg, *Fortschr. Neur. Psychiat.*, **12**, 1 (1940).
52. P. O. Nicholas, *Lancet*, **2**, 40 (1968).
53. T. G. Villar, *Amer. Rev. Resp. Dis.*, **110**, 545 (1974).
54. J. C. Pimental and A. P. Menezes, *Gastroenterology*, **72**, 275 (1977).
55. B. J. Matter, J. Pederson, G. Psimenos and R. D. Lindeman, *Trans. Amer. Soc. Artif. Int. Organs*, **15**, 309 (1969).
56. P. Ivanovich, A. Manzler and P. Drake, *Trans. Amer. Soc. Artif. Int. Organs*, **15**, 316 (1969).
57. A. D. Manzler and A. W. Schreiner, *Ann. Int. Med.*, **73**, 409 (1970).
58. S. R. Mutal, *Forensic Sci.*, **1**, 245 (1972).
59. S. R. Cohen, *J. Occup. Med.*, **16**, 621 (1974).
60. J. Thompson, D. D. Jones and W. H. Beaseley, *Brit. J. Ind. Med.*, **34**, 32 (1977).
61. D. B. Harker, *The Veterinary Record*, **99**, 78 (1976).
62. I. H. Scheinberg and I. Sternlieb, *Pharmacol. Rev.*, **12**, 255 (1960).
63. M. S. Seelig, *Amer. J. Clin. Nutr.*, **25**, 1022 (1972).
64. E. B. Sandell, *Colorimetric Determination of Traces of Metals*, Interscience Publishers, New York, 1959.
65. J. Stary and J. Smizanska, *Anal. Chim. Acta*, **29**, 546 (1963).
66. C. H. R. Gentry and L. G. Sherrington, *Analyst*, **75**, 17 (1950).
67. R. T. McIntyre, E. W. Berg and D. V. Campbell, *Anal. Chem.*, **28**, 1316 (1956).
68. P. Calme, J. P. Gerhard and E. Kraemenge, *Mikrochim. Acta*, **6**, 1276 (1969).
69. A. A. Shilt and P. J. Taylor, *Anal. Chem.*, **42**, 220 (1970).
70. K. Khalifa, H. Doss and R. Awadallah, *Analyst*, **95**, 207 (1970).
71. P. Carter, *Clin. Chim. Acta*, **39**, 497 (1972).
72. K. T. Lee, Y. -F. Chang and S. F. Tan, *Mikrochim. Acta*, **2**, 505 (1976).
73. A. A. Alexiev, P. R. Bontchev and V. Bardarov, *Mikrochim. Acta*, **2**, 535 (1976).
74. A. A. Alexiev, P. R. Bontchev and S. Gantcheva, *Mikrochim. Acta*, **2**, 487 (1976).
75. K. Ritchie and J. Harris, *Anal. Chem.*, **41**, 163 (1969).
76. J. H. Anglin, Jr., W. B. Batten, A. I. Raz and R. M. Sayre, *Photochem. Photobiol.*, **13**, 279 (1971).
77. A. K. Perkons and R. E. Jervis, *J. Forensic Sci.*, **11**, 50 (1966).
78. G. S. Fell, H. Smith and R. A. Howie, *J. Clin. Pathol.*, **21**, 8 (1968).
79. G. C. Battistone, E. Levie and R. Lofberg, *Clin. Chim. Acta*, **30**, 429 (1970).
80. M. L. Leu, G. T. Strickland and S. J. Yeh, *J. Lab. Clin. Med.*, **77**, 438 (1971).
81. M. M. Molokhia and B. Portnoy, *Brit. J. Dermatol.*, **88**, 347 (1973).
82. M. M. Molokhia and B. Portnoy, *Brit. J. Dermatol.*, **83**, 376 (1970).
83. A. P. Grimanis, *Talanta*, **15**, 279 (1968).
84. E. L. Kanabrocki, L. F. Case, L. Graham, I. Fields, Y. T. Oester and E. Kaplan, *J. Nucl. Med.*, **9**, 478 (1968).
85. M. Worwood and D. M. Taylor, *Int. J. Appl. Radiat. Isotopes*, **19**, 753 (1968).
86. K. K. S. Pallay, C. C. Thomas, Jr. and C. M. Hyche, *J. Radioanal. Chem.*, **20**, 597 (1974).
87. T. E. Henzler, R. J. Korda, P. A. Helmke, M. R. Anderson, M. M. Jiminez and L. A. Haskin, *J. Radioanal. Chem.*, **20**, 649 (1974).
88. J. M. Lo, J. C. Wei and S. J. Yeh, *Anal. Chem.*, **49**, 1146 (1977).
89. B. J. Maziere, J. Gros and D. Comar, *J. Radioanal. Chem.*, **24**, 279 (1975).
90. J. Smeyers-Verbeke, D. L. Massart, J. Versieck and A. Speecke, *Clin. Chim. Acta*, **44**, 243 (1973).
91. J. P. Franke and R. A. de Zeeuw, *J. Anal. Toxicol.*, **1**, 291 (1977).
92. J. W. Jones, R. J. Gajan, K. W. Boyer and J. A. Fiorino, *J. Assoc. Offic. Anal. Chemists*, **60**, 826 (1977).

93. J. L. Campbell, B. H. Orr, A. W. Herman, L. A. McNelles, J. A. Thomson and W. B. Coor, *Anal. Chem.*, **47**, 1542 (1975).
94. D. Kolibova and V. Sychra, *Anal. Chim. Acta*, **63**, 479 (1973).
95. A. Montaser, S. R. Goode and S. R. Crouch, *Anal. Chem.*, **46**, 599 (1974).
96. A. Montaser and S. R. Crouch, *Anal. Chem.*, **46**, 1817 (1974).
97. R. Herrmann and W. Lang, *Z Klin. Chem.*, **1**, 182 (1963).
98. E. Berman, *At. Abs. Newsletter*, **4**, 296 (1965).
99. S. Sprague and W. Slavin, *At. Abs. Newsletter*, **4**, 351 (1965).
100. J. B. Dawson, D. J. Ellis and H. Newton-John, *Clin. Chim. Acta*, **21**, 33 (1968).
101. H. T. Delves, *Clin. Chim. Acta*, **71**, 495 (1976).

CHAPTER 13

GOLD

There is ample evidence that gold was being used for ornamental purposes prior to any human record.[1] Quantities of the metal were present in the native form. It is possible to work gold cold. Egyptians were probably the earliest metallurgists involved with the metal.

From the biochemical viewpoint, gold is not considered an essential element. It is considered a substance with toxic potential, however.

The Chinese were probably the first to use gold in medicine.[2] At various periods throughout recorded time, it has been employed as a panacea for all diseases. Alchemists attempted to prepare 'an elixir of life' from gold to cure all ills and confer eternal youth. For centuries the metal has been used quite effectively as an antipruritic to relieve the itching palm. However, untoward reactions, including habituation, from that specific use have been reported. To quote Virgil:[3] 'To what will you not drive mortal hearts you accursed hunger for gold?'

Interest in the medicinal use of gold in more modern times was stimulated by Robert Koch, who in 1890 observed that gold cyanide arrested the growth of tubercle bacilli *in vitro*. Though he was unable to demonstrate its effectiveness in infected animals, gold began being used sporadically in treating tuberculosis. Gold therapy was extended to the treatment of arthritis because some clinicians early in the 1900s believed rheumatoid arthritis to be a form of tuberculous joint disease. At present gold salts are employed widely in treating rheumatoid arthritis.

The pharmacology of gold in patients and experimental animals has been extensively studied in the past.[4-9] The pharmacodynamics of the metal, its tissue deposition etc. are a great source of interest to current investigators.[10-17] Biochemically, gold is indeed a rare metal.

PHARMACOLOGY—TOXICOLOGY OF GOLD

The sodium thiomalate and sodium thiosulfate salts of gold are water soluble, but very little is absorbed from the gastrointestinal tract. Absorption from intramuscular sites in injection is rapid, however.

Excretion of gold is slow and is primarily via the kidneys. Five percent is excreted within the first 24 hours. Eighty five percent of a dose is still retained after seven days. Obviously bodily stores increase with each successive dose. The element can be detected in the urine for as long as a year following cessation of therapy.

Secretion of gold in the maternal milk of a patient on chrysotherapy has been demonstrated by Bell and Dale,[18] among others. Using neutron activation analysis, they found maternal milk to contain 22 ng·ml^{-1} and infant urine, 0.4 ng ml^{-1}.

The tissue distribution of gold is dependent upon the compound administered. Highest concentrations are found in the kidneys with appreciable quantities in the liver and spleen when water soluble compounds are injected. Since insoluble compounds are apparently taken up by phagocytes, the concentrations found in the liver and spleen are greater than that in the kidney.[6,19,20]

Gold apparently passes the placental barrier. Racker and Henderson[21] discuss an instance where a patient had received gold sodium thiomalate for rheumatoid arthritis prior to conception. Pregnancy was terminated after 20 weeks. Gold deposits were found in the placenta as well as the liver and kidney of the fetus.

Certain distribution studies reported recently (thirty and more years after the studies cited above) have decided points of interest. For example, Hashimoto et al.[22] describe the occurrence of a corneal chrysiasis, a gold deposition in the cornea, seen during the course of gold therapy. The degree of corneal chrysiasis observed in the 40 patients on chrysotherapy was related to the total dose of gold received and the amount of drug injected weekly. Deposits in the cornea disappeared within a short time when therapy was discontinued. Patients did not experience visual disturbances.

In a patient who had received 5 g of gold over an extended period, Gottlieb et al.[23] found the maximum concentrations in the para-aortic lymph nodes, adrenals, and other organs of the reticuloendothelial system. However, because of their small mass, gold storage capacity is limited. Comparatively low concentrations were found in tissues making up the joint structure. The authors report that the quantities of gold found in mg per organ, were: bone marrow, 159; liver, 148; skin, 117; bone, 110; muscle, 33; spleen, 19; lymph node, 12; adrenal, 3; and thyroid, 1.

Vernon-Roberts et al.[24] using neutron activation analysis studied gold distribution in seven patients who had died over a period ranging from 0 to 23 years after therapy. They noted that, during active gold therapy, gold was found to be abundant in the synovial living cells except where a fibrin layer was present on the surface. The metal accumulated progressively in subsynovial connective tissue during therapy. Distribution was not uniform, however. Deposits were found within the macrophages of renal tubular epithelium, semeniferous tubules, hepatocytes, and adrenal cortical cells, as well. The metal seemed selectively concentrated within inflamed synovial tissues during therapy.

Gold persisted in synovial and other tissues for up to 23 years following the cessation of chrysotherapy.

Jeffery et al.[25] determined gold in the skin and plasma of groups of patients on chrysotherapy. The concentration of the metal in skin and nails was higher than in the plasma. Highest skin gold levels were obtained in the group on long-standing therapy who were exhibiting altered skin pigmentation, i.e. chrysiasis. Gold apparently does not accumulate in hair.

Penneys et al.[26] observed a direct correlation between the cumulative gold dosage and skin gold level.

GOLD TOXICITY

According to the literature of the 40s, about one in every four patients is said to suffer a toxic reaction at some time during chrysotherapy. Reactions are severe for one in thirty, and one patient in every two hundred may die.[27] Reactions seem to be more frequent in those with arthritis who have had the disease for more than ten years.[28] Some workers found the incidence of toxic reactions to be unrelated to plasma gold levels.[5]

The most common toxic effects involve the skin and mucous membranes. The reactions of skin are often preceded by pruritis and may vary from simple itching to severe exfoliative dermatitis.

Gold compounds can be toxic to hematopoietic organs. Occurrence of thrombocytopenia, leukopenia, agranulocytosis, and aplastic anemia has resulted. Gold preparations can apparently affect blood coagulation. Sipka et al.[29] reported an inhibition of factors III and XIII in rabbits. At 2 mg kg^{-1} gold induced an overall antithrombotic state despite the increased tendency to platelet aggregation. Sodium aurothiomalate, when added to cultures of human mononuclear cells in concentrations found in sera and tissues, was found to actively inhibit lymphocytic and monocytic functions.[30-32]

Considerable controversy exists in the literature regarding plasma gold levels and the lack of direct correlation with dosage, clinical response and toxicity. Part of the problem may be due to the regimen of dose administration and gold measurement. Lorber et al.[33] prefer adjusting patient doses to plasma gold levels rather than employing a fixed dose schedule. Nevertheless, plasma gold levels in general do not seem to indicate an impending toxic reaction for most clinicians. Davis and Hughes[34] observed that an eosinophilia accompanied or preceded reactions and consider it a useful guide for monitoring patients receiving gold salts.

The nephrotoxicity of gold salts has long been recognized. Yarom *et al.*[35], employing ultrastructural and microprobe techniques, observed recently that selective lesions occurred in the proximal convoluted tubules. Also, the mitochondria appeared to be the target organelle. Some investigators think the nephrotoxicity of gold may be an immune complex disease.[36] Others suggest gold nephropathy may be a combination of both direct toxic and immunologic injury to the glomeruli and tubules.[37]

Occasionally patients may develop acute respiratory distress associated with pulmonary infiltration during treatment with sodium aurothiomalate. This manifestation, recognized only recently, resolves after gold withdrawal.[38]

Gottlieb and Brown[39] reported the occurrence of acute myocardial infarction following gold thiomalate induced vasomotor reactions (nitritoid-like). The two patients described were elderly and with long-standing rheumatoid arthritis.

Rothermich *et al.*,[40] after reviewing chrysotherapy over the years, concluded that it was a valuable and comparatively safe treatment.

MEDICAL USES OF GOLD

Gold metal is employed in dental alloys. Radioactive gold has application in treatment of cancer. Charged gold leaf has been utilized for neurosurgical procedures as a device for closure of the fragile pia mater.[41] Healing of cutaneous ulcers has been accelerated by its application. Gold leaf serves as a light porous, dry dressing and increases the vascularity of the application site.[42] This practice was probably known to the ancients.

INDUSTRIAL USES OF GOLD

Gold has industrial uses other than serving as security for world currencies. The metal is used in electronics for contacts, printed circuits, connections, precision resistances, and semi-conductors. The space industry utilizes gold in glass-to-metal seals, as a coating on bearings, and as a coating against infra-red and high intensity solar radiation. Gold surfaces also resist corrosion.

ANALYSIS OF GOLD

Colorimetry

Colorimetric methods for the determination of gold are fairly sensitive but are subject to interferences. For example, the diethylaminobenzylidene rhodanine–gold complex has a detection sensitivity of 0.01 $\mu g\ ml^{-1}$ and the isopropyl ether extract of rhodamine B, 3 ng ml^{-1}. The detection sensitivities of gold complexes with stannous chloride, hydrobromic acid, and o-tolidine are 5, 40, and 4 ng ml^{-1} respectively.[43] Silver, palladium, platinum, mercury, and copper react with the reagents as well.

Rhodanine (diethylaminobenzylidene rhodanine) has been utilized for the determination of gold in biological materials.[44,45].

o-Dianisidine forms a red complex with auric gold. The reaction has been applied in the analysis of gold in biological fluids after the destruction of organic matter with sulfuric acid and hydrogen peroxide. Gold is then oxidized with aqua regia prior to complexing with o-dianisidine.[46]

Dithizone reacts with gold in dilute mineral acids.[47] Copper is a major interference.

Gold can be extracted as a diethyldithiocarbamate.[48]

Gold as chloroauric acid or bromoauric acid can be extracted from aqueous media by organic solvents and measured colorimetrically.[43] Extraction of gold from an acidic solution could be utilized in isolating the element for analysis by various instrumental means.

Fluorimetry

The fluorescence of the gold–Rhodamine B complex has been utilized in gold determinations. A detection sensitivity of 0.5 μg ml^{-1} has been reported for the complex.[49] There are intereferences, however, since rhodamine forms fluorescent complexes with cobalt, iron, mercury, manganese, tin, antimony and thallium, among others.[50]

Gold forms fluorescing complexes with acridine.[51] Chromium, selenium, tellurium react with this reagent as well.

Chromatography

Paper chromatography has been employed to separate gold from platinum metals.[51] Thin layer chromatography could also be applied.

Charcoal has been used as an adsorbent in determining gold in stream waters.[52] After adsorption the charcoal was ashed, brought into solution, and the gold determined by complexation with rhodanine. Ethyl cellulose was found to be a useful adsorbent in gold analysis.[53]

Emission spectroscopy

Spectrographic methods were the earliest of instrumental techniques applied in measuring gold concentrations in biological materials. The Gerlachs studied tissues of tuberculous subjects on gold therapy.[54] Gaul and Staud[55] examined biopsy material from patients receiving gold sodium thiosulfate.

Few applications of neutron activation analysis to the determination of gold in biological materials have been used,[56] but the technique has been utilized in analyzing rocks and ores for the metal.[57,58]

Recently, Purdham et al.[59] described an X-ray fluorescence spectrometric procedure for gold analysis in biological fluids. Gold was initially precipitated as a telluride then solubilized in aqua regia, evaporated to dryness, solubilized in hydrochloric acid and finally re-precipitated with hydrogen sulfide prior to instrumental analysis.

Atomic absorption spectrometry

Detection limits for gold at the 243.0 nm resonance line are said to be about 0.05 μg ml^{-1}, very favorable detection limits indeed for analyses of biological materials from patients receiving chrysotherapy. Many reports of flame atomic absorption analyses of serum and urine have appeared since the instrumentation became available.[60-65] Some procedures involved simple, direct aspiration of diluted sera and urine. Others isolated gold by chelation–extraction prior to analysis.[66] The latter methods seem the more valid.

Certain components in biological materials, primarily potassium, phosphate, and proteins, markedly enhance the absorbance signal from gold. As an example, Figure 13.1 illustrates the effect of potassium concentration.

Tissue and plasma samples must first be wet or dry-ashed. Trichloroacetic acid precipitation will not release gold bound to plasma proteins, either in vivo or in vitro. Approximately 5% is recovered following trichloroacetic acid precipitation. Recovery is total after acid digestion, however.[67,68]

Less than 5% of the circulating gold is unbound. The major portion

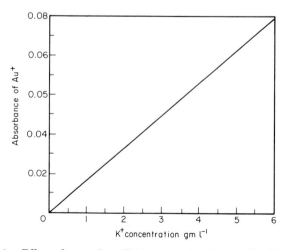

Fig. 13.1.   Effect of potassium (K$^+$) on the absorbance of gold (Au$^+$)

is tightly bound to the fibrinogen and α_1-globulins. We have isolated the fibrinogen fraction from plasma samples by the addition of 1M calcium chloride and found approximately 70% of the plasma gold content, in a blood sample from a patient on chrysotherapy, to be trapped in that fibrinous film. Gold was not detected in erythrocytes.[65]

Urine specimens can be chelated and extracted directly after adjusting the pH to 5–6. Diethyldithiocarbamate is an effective chelate. A lean oxidizing flame is used in gold analysis.

Balasz[65] prefers to extract the gold halide directly after ashing.

Flameless atomic absorption

Methods describing applications of flameless atomization have been reported.[69-72].

The procedure of Dunckley and Staynes[73] is a clever application of chemistry and instrumentation. Gold in urine is precipitated out as a telluride, dissolved in aqua regia and extracted into methylisobutyl ketone prior to analysis in the graphite furnace.

REFERENCES

1. G. Agricola, *De Re Metallica*, trans. by H. C. Hoover and L. H. Hoover, Dover Publishing Inc., New York, 1952.
2. W. D. Block and K. van Goor, *Metabolism, Pharmacology, and Therapeutic Uses of Gold Compounds*, Charles C. Thomas, Springfield, 1956.
3. Virgil, *Aeneid*, III, 1–55, 19 BC.
4. R. H. Freyberg, W. D. Block and S. Levey, *J. Clin. Invest.*, **20**, 401 (1941).
5. R. H. Freyberg, W. D. Block and G. S. Wells, *Clinics*, **1**, 537 (1942).
6. W. D. Block, O. H. Buchanan and R. H. Freyberg, *J. Pharmacol. Exptl. Therap.*, **73**, 200 (1941).
7. W. D. Block, O. H. Buchanan and R. H. Freyberg, *J. Pharmacol. Exptl. Therap.*, **76**, 355 (1942).
8. W. D. Block, O. H. Buchanan and R. H. Freyberg, *J. Pharmacol. Exptl. Therap.*, **82**, 391 (1944).
9. W. D. Block and E. L. Knapp, *J. Pharmacol. Exptl. Therap.*, **83**, 275 (1945).
10. N. L. Gottlieb, P. M. Smith and E. M. Smith, *Arth. Rheum.*, **15**, 582 (1972).
11. B. R. Mascarenhas, J. L. Granda and R. H. Freyberg, *Arth. Rheum.*, **15**, 391 (1972).
12. M. Harth, *Clin. Pharmacol. Therap.*, **15**, 354 (1974).
13. N. L. Gottlieb, P. M. Smith, N. S. Penneys and E. M. Smith, *Arth. Rheum.*, **17**, 56 (1974).
14. Y. S. Katayama, H. Korski and H. Danbara, *J. Nutr.*, **105**, 957 (1975).
15. N. S. Penneys, S. McCreary and N. L. Gottlieb, *Arth. Rheum.*, **19**, 927 (1976).
16. A. F. Oryshak and F. N. Ghadially, *J. Pathol.*, **119**, 183 (1976).
17. N. S. Penneys, S. McCreary and N. L. Gottlieb, *Arth. Rheum.*, **19**, 927 (1976).
18. R. A. F. Bell and I. M. Dale, *Arth. Rheum.*, **19**, 1374 (1976).
19. H. Elftman, A. G. Elftman and R. L. Zwemer, *Anat. Rec.*, **96**, 341 (1946).

20. C. W. Sheppard, E. B. Wells, P. F. Hahn and J. P. B. Goodell, *J. Lab. Clin. Med.*, **32**, 274 (1947).
21. I. Rocker and W. J. Henderson, *Lancet*, **2**, 1246 (1976).
22. A. Hashimoto, Y. Maeda, H. Ito, M. Okazaki and I. Hara, *Arth. Rheum.*, **15**, 309 (1972).
23. N. L. Gottlieb, P. M. Smith and E. M. Smith, *Arth. Rheum.*, **15**, 16 (1972).
24. B. Vernon-Roberts, J. L. Dore, J. D. Jessop and W. J. Henderson, *Ann. Rheum. Dis.*, **35**, 477 (1976).
25. D. A. Jeffery, D. F. Biggs, J. S. Percy and A. S. Russell, *J. Rheum.*, **2**, 29 (1975).
26. N. S. Penneys, K. Kramer and N. L. Gottlieb, *J. Invest. Derm.*, **65**, 331 (1975).
27. C. L. Short, W. W. Beckman and W. Bauer, *New Eng. J. Med.*, **235**, 362 (1946).
28. C. Ragan and T. L. Tyson, *J. Amer. Med. Assoc.*, **133**, 752 (1947).
29. S. Sipka, E. Padar and T. Szilagyi, *Haematologia*, **10**, 361 (1976).
30. K. E. Viken and J. O. Lamvik, *Acta Pathol. Microbiol. Scand. Section C*, **84**, 419 (1976).
31. P. E. Lipsky and M. Ziff, *J. Clin. Invest.*, **59**, 455 (1977).
32. M. Harth, C. R. Steller, N. R. St. C. Sinclair, J. Evans, D. McGerr and R. Zubere, *Clin. Exptl. Immunol.*, **27**, 357 (1977).
33. A. Lorber, T. M. Simon, J. Leeh and P. E. Carroll, Jr., *J. Rheumatol.*, **2**, 401 (1975).
34. P. Davis and G. R. V. Hughes, *Arth. Rheum.*, **17**, 964 (1974).
35. R. Yarom, H. Stein, P. D. Peters, S. Slavin and T. A. Hall, *Arch. Pathol.*, **99**, 36 (1975).
36. T. Palosuo, T. T. Provost and F. Milgrom, *Clin. Exptl. Immunol.*, **25**, 311 (1976).
37. I. Watanabe, F. C. Whittier, J. Moore and F. E. Cuppage, *Arch. Pathol. Lab. Med.*, **100**, 632 (1976).
38. P. W. Gould, P. L. McCormack and D. C. Palmer, *J. Rheumatol.*, **4**, 252 (1977).
39. N. L. Gottlieb and H. E. Brown, *Arth. Rheum.*, **20**, 4 (1977).
40. N. O. Rothermich, V. K. Philips, W. Bergen and M. H. Thomas, *Arth. Rheum.*, **19**, 1321 (1976).
41. J. P. Gallagher and C. G. Geschickter, *J. Amer. Med. Assoc.*, **189**, 928 (1964).
42. N. M. Kanof, *Cutis*, **18**, 395 (1976).
43. E. B. Sandell, *Colorimetric Determination of Traces of Metals*, Interscience Publishers, New York, 1959.
44. B. K. Merejkovsky, *Bull. Soc. Chim. Biol.*, **15**, 1336 (1933).
45. S. Natelson and J. L. Zuckerman, *Anal. Chem.*, **23**, 653 (1951).
46. W. D. Block and O. H. Buchanan, *J. Biol. Chem.*, **136**, 379 (1940).
47. L. Erdey and G. Rady, *Z. Anal. Chem.*, **135**, 1 (1952).
48. H. Bode, *Z. Anal. Chem.*, **143**, 182 (1954).
49. J. Marienko and I. May, *Anal. Chem.*, **40**, 1137 (1968).
50. C. E. White and J. Argauer, *Fluorescence Analysis: A Practical Approach*, Marcel Dekker, New York, 1970.
51. K. P. Stolyarov, N. N. Grigoriev and G. A. Khomenok, *Chem. Abst.*, **78**, 78233V (1973).
52. E. Bauer, *Helv. Chim. Acta*, **25**, 1207 (1942).
53. J. A. Lewis and P. A. Serin, *Analyst*, **78**, 385 (1953).
54. W. Gerlach and W. Gerlach, *Virchow's Arch. f. Path. Anat.*, **28**, 209 (1931).
55. L. E. Gaul and A. H. Staud, *Arch. Dermat. Syph.*, **28**, 790 (1933).
56. D. Gibbons, *Int. J. Appl. Radiation Isotopes*, **4**, 45 (1958).
57. A. R. De Grazia and L. Haskin, *Geochim. Cosmochim. Acta*, **28**, 559 (1964).
58. F. O. Simon and H. T. Millard, *Anal. Chem.*, **40**, 1150 (1968).

59. J. T. Purdham, O. P. Strausz and K. I. Strausz, *Anal. Chem.*, **47**, 2030 (1975).
60. W. T. Frajola and P. B. Mitchell, *Fed. Proc. (Abstracts)*, **26**, 1967.
61. A. Lorber, R. C. Cohen, C. C. Chang and H. E. Anderson, *Arth. Rheum.*, **11**, 170 (1968).
62. S. L. Tompsett, *Proc. Assoc. Clin. Biochem.*, **5**, 125 (1968).
63. A. A. Dietz and H. M. Rubenstein, *Clin. Chem.*, **15**, 787 (1969).
64. J. V. Dunckley, *Clin. Chem.*, **17**, 992 (1971).
65. N. D. H. Balazs, D. J. Pole and J. R. Masarel, *Clin. Chem. Acta*, **40**, 213 (1972).
66. M. Harth, D. S. M. Haines and D. C. Bondy, *Amer. J. Clin. Pathol.*, **59**, 423 (1973).
67. E. Berman, *Appl. Spectros*, **29**, 1 (1975).
68. E. Berman, in *Applied Atomic Spectroscopy*, Vol. 2, (ed. E. L. Grove) Plenum Press, New York, 1978.
69. M. P. Bratzel, Jr., C. L. Chakrabarti, R. E. Sturgeon, M. W. McIntyre and H. Agemain, *Anal. Chem.*, **44**, 372 (1972).
70. J. Aggett, *Anal. Chim. Acta*, **63**, 473 (1973).
71. F. J. M. J. Maessen, F. D. Posma and J. Balke, *Anal. Chem.*, **46**, 1445 (1974).
72. H. Kamel, D. H. Brown, J. M. Ottaway and W. E. Smith, *Analyst*, **102**, 645 (1977).
73. J. V. Dunckley and F. A. Staynes, *Ann. Clin. Biochem.*, **14**, 53 (1977).

CHAPTER 14

IRON

Iron was the first metal utilized by ancient man. Being hard, ductile, and malleable, it could be fashioned, fairly easily, into objects such as weapons, tools, and cooking vessels. The element, widely distributed and responsible for the color of most soils, occurs in the earth's crust at a concentration of almost 5%.

Iron is very essential to all organisms, animal and plant. It functions as a catalyst and is present in amounts greater than that of any other trace element. The iron content of an animal body varies with its state of health, nutrition, age, sex and species. For the most part, iron normally exists bound to protein, either as porphyrins or as the heme compounds, hemoglobin and myoglobin, and to a lesser extent, as non-heme protein-bound compounds, ferritin, transferrin and hemosiderin. In certain disease states the last named, an iron-storage compound, is greatly increased.

According to Jacobs and Worwood[1] 57.6% of the body iron in man is contained in hemoglobin and 8.9% in myoglobin. Approximately 33% is contained in non-heme iron complexes, including ferritin and hemosiderin. The cytochrome and catalase enzymes contain about 0.5%.

The heme pigments, hemoglobin in erythrocytes, and myoglobin in muscles, function as oxygen carriers.[2] Heme-containing enzymes such as the cytochromes in mitochondria[3] and catalase in red blood cells,[4] are concerned with electron transport and peroxidase breakdown, respectively.

Included among the non-heme iron-containing proteins, in addition to the storage compounds named, are certain enzymes such as aconitase, involved in the citric acid cycle, the succinic dehydrogenases and NADH dehydrogenase, concerned with iron transport; and xanthine oxidase, involved in uric acid metabolism.[5]

Since ancient times all civilizations, east and west, associated iron with strength. A common remedy for weakness prescribed by Greek physicians, for example, was the drinking of waters in which old swords had been allowed to rust. The ancient Hindus employed a calcined iron preparation compounded from macerated sheets of roasted iron, whey, vinegar, cow's urine and milk. Sydenham, a late-seventeenth century English physician, prescribed a somewhat more palatable preparation, 'steel in substance', 'iron or steel filings steeped in cold, Rhinish wine, for the treatment of chlorosis, an iron-deficiency anemia.[6] About 1 g of iron was ingested daily by the Sydenham regimen. Fortunately, the iron present in the therapy described was a rather poorly absorbed form. Ferrous iron is absorbed more easily than the ferric form. Excess iron can be toxic.

The small intestine, acting as both an absorptive and excretory organ for iron, serves as the primary control of the body iron content.[7-9] Small excesses of iron within the villous epithelial cells are oxidized to the ferric state and combine with the protein apoferritin to yield ferritin. Iron, in the latter form, is stored temporarily in the intestinal mucosa until excreted via the feces. This control mechanism, or 'mucosal block', so called, is overwhelmed when large quantities of absorbable iron are ingested.[10-13]

TOXICITY OF IRON

All iron preparations ingested in excess can be equally toxic per unit of soluble iron. Acute iron poisoning, while a rarity in adults, is not uncommon among young children. The attractively colored, sugar-coated tablets of commercially available preparations are similar in appearance to sugar-coated chocolate candies. Serious acute poisoning in children can occur following ingestion in excess of one gram of iron.[14] The normal iron intake for children is 10–20 mg kg^{-1}.

Iron poisoning can be fatal. According to Herbert,[6] there is an average of at least one infant or child death a month in the US from the ingestion of ferrous sulfate preparations in excess of two grams. Fatalities were more frequent before the availability of desferroxamine, a specific chelating agent for iron. Dimercaprol, incidentally, is contraindicated because it may form a toxic complex with iron.

The education of parents and responsible adults concerning the dangers inherent in iron tablets and 'pregnancy pills', etc. has markedly reduced the incidence of acute iron poisoning in children who are crawling or climbing about and exploring their environment.

Very occasionally one does encounter a rather bizarre use of iron tablets. Recently a three-week-old infant with iron poisoning, the victim of an attempted homicide, was admitted to Cook County Hospital. It seems that a distraught relative living with the victim's family had lost her own child shortly before this infant was brought home. Fully cognizant of the implications of her actions, she proceeded to add ferrous sulphate tablets to the infant's feed formula when the latter was about 10 days old. Bouts of vomiting ensued.

Upon admission the infant's serum iron level was 262 μg%. Urine iron content was 3.7 μg ml^{-1} (3.7 mg l^{-1}!). A sample of the formula submitted for analysis contained 5.1 μg ml^{-1} of iron. The child apparently recovered.

Signs of iron ingestion may occur within 30 min or may be delayed for hours following ingestion. Symptoms initially include nausea and vomiting due to gastrointestinal irritation and necrosis. There may be a pallor or cyanosis, lassitude, drowsiness, hematemisis, and diarrhea. Shock and cardiovascular collapse may follow. Death has been known to occur within six hours. There may be an apparent, though transient, period of recovery, followed by death in 12–48 hours. The findings upon autopsy include hemorrhagic gastroenteritis (iron sulfate is corrosive) and hepatic injury.

The last fatality from iron poisoning seen at Cook County Hospital occurred in late December 1975 in a child transferred from another institution. The patient was badly managed initially. Rather than perform a careful, copious gastric lavage, syrup of ipecac was administered. More of the dose was retained than was regurgitated as evidenced by the ipecac still present in the stomach, along with unabsorbed ferrous sulfate forty-eight hours later. Emetine was detected in the intestine, brain, liver, lungs, kidney and bile. An estimated two grams of unabsorbed iron salts were harvested from the stomach and intestines.

Results of iron analyses performed on various tissues are shown in Table 14.1. Levels obtained from normal tissues by Tipton and Cook[15] are listed alongside for comparison.

The serum iron level was 2.1 mg% with 100% saturation. Normal serum iron levels for the population of that child's origin range from 85 to 150 μg%. Total iron binding capacities vary between 250 and 450 μg%. Iron saturation is 25–38 μg%.

In hemochromatosis, an iron storage disease, serum iron levels of 191–290 μg% total iron binding capacities of 205–330 μg%, and a mean saturation of 96% have been reported.[16]

TABLE 14.1
Iron concentrations–fatal iron poisoning in a 3-year-old (male)

Tissue	Iron poisoning (μg g^{-1} wet weight)	Controls (Ref. 15) (μg g^{-1} wet weight)
Brain	483	41–75
Intestine[a]	1448–2060	14–66
Kidney	982	45–107
Liver	1504	82–308
Lung	1574	143–495
Spleen	1344	126–546
Stomach[a]	4821	—

[a] Intestine and stomach washed free from particulates prior to weighing and ashing.

All iron contents in the case described above were determined by flamless atomic absorption spectrometry. Tissues were wet ashed prior to analysis. Serum proteins were precipitated by 5% trichloroacetic acid as described previously elsewhere.[17]

INDUSTRIAL USES AND EXPOSURE TO IRON

Iron and its compounds enjoy various industrial applications, other than that in the manufacture of steel. Certain salts such as the acetate, chloride and nitrate, are used as mordants in dyeing and printing textiles. The chloride is employed, also, in photoengraving. Ferric chromate and ferrocyanide are utilized as pigments. Ferric hydroxide is important in water purification systems. The oxide is employed as a pigment, a polishing agent for glass and precious metals, as well as being used in magnets and magnetic tape. Ferrous oxide is utilized in steel manufacture, in enamels and in heat absorbing glass.

Untoward reactions from industrial exposure to iron compounds are rare. Hematite dusts (ferric oxide) have been known to cause a benign pneumoconiosis.

ANALYSIS OF IRON

Colorimetric

While many reagents form colored complexes with iron, few are suitable for determination of trace quantities. Methods suitable for the analysis of biological materials have been described using complexing agents such as thiocyanate;[18-20] mercaptoacetic acid.[21,22] Many investigators have employed 1,10-phenanthroline and 2,2-bipyridine and related compounds.[23-29] The use of 4,7-diphenyl-1,10-phenanthroline followed by isoamyl alcohol extraction has been suggested.[30] Thioglycollic acid has been advocated as a reagent.[31]

Thenoyltrifluoroacetone in xylene was found to be a very selective chelate of iron.[32]

Ferric iron is quantitatively extracted as a cupferrate over the pH range 0–12 by various organic solvents.[33]

A procedure employing 4-aminophenazone, following trichloroacetic acid precipitation of serum proteins, was described fairly recently.[34]

The iron analysis method of Peters, Giovanniello et al.,[35] considered the classical reference for the clinical laboratory, uses bathophenanthroline as the complexing agent.

All colorimetric determinations suffer from interferences of some sort. For example, copper, nickel, and cobalt react with 1,10-phenanthroline and 2,2-bipyridine. Salicylates form a colored complex with iron that absorbs in the wavelength range in which most iron complexes formed during analysis are measured. Salicylates present in therapeutic levels, for example 200 μg ml^{-1}, can yield falsely elevated results in serum iron determinations. Acetaminophen seems to inhibit color development somewhat. Versenate (EDTA) used in treating lead poisoning inhibits color formation, often totally. Investigating the effects upon colorimetric iron analysis of commonly employed therapeutic agents, such as the antibiotics diuretics, anti-hypertensives, etc., could indeed prove quite interesting.

Miscellaneous Instrumental Techniques

Iron forms measurable fluorescent complexes with luminol and peroxide, with carminic acid, and with Rhodamine B.[36] However, to date practical methods employing fluorimetry for the determination of iron in biological materials have not been described.

Neutron activation analysis has been employed in measuring iron in foods.[37]

Campbell *et al.*[38] utilized X-ray fluorescence in deterimining the iron content of plants and other biological materials, including wines.

Hambidge[39] employed a DC arc in argon for the analysis of iron in serum, urine and hair. Samples were dry ashed in a low temperature asher, and then dissolved in hydrocholoric acid prior to instrumental analysis.

Atomic fluorescence has been used in measuring serum iron. A detection sensitivity of 0.5 μg ml^{-1} was reported. Since the normal range for serum iron is 0.85–1.5 μg ml^{-1}, samples must be concentrated for clinical usefulness.

Atomic Absorption Spectrometry

Detection limits for iron by flame atomic absorption spectrometry are listed as 0.01 μg ml^{-1} using the 248.3 nm resonance line.[40]

Possibly the first application of the instrumentation to iron measurement in a clinical laboratory situation was that of Herrmann[41] and co-workers, who determined hemoglobin by measuring iron in blood by atomic absorption spectrometry.

Many further procedures employing either flame or flameless atomization have been described.[42-46]

Despite the favorable detection limits, direct analysis methods are of questionable validity. Non-specific molecular absorbance generated by the biological matrix is not completely eliminated by employing background correction. Results are falsely elevated as a consequence. Proteins should be removed by trichloroacetic acid precipitation prior to analysis. Urines are chelated and extracted.

A method we found quite applicable to true clinical situations in the field of flame atomization is that of Zettner and co-workers.[47] Serum proteins are precipitated with trichloroacetic acid. Iron is chelated by bathophenanthroline and extracted into methylisobutyl ketone prior to analysis. Other chelating agents, for example cupferron or ferrozine, could be utilized similarly.

Since detection limits with flameless atomization devices are extended about tenfold, iron in serum samples can be measured after protein precipitation. Standards are diluted with trichloroacetic acid in a like ratio (1 : 5). Background correction is not necessary since iron is atomized a few seconds after the signal from non-specific molecular absorption has disappeared.

When determining iron in urine or tissue digests by flameless atomization, however, background correction is necessary. The greater sodium signal emanating from these matrices overwhelms the iron signal. Urine samples normally contain 0.1 μg ml^{-1} of iron.

Iron is a ubiquitous metal. Extraneous iron contaminants in laboratory vessels and sampling equipment present a bit of a challenge in controlling the quality of iron analyses, irrespective of the method employed. The problem becomes exquisite when applying micro and flameless procedures. Iron residues in pipet tips, for example, can be many fold greater than the total iron in an aliquot taken for analysis. All equipment used in iron analysis should be acid washed and rinsed in deionized water.

REFERENCES

1. A. Jacobs and M. Worwood, in *Blood and Its Disorders* (eds R. M. Hardesty and D. J. Weatherall) Oxford, Blackwell, 1974.
2. M. Worwood, *Seminars in Hematology*, **14**, 3 (1977).
3. R. Lembert and J. Barrett, *Cytochromes*, Academic Press, New York, 1973.
4. A. S. Brill, in *Comprehensive Biochemistry* (eds M. Florkin and E. H. Stotz) Elsevier, 1966.
5. D. O. Hall, R. Cammack and K. K. Rao, in *Iron in Biochemistry and Medicine*, (eds A. Jacobs and M. Worwood), Academic Press, New York, 1974.
6. V. Herbert in *Pharmacological Basis of Therapeutics*, 15th edn (eds L. S. Goodman and A. Gelman) Macmillan, New York, 1977.
7. M. E. Conrad, L. R. Weintraub and W. H. Crosby, *J. Clin. Invest.*, **43**, 963 1964.
8. T. H. Bothwell, *Brit. J. Haemat.*, **14**, 453 (1968b).
9. A. Jacobs, in *Symposium on Iron Deficiency and Iron Overload*, (ed. S. T. Callender) see also *Clin. Haemat.*, **2**, 241 (1973).
10. W. H. Crosby, *Ser. Haemat.*, **6**, 66 (1965).
11. W. H. Crosby, *Amer. J. Clin. Nutr.*, **21**, 1189 (1968).
12. M. D. Smith and I. M. Pannacciulli, *Brit. J. Haemat.*, **4**, 428 (1958).
13. T. H. Bothwell and C. Isaacson, *Brit. Med. J.*, **1**, 522 (1962).
14. G. Forbes, *Brit. Med. J.*, **1**, 367 (1947).
15. I. H. Tipton and M. J. Cook, *Health Phys.*, **9**, 103 (1963).
16. G. E. Cartwright and M. M. Wintrobe, *J. Clin. Invest.*, **33**, 685 (1954).
17. E. Berman, *Appl. Spectros.*, **29**, 1 (1975).
18. R. Stugart, *Ind. Eng. Chem. (Anal. Ed.)*, **3**, 390 (1931).
19. H. L. Roberts, C. L. Beardsley and L. V. Taylor, *Ind. Eng. Chem. (Anal. Ed.)*, **12**, 365 (1940).
20. C. Hoffman, T. R. Schweitzer and G. Dalby, *Ind. Eng. Chem. (Anal. Ed.)*, **12**, 454 (1940).

21. G. Leavell and R. R. Ellis, *Ind. Eng. Chem.*, (*Anal. Ed.*), **6**, 46 (1934).
22. R. A. Koenig and C. R. Johnson, *J. Biol. Chem.*, **142**, 233 (1942).
23. L. G. Saywell and B. B. Cunningham., *Ind. Eng. Chem.* (*Anal. Ed.*), **9**, 67 (1937).
24. F. C. Hummel and H. H. Willard, *Ind. Eng. Chem.* (*Anal. Ed.*), **10**, 14 (1938).
25. S. H. Jackson, *Ind. Eng. Chem.* (*Anal. Ed.*), **10**, 302 (1938).
26. R. H. Schaefer, *Biochem. Z.*, **304**, 417 (1940).
27. R. H. Thorpe, *Biochem. J.*, **35**, 672 (1941).
28. R. A. Koenig and C. R. Johnson, *J. Biol. Chem.*, **143**, 159 (1942).
29. W. S. Pringle, *Analyst*, **71**, 491 (1946).
30. R. E. Peterson, *Anal. Chem.*, **25**, 1337 (1953).
31. E. Lyons, *J. Amer. Chem. Soc.*, **49**, 1916 (1927).
32. F. L. Moore, W. D. Furman, J. G. Ganchoff and J. G. Siviak, *Anal. Chem.*, **31**, 1148 (1959).
33. J. K. Foreman, C. J. Riley and T. D. Smith, *Analyst*, **82**, 89 (1957).
34. A. Kyaw, *Clin. Chim. Acta*, **69**, 351 (1976).
35. T. Peters, T. J. Giovanniello, L. Apt and F. Ross, *J. Lab. Clin. Med.*, **48**, 274 (1956).
36. P. A. St. John, in *Trace Analysis: Spectroscopic Methods for Elements*, John Wiley & Sons, New York, 1976.
37. R. R. Becker, A. Vaglia and E. R. Schmid., *Radiochem., Radioanal. Letters*, **19**, 343 (1974).
38. J. L. Campbell, B. H. Orr, A. W. Herman, L. A. McNelles, J. A. Thomson and W. B. Coor, *Anal. Chem.*, **47**, 1542 (1975).
39. K. M. Hambidge, *Anal. Chem.*, **43**, 103 (1971).
40. W. E. Rippetoe, V. I. Muscat and T. J. Vickers., *Anal. Chem.*, **46**, 796 (1974).
41. R. Herrmann, W. Lang and D. Stamm, *Blut* **11**, 135 (1965).
42. S. Sprague and W. Slavin, *At. Abs. Newsletter*, **4**, 228 (1965).
43. E. D. Olson and W. B. Hamlin. *Clin. Chem.*, **15**, 438 (1969).
44. E. D. Olson, P. I. Jatlow, F. J. Hernandez and H. L. Kahn, *Clin. Chem.*, **19**, 326 (1973).
45. M. I. Glenn, J. Savory, S. A. Feen, R. D. Reeves, C. J. Molnar and J. D. Winefordner, *Anal. Chem.*, **45**, 203 (1973).
46. Y. Yeh. and F. Zee, *Clin. Chem.*, **20**, 360 (1974).
47. A. Zettner, L. C. Sylvia and L. Capacho-Delgado, *Amer. J. Clin. Pathol.*, **45**, 533 (1966).

CHAPTER 15

LEAD

Lead, one of the seven metals of antiquity, has accompanied all civilizations since their beginnings. It was in use before the time of the Hebrew exodus from Egypt. Lead does not occur in the elemental state in nature but as its sulfide ore or galena. Discovery of metallic lead may have resulted from the accidental dropping of galena into a campfire.

There are numerous references to the metal in the scripture. For example, Moses, in Numbers XXXI, 22–23, directs the Israelites regarding purification of gold, silver, brass, iron, tin, and lead objects acquired with their plunder from the Midianites. The book of Jeremiah contains an allusion to the process of cupellation, the procedure whereby silver is separated from argentiferrous galena: 'The bellows are burned, the lead is consumed of the fire; the founder melteth in vain, for, the wicked are not plucked away. Reprobate silver shall men call them . . .'

Judging by his many metaphors in metallurgical terms, Jeremiah probably had considerable metallurgical experience. Exposure to lead may account, in part, for his rather saturnine comments.

Lead is mentioned in the tribute lists of the Pharoah Thotmes III (c. 1500 BC).

Ancient peoples used lead for many purposes, including amulets, rings and other ornaments; dishes and trays; as a core for bronze statuettes and figures; sinkers for fishing nets, etc. Probably the first use of galena was as an eyepaint.

Lead poisoning was recognized in antiquity. Hippocrates (370 BC) was said to have recognized lead colic in a lead worker.[1] Nicander in his book on poisons (first century BC) mentions litharge (lead oxide) and the symptoms of intoxication following its use.[2] Effects of lead were discussed by Dioscorides, Vitruvius and Pliny.

The Greeks and Romans are said to have used lead extensively in cooking. Bronze pots gave food a bitter taste, but lead sweetened it. Olive oil was often stored in lead-lined vessels. Wine was prepared in lead-lined pots. The systems of Roman aqueducts were lead lined. Horace, about 25 BC, questioned the purity of water in relation to lead pipes and rather wisely prefers that from a murmuring book!

According to Patterson,[3] the estimated lead consumption in the Roman Empire during the first centuries of the present era approximated four kilograms per capita per year, an amount approaching the current consumption in the US of six kilograms per capita per year. Those Romans were an agrarian population without motor vehicles. Probably the greatest portion of the lead processed in the US is used to make battery oxides and lead alkyls.

Hofman[4] in 1885, and Gilfallan,[5] eighty years later advanced the theory that the fall of the Roman Empire was largely due to endemic lead poisoning. Perhaps overt lead exposure may have played a role. However, some credit is due the ravages of malaria, tuberculosis, and the greater vigor and war skills of the invading hordes from the north.

Following the fall of Rome in the fourth century AD, lead use declined and remained at a comparatively low level for about 600 years. Around the tenth century, lead ores began being mined in eastern Germany. Lead exposure escalated.

Sweetening of wine with lead salts, a common practice in continental Europe despite the imperial edicts, made lead colic an endemic 'disease' over the centuries. 'Devonshire colic', discussed by Sir George Baker in 1767, was thought to be due to the lead introduced during the manufacture and storage of cider.[6] Lead colic, very common in colonial America, was also associated with the distilling of rum in lead heads and pipes.[7]

The use of pewter and earthenware with high lead contents for food preparation and storage added to the body burden of lead among inhabitants of early America. Galena, along with iron ore, began being mined near Jamestown in 1621, fourteen years after the founding of the colony.[8] Soils near lead smelters acquire an increased lead content.

TOXICOLOGY OF LEAD

Lead is a general protoplasmic poison that is cumulative, slow acting and subtle, and produces a variety of symptoms. Like other heavy metals, it has an affinity for sulfur. Though it exerts much of its activity through sulfhydryl inhibition, lead interacts with carboxyl and phosphoryl groups also. The element interferes with heme synthesis.

Lead may be absorbed into the body by ingestion, inhalation and through the skin. Absorption is governed by chemical structure.

Inorganic lead salts do not penetrate the intact skin but can be absorbed through cuts and abrasions. Significant quantities can also be absorbed from a bullet or shot wound. Lead shot (buckshot) is considered particularly dangerous, because of its larger surface area permitting greater absorption. There have been reports of lead poisoning occurring within a month after a bullet wound.

It is quite likely that Andrew Jackson's many bouts with the ague were due, in part, to lead poisoning. His anatomy carried a few bullets, mementoes from the duels he fought.

Metabolism of lead

Organic lead compounds, for example tetraethyl lead, rapidly penetrate the intact skin. Absorption into bodily tissues is more rapid than with inorganic compounds. Because of greater lipid solubility large amounts of organic lead gain access to nerve tissue.

Major routes of lead absorption are the respiratory and gastrointestinal tracts. Most industrial exposures in foundries, smelters, or areas where lead salts are processed follow inhalation of lead dusts or fumes. Absorption is dependent upon particle size and solubility. Blumgart[9] and Minot,[10] more than 50 years ago, had demonstrated that lead can be absorbed from all portions of the respiratory tract, including the nasal passages. The ancients appreciated the hazards in smelter fumes. Strabo (63 BC–24 AD) wrote: 'The furnaces for silver are constructed lofty in order that the vapour, which is dense and pestilent, may be raised and carried off.'[11]

Burning of leaded battery casings in space heaters, occasionally practiced in certain depressed regions, resulted in incidents of lead poisoning among both children and adults so exposed. A fairly common practice in some ghetto areas in Chicago is the burning of cheap coal (coal dust) that has been moistened with drained crankcase oil.

Most of the lead exposure that occurs among non-industrial populations results from the food and drink consumed. About 5–10% of the lead ingested is absorbed into the body; the remainder is excreted via the feces and is of little concern. The classic studies by Kehoe et al.[12] indicate that under normal circumstances, 300 μg are ingested daily by an adult. Ingestion of 600 μg of lead daily can result in lead intoxication. The World Health Organization suggests that the daily lead intake should not exceed 5 μg kg (body weight)$^{-1}$.

Following absorption into the blood stream, 95% of the lead circulates loosely bound to the erythrocytes. Only 5% is in the plasma. Previously, it was thought that erythrocyte lead was bound primarily to the cell membrane. Recently, however, it has been observed that the lead is bound primarily to the hemoglobin in the cell.[13] Raghavan and Gonick,[14] while examining the lead-binding properties of erythrocytes obtained from both normal and lead-exposed adults, discovered the existence of a low molecular weight protein (c. 10000) in the latter. This protein was not found in the erythrocytes of unexposed individuals. It was found, further, that while lead binds primarily to hemoglobin in normal cells, appreciable binding of lead to the low molecular weight protein present in the erythrocytes of lead-exposed individuals occurs. The elucidation of the characteristics of the protein may assist in explaining the vagaries of different methods of blood lead analysis.

Lead disappears from blood into tissues at a rate which follows first-order kinetics. Lead diffuses from red cell to plasma, then to the extracellular space, and then into the intracellular spaces or tissue cells. The liver and kidney, among the soft tissues, contain the highest concentrations. Lead diffuses from the soft tissues; part is stored in the bone, part is excreted.

During the early period of lead deposition, expecially in growing bone, concentration is greatest in the epiphyseal portion. Skeletal lead is fairly inert. It leaves the skeleton very slowly. Under certain conditions, however, such as a disturbance in acid–base fluid balance or a respiratory infection, lead may suddenly be mobilized from the bones, and symptoms of an acute lead intoxication may be precipitated.

In the stable state more than 90% of the total body lead is stored in the skeleton. The form in which it is stored is becoming more clearly defined. X-ray diffraction studies indicate that lead assumes a position within bone crystal either by displacing other cations or by occupying lattice interstices.[15] However, the possibility that lead may also be bound to organic components or be deposited as a discrete crystal of an insoluble compound are not excluded.

At very low concentrations lead is an effective nucleating agent for inducing the formation of calcium phosphate crystals. This mechanism may be important in trapping lead at the surface of bone crystal.[16] Lead follows the metabolic pathways of calcium.

Deposits in bone and tissue comprise the body burden of lead and represent the cumulative difference between the quantities absorbed and excreted. The total body burden rises with age. Barry and Mossman[17] have shown that the burden by the age of 60–69 years may have increased 100-fold over that at 0–9 years.

A survey conducted by Weiss *et al.*[18] indicates that lead exposure among the general population in America may be less than it was a century ago. Adult hair samples dating from 1870 to 1923 were found to contain 144–184 μg g^{-1} of lead compared to the present day value of 15–17 μg g^{-1}. Lead contents of children's hair samples from these two periods were 77–109 μg g^{-1} and 5.5–7.6 μg g^{-1}, respectively.

The difference between the two groups can be explained in part by a decrease in dietary lead intake. However, atmospheric lead originating from smelting and other industrial processes has decreased despite the increased industrialization. During the early years of the industrial revolution, lead smelters, for example, contributed much lead to the atmosphere because of inefficient smelting processes and poor design of furnaces. Prior to 1880 the loss was 2% compared to 0.5% before 1920 and about 0.06%, currently.

The lead content in foods varies with the country of origin; the degree and type of industrialization are factors which influence this.

Present sources of lead exposure

The hazards of leaded paints to young children who explore their environment with their mouths has been a point of discussion, often rather heated, for years. Very wisely, lead contents in paint have been reduced by government regulations. Unfortunately the plaster beneath paint and the putty on the window frames have been ignored. Putty, with its lead content of approximately 30%, is quite palatable to children.

Glazed earthenware is notorious as a source of lead poisoning. Many incidents, including fatalities, have been reported in the US and the UK following the use of glazed earthenware in preparing and storing foods, especially acidic foods. Ceramic dishes and cups, irrespective of their country of origin, are suspect until they are shown by analysis to be safe.

Storing spirits in decanters made from leaded glass may enhance the flavour but could present somewhat of a hazard. A recent import in one of the better stores in Chicago bore the label, 'contains 30% PbO'!

Pewter ware, even modern pewter, is more wisely used for display purposes or for serving dry foods. Foods with a pH below 6 leach lead, antimony and other metals from this alloy.

The increased use of pewter during the bi-centennial in the US prompted examination of representative products available locally. Four percent acetic acid was left in contact with the metal cups for two hours, and the quantities of lead, antimony, cadmium, copper and zinc leached out were determined. The findings are listed in Table 15.1.

Not all candles contain leaded wicks, but candles made for use under

TABLE 15.1
Concentrations of metals extracted from pewter after two hours contact with 4% acetic acid (μg%)

Item	Lead	Antimony	Cadmium	Copper	Zinc
American A	248	36	6	9	4
American B	52	16	3	3	1
American C	79	18	7	4	3
American D	232	33	21	4	1
American E	148	26	44	6	1
English	29	17	12	4	1
French	256	45	4	2	1
Armetale[a]	12	7	1	395	12500

[a] Armetale is not a pewter, but is an alloy rather favored by the 'younger set'.

chafing dishes invariably contain a wick with a leaded core. Wicks in candle-making kits are also leaded. Most decorative candles contain a leaded wick. The lead fall-out on foods, when these candles are burned during a meal, can approximate milligram quantities.

Hair dyes usually contain lead compounds, as do eyepaints. These cosmetic preparations can be potentially hazardous for the user as well as for the young children in the family.

Toys may contain potential hazards in their painted and metal parts. Highly colored pages in magazines contain leaded pigments.

Different canned foods and packaged preparations marketed in the US and Britain have been found to contain questionable amounts of lead. Waldron and Stöfen[19] suggest that it might be in the public interest to indicate the lead content in packaged foods.

Seafoods, such as shrimps, clams, scallops, oysters, carry rather interesting metal concentrations. In all probability the Hebrew fathers were not thinking of lead, cadmium, or mercury hazards when they banned consumption of the unclean creatures of the sea, but after examining the trace metal contents in these little creatures, it seems quite a wise rule.

Diagnosis of lead poisoning

Without a doubt the best account of lead poisoning in the medical literature, past and present, is that of Tanquerel des Planches.[20] Little regarding its physical diagnosis, sources of exposure and modes of absorption have been added since the treatise appeared in 1838. Means of laboratory diagnosis have advanced, however.

Tanquerel, a Paris physician, began systematic inquiries concerning lead in 1831. His entire philosophy is explained in the following quote. 'At the hospital I could only study the morbid phenomena which lead diseases manifest during life and the traces that they leave after death, but the different influences which are present at the development of these affections are to be studied only in the midst of labors which cause them.' In addition to treating about 2171 patients with lead poisoning, he also carried out animal experiments. Chemical analyses were performed by his colleagues, Chevalier, Devergie, and Guibort.

In addition to lead colic (of which he saw 1217 cases), arthralgias, paralysis, and encephalopathy of lead poisoning, Tanquerel discusses characteristic phenomena observed in those with the disease:

(1) A leaden discoloration of the teeth and mucous membranes of the mouth due to sulphuret of lead. A portion of the gums nearest the teeth assumes a bluish or slate-grey color.
(2) A lead taste, generally described as sugary, stypzic and astringent.
(3) A lead jaundice or 'peculiar discoloration of solids and liquids in the

system'. The skin is of an earthy yellow tint. Whites of eyes are yellow
too. Lead was found in the blood of those with lead jaundice.
(4) A lead emaciation or wasting of the flesh is found only in those who
absorb large quantities of lead. Tanquerel considers this 'wasting of
flesh' due to changes of the blood produced by lead!

Incidentally, metals other than lead can combine with sulfur and leave
deposits at the gum line. The lead line so called is not observed in edentulous
individuals and in those who practice good habits of oral hygiene.

Tanquerel and others in his day described modes of lead absorption uncon-
nected with the lead trades. Numerous incidents following cutaneous absorption, for example, were described. Use of 'fomentations', i.e. hot packs
or simple external applications by surgeons of Goulard's extract, a solution
of lead subacetate, was common. Ointments for friction rubs and cosmetics
contained litharge. Orfila[21] reported a case of lead disease exhibiting colic;
paralysis of the eye, muscles of the neck, and hands; great emaciation, and
other symptoms 'peculiar to lead' that followed the use of a cosmetic preparation containing lead.

Cooking utensils made of copper and lined with an alloy composed of
a great quantity of tin and very small portions of lead were cited as causing
lead disease. Bonbons colored yellow, green, blue and red were known
to cause lead colic in adults and children. Lead chromate was used to
color bonbons yellow, green and blue. Minium (lead tetroxide—red lead)
was the red coloring agent. Almonds and colored sugar plums of the day
contained quantities of lead. Papers used for wrapping bonbons contained
lead chromate or carbonate. Chocolates were covered with lead foil.

Children's playthings, for example soldiers and trumpets, were recognized
as a source of lead exposure.

Poisonings resulting from absorption by respiratory organs were mentioned,
for example incidents that occurred after periods of sleeping in rooms newly
painted with leaded paints.

A few cases of lead poisoning, two ending fatally, were traced to snuff
being sold at that time. Chevallier in 1831 reported samples he had analyzed
contained acetates, carbonates and hydrochlorates of lead in quantities of
6–30 grains per pound. Lead absorbed through the nasal mucosa produced
rather deleterious effects. Otto reported two cases that arose from snuff
sold in Copenhagen in 1842. The snuff contained 16–20% red lead. Subjects
involved, rather prominent, apparently succumbed in about four months
from lead disease.

Tanquerel observed that 'want of cleanliness' contributes to the development of lead colic among workmen. He also noted the seasonal variations
in outbreak of lead colic.

More cases of lead colic occurred in the summer months; May, June,
July and August. Their distribution is listed in Table 15.2.

TABLE 15.2
Occurrence of lead colic as observed by Tanquerel des Planches[20]

Month	No. of cases	Month	No. of cases
January	67	July	190
February	77	August	127
March	95	September	92
April	99	October	81
May	115	November	78
June	137	December	59

Laboratory diagnosis of lead poisoning

Criteria for the laboratory diagnosis of lead poisoning have evolved over the years. These are still in a state of flux. Instrumental developments and the resulting refinement in chemical–instrumental methods are the primary factors.

Various approaches including measurements of urinary coproporphyrin and δ-aminolaevulinic acid (ALA) excretion, examining blood films for basophilic stippling of erythrocytes, measuring free erythrocyte protoporphyrin (FEP) and activity of the enzyme ALA dehydrase in blood, determining the lead contents of blood, urine, hair, teeth, tissues etc. have all been applied in assessing the degree of lead exposure and effects thereof.

Porphyrinuria or increased urinary porphyrin excretion was first demonstrated in human cases of lead poisoning as early as 1880.[22] Though urinary coproporphyrins are increased in other states, their determination is still considered a fair index of excess lead absorption. Urinary coproporphyrin, as a screening tool, suffers only from the number of false positives obtained. When properly performed, this test does not yield false negative results. Urinary coproporphyrins are negative during treatment of lead intoxication with EDTA. Whether the chelate inhibits excess excretion of coproporphyrins or merely inhibits the reaction in the test tube is not known.

Excretion of ALA can be an index of long-standing, severe exposure to lead. Animal experiments have yielded interesting results. As with many animal studies, findings cannot always be extrapolated successfully to human subjects. Many children at high risk would be missed. We were fascinated to discover that only 1 of the 20 children with lead encephalopathies seen at Cook County Hospital during the summer of 1966 showed a positive ALA. The other 19 were negative!

Basophilic stippling of erythrocytes, for many years considered pathognomonic of lead poisoning, was first demonstrated in the early 1900s. Studies have shown, however, that stippled red cells are seen in but 25% of the subjects with severe lead poisoning. Furthermore, other conditions which cause increased destruction of erythrocytes, a massive streptococcal infection, for example, or exposure to some blood hemolysin, can produce basophilic stippling. Examining a blood film for stippled red blood cells is but a curiosity, merely an additional test in the 'workup' of lead poisoning.

Measuring inhibition of the enzyme ALA dehydrase has definite points of interest. It is considered a very sensitive test.[23] However, by no means is it a specific indicator of blood lead concentration. Inhibition of ALA dehydrase *in vitro* by copper and mercury has been demonstrated.[24] Some correlation was found between urinary mercury excretion in workers exposed to mercury and inhibition of blood (ALA dehydrase).[25] Thompson *et al.*[26] found ALA dehydrase activity to be an indicator of total metal ion concentration in blood and considered its determination to be of doubtful value in screening large populations for increased lead absorption. Ethanol has also been shown to inhibit ALA dehydrase activity in man.[27]

Free erythrocyte protoporphyrin (FEP) levels in red blood cells rise following increased lead exposure; the rise is not immediate, however. FEP determination seems a promising means of observing the effects of lead on hemoglobin synthesis. Nevertheless to employ FEP determinations as the primary test, in surveying children in risk populations for evidence of increased lead absorption, is less than wise. Studies have reported that 37% of children with elevated blood lead levels had normal FEP levels.[28,29] The quality of an investigation should not be sacrificed for expediency. It would be a disservice to the children screened to determine only FEP. To do both an FEP and a blood lead analysis would be ideal.[30]

The lead content in blood is probably the best single index of the level of current and recent absorption.[31] Analyzing hair, teeth etc., for lead content may yield fascinating information but this is of historical value only.

Lead, a general protoplasmic poison, can affect kidney function. Hence measuring urinary lead excretion has definite limitations. Urine lead levels may not correlate with symptoms or with blood lead levels. Often in the face of good physical evidence of lead poisoning plus a significantly elevated blood lead level, urinary lead contents can fall within normal limits.

A series of blood and urine lead concentrations found in representative cases diagnosed as lead poisoning are listed in Table 15.3.

The last subject listed was a foundry worker who exhibited classical symptoms of lead intoxication. Following the administration of versenate, he began to excrete 6 mg l^{-1} of lead.

Normal blood lead levels are considered to be 20 μg% or less. Levels above 40 μg% indicate increased lead exposures. American cigarette smokers normally show levels around 30 μg%. Blood levels above 60 μg% are considered diagnostic. Hospitalization is suggested for subjects with blood lead levels of 80 μg% or higher.

Normal urinary excretion of lead is less than 80 μg l^{-1} in children, and 150 μg l^{-1} in adults.

TABLE 15.3
Blood and urine lead levels in diagnosed cases of lead poisoning

Blood Pb ($\mu g \%$)	Urine Pb ($\mu g\ l^{-1}$)
48	180
57	110
60	350
64	180
84	350
88	200
88	135
108	1800
110	1000
350	60

Lead content in human tissues

As a general rule, tissue lead contents determined prior to 1940 are considerably higher than values obtained later. Improvements in analytical technology and in the purity of reagents are factors contributing to this.

In a compilation by Cumings,[6] of lead contents in fresh tissues reported from approximately 1940 to 1959 brain was found to contain up to 0.11 mg%; liver, up to 0.5 mg%; kidney, up to 2.2 mg%; and rib, up to 2.8 mg%. Possibly some of the 'controls' may have been industrial workers who succumbed from other causes. In the compilation of levels obtained from cases of lead encephalopathy, brain was reported to contain up to 1.7 mg%;

TABLE 15.4
Comparison of tissue lead in poisoned and normal subjects

Tissue	Lead content (mg% wet weight)	
	Poisoned	Normal
Cerebrum (white)	1.843	0.04 (average)
Cerebrum (grey)	0.666	0.04 (average)
Cerebellum	3.175	0.04 (average)
Caudate nucleus	1.177	0.04 (average)
External capsule	2.975	0.04 (average)
Globus pallidum	7.427	0.04 (average)
Putamen	2.697	0.04 (average)
Liver	3.41	0.04–0.28
Kidney	2.39	0.04–0.16
Adrenal	3.14	0.04–0.16
Spleen	1.95	0.04–0.16
Femur	3.2	0.67–1
Bone marrow	5.75	—
Blood	0.2	<0.02

TABLE 15.5
Lead content in tissues of lead
encephalopathy (female of $2\frac{1}{2}$ years)

Tissue	Lead content (mg% wet weight)
Bone	7.876
Kidney	0.847
Liver	0.806
Spleen	0.741
Frontal lobe L (white)	1.814
Frontal lobe L (grey)	1.015
Frontal lobe R (white)	2.263
Frontal lobe R (grey)	1.779
Anterior L basal ganglion	1.868
Basal ganglion R	1.014
Cerebellum R	1.183
Dentate nucleus	1.864
Globus pallidum	4.542
Thalamus	1.094
Lentiform nucleus	1.681
Putamen	2.789
Thoracic segment	3.269
Parietal lobe R	2.18
Occipital lobe R	3.165
Temporal lobe R	1.585

liver up to 26.1 mg%; and kidney, up to 9.5 mg%. Lesser values for normals were reported later by Kehoe.[32]

A comparison of lead contents in tissue obtained from a lead poisoned child and from normal adults in the Chicago area is listed in Table 15.4. The child in question survived for only five days following diagnosis.

We had occasion to examine different segments of the brain and other tissues obtained from a $2\frac{1}{2}$ year old female who succumbed from lead encephalopathy. Findings are listed in Table 15.5. Note that the lead contents in all segments of the brain are greater that those of the liver and kidney. In addition, the white matter in the frontal lobes had higher lead contents than did the grey.

The child's blood lead level upon admission was 385 μg% and two days later, 345 μg%. An initial urine sample for lead analysis was not obtained, but after initiating chelation therapy (BAL-EDTA), the urine lead content was 15.6 mg 1^{-1}. Shortly before death, the blood lead level was 112 μg% and the urine lead, 6 mg 1^{-1}.

ANALYSIS OF LEAD

Colorimetric

Of all the reagents which form colored complexes with lead, dithizone is considered the best reagent for determination of trace quantities of that metal. According to Cholak *et al.*[33] the results obtained are comparable to those determined by spectrographic analysis. There are exquisite sources of error in the dithizone method when analyzing biological materials in clinical situations. Dithizone forms colored complexes with 17 metals including mercury, bismuth, copper, gold, thallium and cadmium. The latter two, not masked by cyanide, are co-extracted with lead. Needless to say, a colorimeter cannot differentiate between the dithizonates of lead, thallium and cadmium. Nevertheless many modifications of the dithizone method for lead analysis can be found in the literature.[34-45]

Lead can be quantitatively extracted as a benzoyl acetonate at pH 7–10.[46] Hydroxyquinoline forms a complex extractable over the pH range of 6–10.[47] At a pH of approximately 5, lead can be extracted as a thenoyltrifluoroacetonate complex.[48] Cupferron extracts lead completely over the pH range of 3–9.[49]

Vesterberg and Sjoholm[50] employed β-naphthyl thiocarbazone in the analysis of lead in biological materials. The complex is completely extractable at pH 9.8.

According to Bode,[51] lead is quantitatively extractable as a diethyldithiocarbamate over to pH range 4–11. We found a pH of 5.5–6.5 optimal. Recovery was considerably less below 5 and above 7. Versenate interferes with chelation by dithiocarbamate.

Ammonium pyrollidine dithiocarbamate is a fairly effective lead chelate at a narrow pH range of 2.2–2.8.[52]

Emission Spectroscopy

With a Hilger quartz spectrograph and an oxyacetylene flame, Sheldon and Ramage[53] in 1931 were able to identify and determine lead, among other metals, in small pieces of tissue and in 0.1 ml aliquots of blood. Prior to instrumental analysis, tissues were washed, dried to constant weight, and ground into powder; blood specimens were soaked into filter paper and dried to constant weight.

Somewhat later, Gerlach *et al.*[54] analyzed blood and tissues for lead and thallium. These investigators wet ashed specimens before subjecting them to spectrographic analysis.

Perhaps the first application of spectrographic analysis which resembled modern clinical laboratory practice was that of Shipley, Scott and Blumberg.[55] Faced with an epidemic of childhood lead poisoning in 1932, they developed a semiquantitative spectrographic method for the detection of lead in blood. The analytical procedure required about 24h.

Yaokum *et al.*[56] recently employed a DC arc spectrograph in a controlled argon–oxygen atmosphere for measuring lead, platinum and manganese in tissues.

X-Ray spectroscopy

X-ray emission spectroscopy is a highly useful tool for the direct, nondestructive measurement of elements in biological material. Small-sized samples suffice.

Grebe and Esser[57] investigated concentrations of lead and other metals in tissues as early as 1936. The first applications of the technique to clinical laboratory samples were made in the 1960s by Natelson et al.[58-61] They determined lead, thallium, gold, mercury, bismuth etc. in microliter quantities of serum.

X-Ray fluorescence

X-ray fluorescence techniques have been utilized for the determination of toxic elements such as lead, arsenic, and selenium. A few procedures concerning applications to blood lead analysis have been described.[62-63]

Agarwal et al.[64] determined lead and other toxic metals in urine by X-ray fluorescence after extracting samples on ion exchange resin columns.

Dried blood plasma has been analyzed for numerous elements, lead included, by spark source mass spectrometry.[65]

Polarography

Isolation of lead by electro-deposition of the metal from a solution of decomposed biological materials was recommended as early as 1941 by Bambach and Cholak.[66] Microgram quantities of lead could be recovered. Iron and copper were interferences, but their effects apparently could be minimized by the addition of potassium cyanide. Electrodeposition techniques found little application in the analysis of biological materials for a period of time. Kublik[67] described the use of a hanging mercury electrode for the analysis of copper, lead and cadmium in waters. Other applications followed.[68,69]

An interesting method for determination of lead in blood by an anodic stripping (ASV) technique was reported by Voloder et al.,[70] in which 0.2 ml of blood was ashed in nitric acid and peroxide before determining the lead content. Supposedly, interferences from other trace metals or elements in blood were not noted. However, any trace of nitric acid was found to interfere with the procedure.

Applications of ASV to the analysis of lead in clinical samples has escalated,[71-76] some procedures being quite useful. However, methods describing the direct analysis of blood samples are of questionable validity, as blood requires destruction before ASV determinations.[77] Copper can interfere, and penicillamine, used in treating chronic lead poisoning in some centers, also interferes with blood lead determination by direct ASV techniques. Wet ashing eliminates the problem.

Franke and de Zeeuw[78,79] employ anodic stripping voltammetry in the differential mode (DPSAV) as a screening as well as a quantifying tool in toxic metal analysis.

Gas chromatography

Few applications of gas chromatography to lead analysis have been reported. Mushak[80] briefly investigated methods for measuring tri- and tetraalkyl lead. Tetraethyl lead was measured as such or converted to a disodium hexachloroplumbate.

Chau and co-workers,[81] by coupling an atomic absorption spectrometer, equipped with a silica furnace, to a gas chromatograph, were able to detect tetraethyl lead at the 0.1 ng level. The various solvents employed in extractions (chloroform, carbon tetrachloride, hexane, and benzene) produced nonspecific molecular absorption signals, but these emerged well before the lead compounds of interest. Nevertheless, the authors suggested using background correction to eliminate potential interferences.

Atomic absorption spectrometry

Detection limits for lead by flame atomic absorption are 0.01 μg ml^{-1} at the 217 nm resonance line and 0.03 μg ml^{-1} at the 283.3 nm line. Electronic noise at the former line makes it a less practical analytical line on many instruments.

Constituents in biological materials, such as blood, urine, hair, tissues, teeth etc., markedly affect the absorption signal of lead. Background correction is only partially effective in minimizing the nonspecific molecular absorption signal. Chelation–extraction methods can be useful both in eliminating matrix interferences and in concentrating the lead present in a sample.

Lead adsorbed on blood cells can be released for chelation–extraction by precipitating proteins with trichloroacetic acid.[82-84] Some investigators prefer acid digestion[85] or hemolysis[86] prior to chelation extractions.

Versenate (EDTA) is destroyed by wet digestion. However, when blood samples are either hemolyzed or precipitated with trichloroacetate acid, versenate will interfere with the chelation–extraction of any lead which is present. EDTA is a more avid chelate of lead than is dithiocarbamate. It has been claimed that addition of a calcium salt can eliminate interference by versenate.[87]

One should bear in mind, however, that the effectiveness of calcium additions in eliminating versenate interference is dependent upon the concentrations of calcium and EDTA present. The latter is an unknown quantity. Consequently, analytical findings under these conditions prove capricious to say the least.

Because the majority of blood analyses performed in clinical areas involve children in the 1–4 year old age group, there has been an impetus to develop micro-analytical techniques. Delves[88] proposed a technique for blood lead determination using 20 μl of sample. Blood proteins are denatured by heat and peroxide prior to atomizing the sample in the flame of a Boling-type burner. Background correction is necessary. About 38 modifications of the system have been proposed since the original paper appeared.[89,90]

Urinary lead can be chelated and extracted directly after pH adjustment. Kopito and Shwachman[91] prefer to coprecipitate the lead with bismuth. Zurlo et al.[92] determined lead in urine after coprecipitation with thorium.

Tissue samples are wet ashed in nitric–perchloric acids and, after pH adjustment, any lead which is present is chelated and extracted. Aqueous standards are carried through the entire analytical procedure.

Flameless atomization devices do permit lead determinations of sorts, directly, on blood aliquots of 20 μl or less.[93-97] Despite various claims to the contrary some sort of chemical pretreatment is necessary if one requires more than a semiquantitative result. True, smaller background signals are generated by microliter samples, but, the signal from the analyte is also correspondingly reduced. A signal emanating from the matrix constituents of an untreated blood or urine specimen will overwhelm that of the lead even when background correction is employed. Chelation–extraction methods have been applied to advantage with flameless atomization techniques.[98-100] Some investigators prefer partial digestion of blood samples with equal volumes of nitric acid prior to analysis in the graphite furnace.[101] Rains[102] employs a 'stripping procedure', i.e. precipitation of proteins in 5% nitric acid solution; for which only 100 μl of blood are required. Background correction was used in all the methods cited above.

When normal blood samples and the usual control material were analyzed by both the stripping method and by a routine chelation–extraction flameless procedure,[101] the results obtained agreed within a microgram or two per 100 ml. However, when applied to the analysis of freshly drawn blood, from children exhibiting definite clinical symptoms of acute lead intoxication, or to samples drawn from various lead-exposed adults with symptoms suggestive of lead poisoning, the two procedures did not yield comparable results. Values obtained by the stripping method were one-third to one-half of those obtained by the chelation–extraction procedure. A comparison of representative values will be found in Table 15.6.

Similar discrepancies have been noted when two chelation–extraction procedures, one chelating lead at pH 5.5–6.5, and the other chelating lead at pH 2.2–2.8 are compared. Lower results are obtained with the latter method.

Since the factor of incidental analytical error has been ruled out fairly well, one can question what possible role the low molecular weight protein described previously[14] may play in the observed, apparently analytical, phenomenon. It could perhaps be postulated that at the more acidic pH any lead bound to the low molecular weight protein may be carried down with the precipitate and not released into the supernatant liquid.

TABLE 15.6
A comparison of blood lead values obtained by chelation-extraction and nitric acid precipitation, both with flameless atomization

Subject	Status	Lead found (μg%)	
		Chelation–Extraction	Nitric acid
Child	asymptomatic	36	34
Child	asymptomatic	24	26
Child	asymptomatic	17	14
Child	asymptomatic	27	25
Child	asymptomatic	29	29
Child	asymptomatic	58	58
Child	asymptomatic	61	58
Child	acute symptoms	143	93
Child	symptoms	56	34
Child	symptoms	56	29
Child	symptoms	65	45
Adult	asymptomatic	24	23
Adult	asymptomatic	39	38
Adult	asymptomatic	45	44
Adult	asymptomatic	57	58
Adult	asymptomatic	61	61
Adult	symptomatic	69	49
Adult	symptomatic	67	48
Adult	symptomatic	57	44
Adult	symptomatic	44	30

Lead analysis in this age of instrumentation is not technically difficult. However, analyses must be approached with understanding of the analyte and its respective matrix, as well as of the principles involved in the specific instrumentation employed.

The ubiquity of lead does present certain hazards, or should one say, challenges, for the analyst. Suitable measures must be taken to eliminate and control lead contaminants—beginning with the obtaining of the sample itself. Children and lead workers can be liberally dusted with lead containing particles.

REFERENCES

1. J. C. Aub, L. T. Fairhall, A. S. Minot and P. Reznikoff, *Medicine*, **4**, 1 (1925).
2. J. Alderson, *Lancet*, **2**, 73 (1852) cited in J. N. Cumings, *Heavy Metals and the Brain*, Charles C. Thomas, Springfield, 1959.
3. C. C. Patterson, *Archives Environmental Health*, **11**, 344 (1965).
4. K. B. Hofman, *Das Blei bie den Volkern des Alterhums*, Hagel, Berlin, 1885.
5. S. C. Gilfallan, *J. Occup. Med.*, **7**, 53 (1965).
6. J. N. Cumings, *Heavy Metals and the Brain*, Charles C. Thomas, Springfield, 1959.

7. C. P. McCord, *Indust. Med. Surg.*, **22**, 393 (1953).
8. C. P. McCord, *Indust. Med. Surg.*, **22**, 534 (1953).
9. H. L. Blumgart, *J. Indust. Hyg.*, **5**, 153 (1923).
10. A. S. Minot, *J. Indust. Hyg.*, **6**, 125 (1924).
11. G. Agricola, *De Re Metallica*, trans. by H. L. Hoover and L. H. Hoover, Dover, New York, 1950.
12. R. A. Kehoe, F. Thamann and J. Cholak, *J. Indust. Hyg. Toxicol.*, **15**, 257 (1933).
13. F. W. Bruenger, W. Stevens and B. J. Stover, *Health Physics*, **25**, 37 (1973).
14. S. R. V. Raghavan and H. C. Gonick, *Proc. Soc. Exptl. Biol. Med.*, **155**, 164 (1977).
15. N. S. McDonald, F. Ergimilian, P. Spain and C. McArthur, *J. Biol. Chem.*, **189**, 387 (1951).
16. H. Fleisch, S. Bisaz and R. Russel, *Proc. Soc. Exptl. Biol. Med.*, **118**, 882 (1965).
17. P. S. I. Barry and D. B. Mossman, *Brit. J. Indust. Med.*, **27**, 339 (1970).
18. D. Weiss, B. Whitten and D. Leddy, *Science*, **178**, 69 (1972).
19. H. A. Waldron and D. Stöfen, *Sub-Clinical Lead Poisoning*, Academic Press, New York, 1974.
20. S. L. Dana, *Lead Disease: A Treatise from the French of L. Tanquerel des Planches*, Daniel Bixby and Company, Lowell, Mass., 1848.
21. Orfila, *Toxicol. t 1 edition quatr.*, 1843, cited in *Lead Disease: A Treatise from the French of L. Tanquerel des Planches*, Daniel Bixby and Company, Lowell, Mass., 1848.
22. J. Binniendijk, cited by B. J. Stokvis in *Z. Klin. Med.*, **28**, 1 (1895).
23. B. Haeger-Aronson, *Brit. J. Ind. Med.*, **28**, 52 (1971).
24. K. D. Gibson, A. Neuberger and J. J. Scott, *Biochem. J.*, **61**, 618 (1955).
25. O. Wada, K. Toyokawa, T. Suzuki, S. Suzuki, Y. Yano and K. Nakao, *Arch. Environ. Health*, **19**, 485 (1969).
26. J. Thompson, D. D. Jones and W. H. Beasley, *Brit. J. Ind. Med.*, **34**, 32 (1977).
27. M. R. Moore, A. D. Beattie, G. G. Thompson and A. Goldberg, *Clin. Sci.*, **40**, 81 (1971).
28. S. Piomelli, *Pediatrics*, **51**, 254 (1973).
29. S. Piomelli, *Pediatrics*, **53**, 303 (1973).
30. R. Fischer, *Pediatrics*, **52**, 467 (1973).
31. J. J. Chisholm, *J. Pediatrics*, **79**, 719 (1971).
32. R. Kehoe, *Arch. Environ. Health*, **8**, 232 (1964).
33. J. Cholak, D. M. Hubbard, R. R. McNary and R. V. Story, *Ind. Eng. Chem. (Anal. Ed.)*, **9**, 488 (1937).
34. H. Fischer and G. Leopoldi, *Z. Angew Chem.*, **47**, 90 (1934).
35. D. M. Hubbard, *Ind. Eng. Chem. (Anal. Ed.)*, **9**, 493 (1937).
36. K. Bambach, *Ind. Eng. Chem. (Anal. Ed.)*, **11**, 400 (1939).
37. K. Bambach and R. E. Burkey, *Ind. Eng. Chem. (Anal. Ed.)*, **14**, 904 (1942).
38. L. J. Snyder, *Anal. Chem.*, **19**, 684 (1947).
39. H. V. Hart, *Analyst*, **76**, 691 (1951).
40. H. C. Lockwood, *Analyst*, **79**, 143 (1954).
41. W. M. McCord and J. W. Zemp, *Anal. Chem.*, **27**, 1171 (1955).
42. S. L. Tompsett, *Analyst*, **81**, 330 (1956).
43. P. B. Hammond, H. N. Wright and M. H. Roepke, *University of Minnesota Agricultural Experimental Station Bulletin No. 221*, 1956.
44. D. G. M. Diaper and A. Kuksis, *Canad. J. Chem.*, **35**, 1278 (1957).
45. E. Berman, *Amer. J. Clin. Pathol.*, **36**, 549 (1961).
46. J. Stary and E. Hlady, *Anal. Chim. Acta*, **28**, 227 (1963).
47. J. Stary, *Anal. Chim. Acta*, **28**, 132 (1963).

48. F. Hagemann, *J. Amer. Chem. Soc.*, **72**, 768 (1950).
49. J. S. Fritz, M. J. Richard, A. S. Bystroff, *Anal. Chem.*, **29**, 577 (1957).
50. R. Vesterberg and O. Sjoholm, *Mikrochem.*, **38**, 81 (1951).
51. H. Bode. *Z. Anal. Chem.*, **143**, 182 (1954).
52. H. Malissa and E. Schoffman, *Mikrochim. Acta*, **1**, 187 (1955).
53. J. H. Sheldon and H. Ramage, *Biochem. J.*, **25**, 1608 (1931).
54. W. Gerlach, W. Rollwaggen and R. Intoni, *Virchow's Arch. f. Path. Anat.*, **301**, 588 (1938).
55. P. G. Shipley, F. T. M. Scott, and H. Blumberg, *Bull. Johns Hopkins Hosp.*, **51**, 327 (1932).
56. A. M. Yaokum, P. L. Stewart and J. E. Sterrett, *Environ. Health Perspect.*, **10**, 85 (1975).
57. V. L. Grebe and F. Esser, *Fortschr. Gebiete Roentgenstrahlen*, **54**, 185 (1936).
58. S. Natelson, M. R. Richelson, B. Shield and S. L. Bender, *Clin. Chem.*, **5**, 579 (1959).
59. S. Natelson and B. Shield, *Clin. Chem.*, **6**, 299 (1960).
60. S. Natelson and B. Shield, *Clin. Chem.*, **8**, 17 (1962).
61. S. Natelson, *Clin. Chem. (Suppl.)*, **11**, 290 (1965).
62. M. F. Lubozynski, R. J. Baglan, D. R. Dyer and A. B. Brill, *Int. J. Appl. Radiat. Isotopes*, **23**, 487 (1972).
63. T. J. Kneip and G. R. Laurer, *Anal. Chem.*, **44**, 57A (1972).
64. M. Agarwall, R. B. Bennett, I. G. Stump and J. M. D'Auria, *Anal. Chem.*, **47**, 924 (1975).
65. W. A. Wolstenholme, *Nature (London)*, **203**, 1284 (1964).
66. K. Bambach and J. Cholak, *Ind. Eng. Chem. (Anal. Ed.)*, **13**, 504 (1941).
67. Z. Kublik, *Acta Chim. (Hung.)*, **27**, 79 (1961).
68. W. Kemula, E. Rakovska and Z. Kublik, *J. Electroanal. Chem.*, **1**, 205 (1960).
69. G. C. Whitnack and R. Sasselli, *Anal. Chim. Acta*, **47**, 274 (1969).
70. K. Voloder, M. Braneca, N. Ivicic and J. Eder-Trifunovic, *Proceedings of the International Symposium, Environmental Health Aspects of Lead*, Amsterdam, October 1972.
71. W. R. Matson, R. M. Griffin and G. B. Schreiber, in *Trace Substances in Environmental Health*, Vol. 4, (ed. D. Hemphill) Univ. of Missouri, Columbia, 1971.
72. L. Duic, S. Szechter and J. Srinivasan, *J. Electroanal. Chem.*, **19**, 76 (1973).
73. B. Searle, W. Chan and B. Davidow, *Clin. Chem.*, **19**, 76 (1973).
74. T. R. Copeland and R. K. Skogerbee, *Anal. Chem.*, **46**, 1257A (1974).
75. G. Morrell and G. Giridhar, *Clin. Chem.*, **22**, 221 (1976).
76. M. J. Pinchin and J. Newham, *Anal. Chim. Acta*, **90**, 91 (1977).
77. P. Valenta, H. Rutzel, H. W. Nurnberg and M. Stoeppler, *Z. Anal. Chem.*, **285**, 25 (1977).
78. J. P. Franke and R. A. de Zeeuw, *Arch. Toxicol*, **37**, 45 (1976).
79. J. P. Franke and R. A. de Zeeuw, *J. Anal. Toxicol.*, **1**, 291 (1977).
80. P. Mushak., *J. Anal. Toxicol.*, **1**, 286 (1977).
81. Y. K. Chau, P. T. S. Wong and P. D. Goulden, *Anal. Chim. Acta*, **85**, 421 (1976).
82. E. Berman, *Atomic Abs. Newsletter*, **3**, 111 (1964).
83. S. Selander and K. Cremer, *Brit. J. Ind. Med.*, **25**, 209 (1968).
84. E. Berman, V. Valavanis and A. Dubin, *Clin. Chem.*, **14**, 239 (1968).
85. P. P. Donovan and D. T. Feeley, *Analyst*, **94**, 879 (1969).
86. D. W. Hessel, *Atomic Abs. Newsletter*, **7**, 55 (1968).
87. L. J. M. Zinterhofer, P. I. Jatlow and A. Fappiano, *J. Lab. Clin. Med.*, **78**, 664 (1971).
88. H. T. Delves, *Analyst*, **95**, 431 (1970).

89. H. T. Delves, *J. Anal. Toxicol.*, **1**, 261 (1977).
90. *Annual Reports on Analytical Atomic Spectroscopy*, Vols 1–6, Society for Analytical Chemistry, London, 1971–1976.
91. L. Kopito and H. Shwachman, *J. Lab. Clin. Med.*, **70**, 326 (1967).
92. N. Zurlo, A. M. Griffin and G. Colombo, *Anal. Chim. Acta*, **47**, 203 (1969).
93. J. F. Rosen, *J. Lab. Clin. Med.*, **80**, 567 (1972).
94. J. Y. Hwang, P. A. Ulluci and C. J. Mokeler, *Anal. Chem.*, **45**, 795 (1973).
95. M. A. Evenson and D. D. Pendergast, *Clin. Chem.*, **20**, 163 (1974).
96. N. P. Kubasik and M. T. Volosin, *Clin. Chem.*, **20**, 300 (1974).
97. F. J. Fernandez, *Clin. Chem.*, **21**, 558 (1975).
98. J. P. Hwang, P. A. Ulluci, S. B. Smith and A. L. Malenfant, *Anal. Chem.*, **43**, 1319 (1971).
99. N. P. Kubasik and M. T. Volosin, *Clin. Chem.*, **19**, 954 (1973).
100. E. Berman, *Appl. Spectros.*, **29**, 1 (1975).
101. K. G. Brodie and B. J. Stevens, *J. Anal. Toxicol.*, **1**, 282 (1977).
102. T. Rains, personal communication.

CHAPTER 16

LITHIUM

Lithium, the lightest of all metallic elements, was discovered in 1817 by Johann Auguste Arfvedson, a Swedish investigator working with Berzelius. Emil Rosenberg, a Philadelphia physician, circa 1876, demonstrated the element's presence, along with that of sodium, potassium, and calcium, in tissues and other biological materials.[1] Its occurrence is rather widespread.

Lithium is found is most rocks of igneous origin. Lepidolite, also called lithium mica, and petalite, a milky white or clear aluminium silicate, contain about 4% lithium. Spodumene contains 8%, and amblygonite is 10% lithium.

The concentration of lithium in sea water is approximately 0.1 ppm. However, high concentrations are found in certain brines such as Searle's Lake in California or Great Salt Lake in Utah. The latter is estimated to contain four million tons of lithium chloride.[2]

LITHIUM IN BIOLOGICAL MATERIALS

Lithium is found in plants and in all animal tissues and fluids. Hamilton et al.[3] found the mean lithium concentration in normal tissue to be as follows: lymph nodes, 0.13–0.27 μg g^{-1} wet weight; lung, 0.05–0.07 μg g^{-1}; brain, 3–5 ng g^{-1}; testis, 2–4 ng g^{-1}; and human blood, 4–8 ng g^{-1}.

Bowen[4] estimated the total dietary intake of lithium by an adult to be about 2 mg daily. Hamilton and Minski[5] consider an intake of 54–160 μg per day to be more likely.

The function of lithium is unknown. No biochemical or physiological system has been found to be lithium dependent. However, the element is tolerated in considerable quantity when compared with the limits for other non-essential metals, for example, beryllium, its neighbor in the periodic table.

Lithium enters into various biochemical activities. Enzymes activated by potassium are inhibited by lithium, and the action of calcium and of magnesium are also antagonized. Lithium affects carbohydrate metabolism at several points, namely, hexokinase activity, activation of liver adenyl cyclase and protein kinase, glycogen synthesis, and pyruvate kinase activity.[6]

A lithium ion can replace both sodium and potassium in some active transport mechanisms. *In vivo*, as well as *in vitro*, studies indicate that the passive diffusion of lithium resembles that of sodium.[7] While therapeutic doses do not appear to alter sodium or potassium concentrations very much, the ion does induce changes in calcium metabolism. Serum calcium concentrations are elevated above normal, and urinary calcium is decreased in the presence of lithium ions.

MEDICAL USES OF LITHIUM

Lithium became an important agent in the therapeutics of psychiatry not long after Cade of Australia[8] reported in 1949 that lithium seemed to have a specific effect upon mania. The extent of ever-escalating interest in the agent can be ascertained from the number of publications on the subject which have been generated since Cade's original work appeared. By 1970 five papers weekly were being published.[9]

Prior to its use in treating manic depressives, lithium had been applied sporadically, but not too successfully, in therapeutics. Garrod[10] in 1859 tried it in treating gout. Lithium was used in diabetes. Apparently, the drug had been tried in psychiatry about the 1920s also.[9]

Lithium chloride had been utilized as a salt substitute in the 1940s, but by 1950 the practice was discontinued following reports of serious side effects and fatalities resulting from the salt's use.

Goode[11] as early as 1903 did a fairly detailed study of the effects of large doses of lithium salts upon cats and demonstrated many of its toxic symptoms and side effects.

INDUSTRIAL USES OF LITHIUM

Lithium and its compounds have a variety of industrial applications. The metal itself is employed in the manufacture of alloys used in the aerospace industry; for example, lithium–magnesium, an alloy of great strength and lightness. Lithium is also alloyed with lead, copper, silver and aluminum.

The hydride is used to produce hydrogen. Lubricating greases containing lithium stearate are characterized by an ability to maintain action over a wide temperature range and in the presence of water. Some salts are very hygroscopic and are utilized widely in air conditioning and de-humidifying systems. The chloride and fluorides are employed in metal cleaner and as a flux in soldering and welding aluminum. The fluoride is also used in the manufacture of enamels and glazes. Lithium hydroxide is used in alkaline storage batteries and in photography. The tetracyanoplatinate finds use in X-ray photography.

Polymerization processes in the rubber industry are facilitated by lithium metal in dispersed form. Lithium minerals, particularly spodumene and lepidolite, are used in the ceramic and glass industries for fluxing, and for producing resistance to cracking, warping and sagging.

TOXICOLOGY OF LITHIUM

Ionic lithium is almost completely absorbed from the gastrointestinal tract. Peak concentrations in the plasma are attained within two to four hours following an oral dose; absorption is complete in approximately eight hours. Lithium does not seem to bind to plasma proteins. It passes through the blood-brain barrier slowly and achieves a concentration in the cerebrospinal fluid equivalent to 40% of that in plasma.[12] Lithium levels in saliva are several times greater than those in plasma. The ion is also excreted in the milk of lactating women receiving lithium therapy.

Lithium is primarily eliminated from the body via the kidneys. Approximately 95% of a single dose is excreted in urine 4%, in sweat, and only 1% via feces.

Nausea, vomiting, profuse diarrhea, and lethargy are among the characteristic symptoms of lithium intoxication.[13] Some patients may experience confusion, and exhibit hyperreflexia, tremors and dysarthrias. In acute incidents, seizures may ensue and progress to coma and death.[14]

Other side effects arising from lithium therapy include hypotension, cardiac arrythmias, proteinuria, edema, polyuria and polydipsia. The latter two symptoms may be quite marked in certain patients[15,16] and renal failure can develop.[17] The occurrence of changes in the size and function of the thyroid has been noted and well documented. [13,17]

Plasma or blood lithium levels must be monitored rather closely, for the salts have a low therapeutic index. A dose that may be adequate for one patient may be potentially lethal for another.[18] Furthermore, dose requirements for an individual patient are also dependent upon changes in body physiology.

Plasma levels are best measured eight to twelve hours after the last dose of lithium salts has been ingested. The therapeutic range lies between 0.8 and 1.5 meq l^{-1} (5–11 μg ml^{-1}) approximately. Patients with levels below 0.8 meq l^{-1} respond poorly while levels above 1.4 meq l^{-1} are associated with an increased risk of side effects. Some typical manics fail to respond to treatment despite the attainment of adequate plasma lithium levels.[19]

The mode of action of lithium in manic depressive states is unknown. Whether the various biochemical effects of lithium are relevant to the clinical actions of the drug is yet to be decided.[20]

DETERMINATION OF LITHIUM

Both flame emission and atomic absorption are used routinely in the determination of lithium. Detection limits by flame emission are 3 pg ml^{-1} and by atomic absorption, 5 ng ml^{-1}.

In order to compensate for the marked matrix interferences inherent in flame emission, standards employed in lithium analyses must be prepared in a matrix comparable in composition to that of the sample. Certain investigators use an aqueous diluent similar to serum in electrolyte concentration, others prepare standards in serum pools obtained from normal sources. Analyses of serum by flame emission are usually performed on a 1 : 100 dilution.[21-23]

Matrix interferences are not encountered when determining serum or plasma lithium concentrations by flame atomic absorption. Sera diluted 1 : 25 with distilled water can be compared accurately with aqueous lithium standards treated similarly. A discernible absorbance signal is not elicited from a normal serum or plasma specimen so treated.

Various applications of atomic absorption spectrometry to analyses of lithium in biological material, some needlessly complex, have been described.[24-31]

REFERENCES

1. E. Rosenberg, *The Use of the Spectroscope in Its Applications to Scientific and Practical Medicine*, G. P. Putnam, New York, 1876.
2. F. N. Johnson (ed.) *Lithium Research and Therapy*, Academic Press, London 1975.
3. E. I. Hamilton, M. J. Minski and J. J. Cleary, *Sci. Total. Environ.*, 1, 341 (1972–73).
4. H. J. M. Bowen, *Trace Elements in Biochemistry*, Academic Press, New York, 1966.
5. E. I. Hamilton and M. J. Minski, *Sci. Total Environ.*, 1, 375 (1972–73).
6. E. T. Mellerup and O. J. Rafaelson, in *Lithium Research and Therapy*, (ed. F. N. Johnson) Academic Press, London, 1975.
7. R. P. Huelin, in *Lithium Research and Therapy*, (ed. F. N. Johnson) Academic Press, London, 1975.
8. J. F. J. Cade, *Med. J. Austral.*, 2, 349 (1949).
9. F. N. Johnson and J. F. J. Cade, in *Lithium Research and Therapy*, (ed. F. N. Johnson) Academic Press, London, 1975.
10. A. B. Garrod, *Gout and Rheumatic Gout*, Walton and Maberly, London, 1859.
11. C. A. Goode, *Amer. J. Med. Sci.*, 25, 273 (1903).
12. S. R. Platman and R. R. Fieve, *Archs. Gen. Psychiat.*, 19, 659 (1968).
13. M. Schou, A. Amdisen and K. Thomsen, *Acta Psychiat. Scand., Suppl.*, 203, 153 (1968).
14. B. M. Saron and R. Gaind, *Clin. Toxicol.*, 6, 257 (1973).
15. T. A. Ramsey, J. Mendels, J. W. Stokes and R. G. Fitzgerald, *J. Amer. Med. Assoc.*, 219, 1446 (1972).
16. I. Singer, D. Rotenberg and J. B. Preschett, *J. Clin. Invest.*, 51, 1081 (1972).
17. L. Vacaflor, in *Lithium Research and Therapy*, (ed. F. N. Johnson) Academic Press, London, 1975).

18. The Late H. I. Coombs, R. R. H. Coombs and U. G. Mee, in *Lithium Research and Therapy*, (ed. F. N. Johnson) Academic Press, London, 1975.
19. M. Peet, in *Lithium Research and Therapy*, (ed. F. N. Johnson) Academic Press, London, 1975.
20. M. Schov, *Ann. Rev. Pharmacol. Toxicol.*, **16**, 231 (1976).
21. A. Amdisen, *J. Clin. Lab. Investigation*, **20**, 104 (1967).
22. A. L. Levy and E. M. Katz, *Clin. Chem.*, **16**, 840 (1970).
23. R. Robertson, K. Fritze and P. Grof, *Clin. Chim. Acta*, **45**, 25 (1973).
24. B. R. Little, S. R. Platman and R. R. Fieve, *Clin. Chem.*, **14**, 1211 (1968).
25. J. Pybus and G. N. Bowers, Jr., *Clin. Chem.*, **16**, 139 (1970).
26. J. L. Hansen, *Amer. J. Med. Technol.*, **34**, 11 (1968).
27. K. Richter, *Z. Med. Labortechnik*, **13**, 263 (1972).
28. K. R. Johnson, *Med. Lab. Technol.*, **30**, 61 (1973).
29. F. J. M. J. Maessen, F. D. Posma and J. Balke, *Anal. Chem.*, **46**, 1445 (1974).
30. J. K. Grime and T. J. Vickers, *Anal. Chem.*, **47**, 432 (1975).
31. G. H. Hisayasu, J. L. Cohen and R. W. Nelson, *Clin. Chem.*, **23**, 41 (1977).

CHAPTER 17

MANGANESE

Manganese was first shown to be required by plants and micro-organisms more than 50 years ago.[1] Its essentiality for animals was demonstrated somewhat later.[2,3]

Manganese activates many enzymatic reactions associated with the metabolism of organic acids, carbohydrates, nitrogen and phosphorus. The element is involved in the photosynthesis of plants, being firmly bound, apparently, to lamellae of chloroplasts. More manganese is contained in roots than in leaves.

Manganese is essential for normal growth, skeletal formation, and for normal reproductive function in mammals and poultry. For example, it has been shown to be necessary for the synthesis of chondroitin sulfate in chicks.[4] Deficiency states have been induced in animals by feeding them manganese deficient diets.

An estimated 3–7 mg of manganese are ingested daily with a well-balanced diet. Nuts and cereals are richest in manganese, followed in order by dried fruits, roots, fresh fruits, non-leafy vegetables, animal tissue, poultry and poultry products, fish and sea foods. Diets consisting primarily of milk, sugar refined cereals and little fruits and vegetables could contain insufficient manganese.[5] Tea is rich in manganese, one cup containing approximately 1.3 mg.

Although possible manganese deficiencies in man had not been demonstrated until fairly recently, Schroeder and his co-workers[6] had been of the opinion for a long time, that certain disease states—for example diabetes, pregnancies involving nervous instability and convulsions, disorders of bony and cartilaginous growth in infants and children, rheumatoid arthritis, certain types of sterility in males and females—should be investigated for possible manganese deficiencies. It might also be wise to monitor the manganese status of patients on long-term parenteral therapy.

One case of apparent manganese deficiency occurring in a human volunteer in a vitamin K deficiency study on a metabolic ward was reported by Doisy.[7] Changes in hair and beard color, low hair growth, and weight loss were noted. This clinical picture was duplicated in chicks fed a similar diet. The experiment was not repeated in other humans, however.

Hydralazine, or aprescoline, a commonly used agent in treating hypertension, has been shown to bind manganese *in vitro*[8] and *in vivo*[9] in rats. There is limited evidence that a manganese deficiency may occur in hydralazine disease, a picture simulating systemic lupus erthematosis, which arises in some hypertensive subjects on hydralazine therapy.[10]

DISTRIBUTION OF MANGANESE IN MAMMALS

Manganese tends to be higher in tissues rich in mitochondria. The element is also involved in melanogenesis;[11] darker hair and feathers containing more manganese and melanin. Different tissues seem to contain comparable levels of manganese irrespective of the species from which these are obtained. Table 17.1 lists a comparison of the manganese contents of various tissues.

TABLE 17.1
Concentration of manganese in human and animal tissues ($\mu g\ g^{-1}$)

Tissue	Man[12]	Rabbit[13]	Average for range of species[13]
Bone	—	—	3.3
Adrenals	0.2	0.67	0.40
Brain	0.34	0.36	0.40
Heart	0.23	0.28	0.34
Kidney	0.93	1.2	1.2
Liver	1.68	2.1	2.5
Lung	0.34	—	—
Muscle	0.09	0.13	0.18
Ovaries	0.19	0.60	0.55
Pancreas	1.21	1.6	1.9
Spleen	0.22	0.22	0.40
Testes	0.19	0.36	0.50
Hair	—	0.99	0.80

The manganese content in the liver and lung, unlike that of other trace elements, varies little with age.[6,12] Most of the manganese in mammalian liver is concentrated in the arginase fraction.[13-15]

MANGANESE IN BLOOD

Bowen,[16] using neutron activation analysis, found human blood to contain 16–32 $\mu g\ l^{-1}$ of manganese. Cotzias *et al.*,[17,18] also employing neutron activation, found that whole blood contained 5.7–11.1 μg l^{-1}; serum, 0.4–1.6 μg l^{-1}; and erythrocytes, 22.3–24.7 μg l^{-1}.

It has been reported that manganese levels are invariably elevated following the occurrence of a myocardial infarction. Hedge et al.[19] considered manganese elevation to be a better diagnostic criterion of coronary occlusion than a rise in serum glutamic oxaloacetic transaminase (SGOT).

Cotzias and co-workers[20] found significantly elevated red cell manganese levels in patients with rheumatoid arthritis, while serum manganese remained essentially with the normal range.

Serum manganese is selectively and almost totally bound to the β_1 globulin fraction.[21,22] A manganese compound, thought to be a manganese-porphyrin compound is found in human and rabbit erthrocytes.[23]

Manganese concentrations in cerebrospinal fluid are reported to be equivalent to one half the levels found in plasma.[24]

Changes in the blood manganese concentration have been reported to occur in epilepsy. According to Pippenger,[25] these variations (elevations and decreases) are apparently dependent upon the anti-epileptic drugs employed in therapy.

MANGANESE TOXICITY

Though man and animals can ingest large doses of manganese in their diets without experiencing untoward effects, the metal as encountered through industrial exposures does induce toxicities. Symptoms observed in man include psychomotor instability and hallucinations. The rigidity and movements of the limbs observed are akin to those of Parkinsonism and also suggest a similarity to Wilson's disease.[26]

Manganese occurs as an oxide in minerals (pyrolusite, braunite, manganite). The metal itself is employed in the manufacture of steel, and of alloys, such as ferromanganese and copper manganese. The acetate, borate, oleate, and carbonate salts find use as drying compounds in varnishes and oils. The acetate is also utilized as a mordant in dyeing, and the carbonate as a pigment 'manganese white'. The chloride is used in batteries. Manganese dioxide has various uses in making amethyst glass, painting on porcelain, etc.

Exposures arise from inhaling and swallowing manganese dusts and fumes. The mining and smelting of manganese ores are the most hazardous processes. Harmful levels are present in plants manufacturing steel and maganese alloys, but exposure is less in dry battery manufacture and in the production of paints, varnishes and fireworks.

The earliest report of neurological lesions occurring in man from exposure to manganese dusts is that of Couper[27] in 1837. He described five cases occurring in a pyrolusite mill.

Fifty years ago von Oettingen[28] described the known pharmacology of manganese in depth. Normally, the element is absorbed slowly and incompletely from the gastrointestinal tract. Manganese is excreted primarily by the intestine.

In humans, poisoning is due to massive inhalation of manganese dusts and fumes; only the lungs and blood seem to constitute sites of high manganese concentrations. The distribution in other organs seems to remain normal.[29]

Monitoring of blood and serum manganese levels is considered to be the best means of assessing exposure. The determination of urinary manganese concentrations has little value since the colon is the major excretory route. Blood levels of 1–47 μg% have been reported in workers exhibiting signs of illness related to overt manganese exposures.[30]

Information concerning tissue deposition in human manganese toxicities is sparse, indeed,[27,31] as is similar information in Parkinsonism, an entity with very similar symptomatology. The status of analytical technology in clinical areas where such cases are seen, generally represents the major factor in this lack of data. Until the advent of atomic absorption spectrometry, neutron activation analysis was the only instrumental technique with adequate sensitivity for manganese analysis on realistic human biological samples.

As mentioned previously (Table 12.3), more than a decade ago we were fortunate in being presented with an opportunity to examine various sections of normal brains, together with a brain from a victim of Wilson's disease, and a brain from a victim of Parkinson's disease. Except for the substantia nigra in the latter, wherein a manganese content of 100 μg g^{-1} wet weight was found, the manganese content in all brain sections examined was not remarkable. Concentrations in 'brain tissue' as reported in the literature range between 0.34 and 0.4 μg g^{-1} wet weight.

The general appearance of the substantia nigra in the Parkinsonian subject was noteworthy—being indurated and markedly pigmented compared to the like sections obtained from the other subjects.

ANALYSIS OF MANGANESE

Colorimetric

Most of the early studies of the manganese content in biological materials applied the Willard-Greathouse periodate method[32] or a modification thereof. Materials are wet or dry ashed. Then the acidic solution of manganous salts is oxidized to permanganate by means of periodic acid or its salts, and the resulting red color is measured.

Constituents such as calcium in the biological matrix interfere, the sensitivity is also inadequate. Skinner and Peterson,[33] employing samples averaging 15 g, analyzed animal tissues by the periodate procedure. As little as '0.01 mg of manganese' was determined satisfactorily.

The formaldoxime method is said to be five times more sensitive than the periodate method. However, its application to biological materials is fraught with difficulties. Trivalent manganese, in an alkaline medium, forms a soluble reddish-brown complex with formaldoxime within a few minutes and this remains stable for hours.[34] Interferences from copper and iron can be prevented by complexing with cyanide. Colored compounds are also formed with nickel and cobalt, and the vanadium formaldoxime complex is also a red color. Bismuth, mercury, and titanium generate a turbidity. Calcium will precipitate as a phosphate in an alkaline medium.

The formaldoxime procedure has been applied in determining manganese contents in plant materials.[35]

Manganese(IV) and (VII) form highly colored complexes with various organic compounds, o-tolidine,[36] tetramethyldiaminodiphenyl methane[37,38] and benzidine[39,40] being among the reagents that have been used.

The reaction of 4,4-tetramethylaminotriphenyl methane and manganese, yielding a yellow oxidation product, has been applied in determining traces of manganese in biological materials;[41] a detection sensitivity of 0.1 ng is claimed. Chloride interferes with the reaction, and chromium and vanadium also produce a yellow color with this reagent. Since the complex is extractable with organic solvents, it could no doubt be applied to advantage in conjunction with a modern instrumental technique.

At pH 9–12 manganese (II) is quantitatively extractable as a benzoyl acetate.[42]

Manganese forms a red-brown complex with 2-nitroso-1-naphthol that is extractable.[43] It also forms an extractable complex with salicylaldoximate.

At pH 6–9 manganese is extractable as a diethyldiammonium diethyl dithiocarbamate[44] and as a diethyldithiocarbamate.[45] We have applied chelation with the latter compound and extraction into methylisobutyl ketone (MIBK) to manganese analysis by both flame and flameless atomic absorption spectrometry.

Manganese forms an extractable complex with 8-hydroxyquinaldine at pH 11–12.[46] Other metals are masked by cyanide. EDTA (Versenate) prevents the extraction of the manganese-8-hydroxyquinaldine complex.

Cupferron over the pH range 4.5–9.5 extracts approximately 15% of the manganese present.[47]

While manganese forms fluorescent complexes with 8-hydroxyquinoline,[48,49] carminic acid,[50] and Rhodamine B,[49] these reactions are not selective. Similar fluorescent complexes are formed by other metals; copper, iron, cobalt, mercury, antimony, among them. Consequently, useful applications of fluorimetry to the determination of manganese in biological material have not been described.

Neutron activation analysis

The work of Bowen in 1956,[16] and later that of Cotzias and his colleagues[17,18,20-24,29], concerning the analysis of manganese in human materials by activation analysis, has already been cited. Olehy *et al.*[51] in 1966 described the activation analysis of manganese, barium, calcium, copper, etc. in human erythrocytes and plasma.

More recently, Henzler and co-workers[52] employed the technique for determining the manganese content in biological materials, after first isolating the element by chelation–extraction. Arsenic, cadmium, cobalt, copper, mercury, antimony, selenium, and zinc are among the trace metals analyzed in a similar manner.

Neutron activation analysis has been applied in measuring manganese contents in chemical reagents[53] and in petroleum.[54]

Emission—spectrography

Manganese was among the metals Sheldon and Ramage[55] were able to identify and quantify in small pieces of tissue and 0.1 ml aliquots of blood. An oxy–acetylene flame and a Hilger quartz spectrograph were utilized in these analyses.

Butt and co-workers[56] studied trace metal patterns in disease states such as hemochromatosis and Laennec's cirrhosis by spectrographic means.

More recently, Hambidge[57] analyzed serum, urine, and hair for their manganese content. Samples were ashed in a low temperature asher and dissolved in hydrochloric acid prior to analysis by a DC arc spectrograph in an argon atmosphere. Manganese has also been determined in plant ash.[58]

The determination of this element as well as others, at the 5–500 ng l^{-1} level by DC arc spectrography[59] has been described. Sample concentration by evaporation is then followed by direct determination.

According to Pinta[60] the most common approach for metal analysis of waters by flame or plasma emission methods is to concentrate the sample by chelation and extraction techniques prior to instrumental analysis.

Atomic absorption spectrometry

Detection limits for manganese by flame absorption spectrometry are of the order of 5 ng ml^{-1} at the 279.5 nm resonance line. Limits are of course extended by flameless atomization.

A number of methods for determination of manganese in blood,[61-66] urine,[67] animal tissues,[68,69] and plant material[70] have been described, some of which seem quite practical. However, one is tempted to take exception to others. For example, certain flameless procedures recommend charring aliquots at 1350 °C, whereas we have noted apparent losses at temperatures above 600 °C. The accuracy of direct analysis procedures is open to question for reasons stated previously (see Chapter 10).

As indicated above, combining chelation of manganese by sodium diethyl dithiocarbamate at pH 6–9 and extraction into methylisobutyl ketone with analysis in the graphite furnace is a practical approach. The extract is dried at 100 °C for 20 s, charred at 300 °C for 40 s, and atomized at 2400 °C for 6 s and background correction is used.

REFERENCES

1. J. S. McHargue, *J. Agr. Res.*, **24**, 781 (1923).
2. A. R. Kemmerer, C. A. Elvehjem and E. B. Hart, *J. Biol. Chem.*, **92**, 623 (1931).
3. J. Waddell, H. Steenbock and E. B. Hart, *J. Nutr.*, **4**, 53 (1931).
4. R. M. Leach, Jr., A. M. Muenster and E. M. Wien, *Arch. Biochem. Biophys.*, **133**, 22 (1969).
5. E. J. Underwood, *Trace Elements in Human and Animal Nutrition*, Academic Press, New York, 1971.
6. H. A. Schroeder, J. J. Balassa and I. H. Tipton, *J. Chronic Dis.*, **19**, 545 (1966).
7. E. A. Doisy, Jr., in *Proceedings of the Missouri 6th Annual Conference on Trace Substances in Environmental Health*, (ed. D. D. Hemphill) University of Missouri Press, Columbia, 1973.
8. H. M. Perry, Jr. and H. A. Schroeder, *Amer. J. Med. Sci.*, **228**, 405 (1954).
9. L. S. Hurley, D. E. Woolley, F. Rosenthal and P. S. Timiras, *Amer. J. Physiol.*, **204**, 493 (1963).
10. P. Comens, *Amer. J. Med.*, **20**, 944 (1956).
11. G. C. Cotzias, P. S. Papavasiliou and S. T. Miller, *Nature (London)*, **201**, 1228 (1964).
12. I. H. Tipton and M. J. Cook, *Health Phys.*, **9**, 103 (1963).
13. H. Fore and R. A. Morton, *Biochem. J.*, **51**, 594 (1952).
14. S. Edelbacher and H. Bauer, *Naturwissenshaften*, **26**, 26 (1938).
15. S. J. Bach and D. B. Whitehouse, *Biochem. J.*, **57**, XXXI, 1954.
16. H. J. M. Bowen, *J. Nucl. Energy*, **3**, 18 (1956).
17. G. C. Cotzias, S. T. Miller and J. Edwards, *J. Lab. Clin. Med.*, **67**, 836 (1966).
18. P. S. Papavasiliou, S. T. Miller and G. C. Cotzias, *Amer. J. Physiol.*, **211**, 211 (1966).
19. B. Hedge, G. C. Griffith and E. M. Butt, *Proc. Soc. Esptl. Biol. Med.*, **107**, 734 (1961).
20. G. C. Cotzias, P. S. Papavasiliou, E. R. Hughes, L. Tang and D. C. Borg, *J. Clin. Invest.*, **47**, 992 (1968).
21. G. C. Cotzias and P. S. Papavasiliou, *Nature (London)*, **195**, 823 (1962).
22. A. C. Foradori, A. J. Bertinchamps, J. M. Gulebon and G. C. Cotzias, *J. Gen. Physiol.*, **50**, 2255 (1967).
23. D. C. Borg and G. C. Cotzias, *Nature (London)*, **182**, 1677 (1958).
24. G. C. Cotzias and P. S. Papavasiliou, *Nature (London)*, **195**, 823 (1962).
25. C. E. Pippenger, personal communication, June 1978.
26. M. M. Canavan, S. Cobb and C. K. Drinker, *Arch. Neurol. Psychiat.*, **32**, 501 (1934).
27. J. N. Cumings in *The Scientific Basis of Medicine Annual Reviews*, The Athlone Press, London, 1965.
28. W. F. von Oettingen, *Physiol. Rev.*, **15**, 175 (1935).

29. G. C. Cotzias, *Physiol. Rev.*, **38**, 503 (1958).
30. R. Penalver, *Ind. Med. Surg.*, **24**, 1 (1955).
31. R. H. Flinn, P. A. Neal, W. H. Reinhart, J. M. Dallaville, W. B. Fulton and A. E. Dooley, U.S. Public Health Service Bulletin No. 247, Washington DC, 1940.
32. H. H. Willard and L. H. Greathouse, *J. Amer. Chem. Soc.*, **39**, 2366 (1917).
33. J. T. Skinner and W. H. Peterson, *J. Biol. Chem.*, **88**, 347 (1930).
34. A. Gottlieb and F. Hecht, *Mikrochemie*, **35**, 337 (1950).
35. E. G. Bradfield, *Analyst*, **82**, 254 (1957).
36. L. Forman, *J. Amer. Water Works Assoc.*, **21**, 1212 (1929).
37. R. G. Harry, *J. Soc. Chem. Ind.*, **50**, 434T (1931).
38. D. J. D. Nicholas, *Nature (London)*, **157**, 696 (1946).
39. R. C. Stratton, J. B. Ficklen and W. A. Hough, *Ind. Eng. Chem. (Anal. Ed.)*, **4**, 2 (1932).
40. A. C. Wiese and B. C. Johnson, *J. Biol. Chem.*, **127**, 203 (1939).
41. E. M. Gates and G. H. Ellis, *J. Biol. Chem.*, **168**, 537 (1947).
42. J. Stary and E. Hladky, *Anal. Chim. Acta.*, **168**, 537 (1947).
43. G. Gorbach and F. Pohl, *Mikrochem.*, **38**, 258 (1951).
44. P. F. Wyatt, *Analyst*, **78**, 656 (1953).
45. H. Bode, *Z. Analyt. Chem.*, **143**, 182 (1954).
46. K. Motojima, H. Yoshida and T. Imanashi, *Japan Analyst*, **11**, 1028 (1962).
47. J. Stary and J. Smizanska, *Anal. Chim. Acta.*, **29**, 546 (1963).
48. B. K. Pal and D. E. Ryan, *Anal. Chim. Acta*, **47**, 35 (1969).
49. C. E. White and R. J. Argauer, *Fluorescence Analysis—A Practical Approach*, Marcel Dekker, New York, 1970.
50. T. S. West, in *Trace Characterization—Chemical and Physical*, (eds W. W. Meinke and B. F. Scribner) NBS monograph 100, US Government Printing Office, Washington DC, 1967.
51. D. A. Olehy, R. A. Schmett and W. F. Bethard, *J. Nucl. Med.*, **7**, 917 (1966).
52. T. E. Henzler, R. J. Korda, P. A. Heimke, M. R. Anderson, M. M. Jimenez and L. A. Hasken, *J. Radioanal. Chem.*, **20**, 649 (1974).
53. L. Kosta and V. Ravnik, *Radiochem. Radioanal. Letter*, **7**, 295 (1971).
54. K. K. Shah, R. H. Filby and W. A. Haller, *J. Radioanal. Chem.*, **6**, 185 (1970).
55. J. H. Sheldon and H. Ramage, *Biochem. J.*, **25**, 1608 (1931).
56. E. M. Butt, R. E. Nusbaum, T. C. Gilmour and S. L. DiDio, in *Metal Binding in Medicine*, (ed. M. J. Seven) Lippincott, Philadelphia, 1960.
57. K. M. Hambidge, *Anal. Chem.*, **43**, 103 (1971).
58. M. H. Chaplin and A. R. Dixon, *Appl. Spectros.*, **28**, 5 (1974).
59. D. Pepin and A. Gardes, *Analusis*, **2**, 549 (1973).
60. M. Pinta, *Modern Methods for Trace Element Analysis*, Ann Arbor Science, Ann Arbor, Mich., 1978.
61. R. T. Ross and J. G. Gonzalez, *Bull. Environ., Contam. Toxicol.*, **12**, 470 (1974).
62. R. A. A. Muzzarelli and R. Rochetti, *Talanta*, **22**, 683 (1975).
63. S. B. Gross and E. S. Parkinson, *Atomic Abs. Newsletter*, **13**, 107 (1974).
64. K. C. Beck, J. H. Reuter and E. M. Perdue, *Geochim. Cosmochim. Acta*, **38**, 341 (1974).
65. D. J. D'Amico and H. L. Klawans, *Anal. Chem.*, **48**, 1469 (1976).
66. J. P. Buchet, R. Lauwerys and H. Roels, *Clin. Chim. Acta*, **73**, 481 (1976).
67. P. G. Van Ormer and W. G. Purdy, *Anal. Chim. Acta*, **64**, 93 (1973).
68. C. J. Pickford and G. Rossi, *Atomic Abs. Newsletter*, **14**, 78 (1975).
69. G. B. Belling and G. B. Jones, *Anal. Chim. Acta*, **80**, 279 (1975).
70. E. B. Bradfield, *Analyst*, **99**, 403 (1974).

CHAPTER 18

MERCURY

Mercury, both as the metallic form (quicksilver) and the sulfide (cinnabar), was known to different people around the world for thousands of years and has played a prominent role in therapeutics, alchemy and folklore. Aristotle mentions the metal in his *Meteorologica*, referring to it as fluid silver.

Procedures for the reduction of mercury from its ores were understood long before 400 BC apparently. Theophratus, a disciple of Aristotle, discusses a method whereby quicksilver is obtained by rubbing cinnabar with vinegar in a 'brass mortar with a brass pestle'.[1] Both Dioscorides and Pliny describe the distillation apparati used to recover quicksilver.

Dangers inherent in mining quicksilver and in the reduction of mercury ores were appreciated by the ancients. Romans initially only used slave labor in their mines. Later these were worked by convicts.

Agricola, in his discussion of the mining and cleaning of quicksilver, suggests means of reducing exposure in the workplace:

'The pots, lest they should be defective, are moulded from the best potters' clay, for, if there are defects, the quicksilver flies out in the fumes. If the fumes give out a very sweet odour, it indicates that the quicksilver is being lost, and since this loosens the teeth, the smelters and others standing by, warned of the evil, turn their backs to the wind which drives the fumes in the opposite direction. For this reason, the building should be open around the front and the sides and exposed to the wind'.

Mercury and its salts enjoyed medical uses in ancient India, China etc., but until the sixteenth century when Paracelsus noted its effectiveness in treating syphillis, they were not used therapeutically in the western world. The Greeks and Romans considered them too toxic for therapeutic use. Following Paracelsus, however, though its hazard potential was appreciated, mercury came to be considered a panacea, a universal antidote for all ills.

149

Woodall,[2] circa 1639, describes mercury as 'the hottest, the coldest, a true healer, a wicked murderer, a precious medicine, and a deadly poison—a friend that can flatter and lie'.

Some of the mixtures employed extensively until fairly recent times are rather fascinating. For example, a cathartic mixture 'mercury mass', also called 'blue pill' or 'blue mass', consisted of 32–34% of metallic mercury mixed with honey, licorice, althea, glycerol and some mercury oleate. It caused systemic mercury poisonings.

Mercurous chloride, or calomel, enjoyed great popularity as a laxative. The compound is still encountered in some proprietary preparations.

Mercuric oxide, salicylate and chloride salts were used as topical antiseptics. Red mercuric sulfide preparations were used in treating syphillis as were mercuric benzoate, mercurous acetate and mercurous iodide.

The importance of mercury and its compounds as therapeutic agents has declined in recent years. Organic mercurials, such as, metaphen and merbromin (mercurochrome) exert a weak bacteriostatic action and are employed as local antiseptic agents. Some compounds like mercurin are still used as diuretics to a limited extent.

TOXIC ACTIONS OF MERCURY

Mercuric ions act as potent enzyme inhibitors, protein precipitants, and corrosives. Mercury, like arsenic, has a great affinity for sulfhydryl groups. It also combines with phosphoryl, carboxyl, amide and amine groups.

The biochemical toxicology of the different compounds varies greatly with the chemical form and entrance route into the body. Divalent inorganic mercurials, being more soluble than the monovalent compounds, are more rapidly absorbed and hence are much more toxic following oral administration. In lipoid vehicles, inorganic mercurials can be absorbed through the intact skin. These compounds can also be absorbed through mucous membranes. One occasionally sees the occurrence of mercurialism, in teething children, traceable to the 'old-wives' practice of rubbing ammoniated mercury ointment on the sore gums.

Elemental mercury, being lipid soluble, can be absorbed through the intact skin. However, negligible amounts of elemental mercury are absorbed through the gastric mucosa. Experiments on rats have shown that the gastrointestinal absorption of elemental mercury is less than 0.01%.[3] Hence mercury, swallowed during the course of breaking a thermometer in one's mouth accidentally, is of lesser concern than are the glass particles which may be swallowed concurrently.

On the other hand mercury vapour, even in minimal concentrations, is hazardous. It has been estimated from animal experiments that 75–100% of the mercury vapour inhaled is taken up by the animal body.[4] Approximately 75% of mercury vapor concentrations between 50 and 350 μg m^{-3} are reported to be retained by exposed man.[5,6] Inhaled mercury rapidly leaves the lungs for the circulatory system.[7,8] Toxic effects are produced after the elemental metal diffusing into the erythrocytes and other tissues is oxidized to the mercuric ion.[9] Clarkson, et al.[10] in their studies on the equilibration of mercury vapor with blood have demonstrated the rapid biotransformation in vitro of mercury from the elemental form to the mercuric ion. Catalase is considered to be the enzyme responsible.[11]

The extent of mercury uptake by tissues is dependent upon the form in which it is initially introduced into the body. For example, Berlin et al[12] showed that uptake by the brain was ten times higher following exposure to mercury vapor than after injection of mercuric salts. This greater penetration and uptake can be explained in part, as observed by Magos,[7,9] that in the early period following exposure to mercury vapor, mercury is partly distributed in the blood as mercury vapor and so penetrates the blood–brain barrier more readily.

The brain is a critical organ in long-term exposure to mercury vapor. Once in the brain, the elemental mercury is oxidized to the mercuric ion. As a consequence, elimination is slow. Mercury tends to accumulate in certain nerve cells.[12-14] Thus, as with other toxic metals such as lead, distribution in the various sections of the brain is uneven.

Uptake of mercury vapor by blood, both in vivo and in vitro, is inhibited by ethyl alcohol. Nielsen-Kudsk[15] demonstrated that a blood alcohol concentration of 40 mg% (0.04%) produces a maximum decrease of 30% in the pulmonary absorption of mercury in vivo. Inhibition of the uptake of mercury by blood in vitro reaches 60% at ethyl alcohol concentrations of 0.2%.

Alkyl mercurials are of great importance as contaminants in the environment, the greatest concern being contamination of the food chain. Chloroethyl mercury is employed as a fungicide for treating seeds.

In addition to agricultural uses, phenyl mercury, the least toxic of the organo-mercurials, is also employed as a preservative in the manufacture of paints and adhesives, and in the paper industry as a slime-mold retardant.

Methyl mercury compounds are the most toxic of the alkyl mercurials used, as well as the most important of the environmental mercury contaminants. The fact that elemental mercury and mercury compounds can be converted to methyl mercury by micro-organisms is a matter of great interest and some concern.[16,17] This biotransformation of mercurials to methyl mercury may be a possibility in humans as well.

The first recorded cases of methyl mercury poisoning were reported by Edwards[18,19] in 1865. Two laboratory technicians at St. Bartholomew's Hospital, using dimethyl mercury whilst researching into the valency of metals and metallic compounds, began complaining of numbness of the hands, deafness, poor vision, sore gums, and ataxia. They both died subsequently although a third technician recovered.

More recent concern with alkyl mercurials stems from repeated outbreaks of methyl mercury poisoning in Japan[20] and elsewhere.[21] Possibly the most famous incidents were those occurring in Minimata, Japan from 1953 to 1960. Minimata disease, so called, was caused by the ingestion of fish and shellfish contaminated by industrial effluents containing methyl mercury. Poisonings in Iraq arising from the consumption of bread made from treated grains (despite warnings) occurred later.[22]

To prevent further situations of a similar nature, the mercury contents of waters, fish, and other foods are monitored fairly regularly in industrialized areas around the world.

INDUSTRIAL MERCURY POISONING

In the past, manufacture of felt (the hatters' trade) was a source of industrial mercurialism, second only to mining of the metal in its severity and frequency of occurrence. Mercury is used, minimally around the world today in hat manufacture and fur processing. Poor housekeeping practices in plants manufacturing mercury thermometers and barometers can also cause chronic poisonings.

Before matters were brought under control a few years ago, employees in a huge plant in the US showed urinary mercury excretions of 1100–2000 μg l^{-1}. Subsequently, most levels fell below 100 μg l^{-1}.

Repeated mercury spills have been implicated as causing elevated urinary mercury levels (up to 330 μg l^{-1}) among persons engaged in amalgam preparation in dental offices.[23] Laboratories, both academic and hospital, were notorious for their *blasé* disregard of the hazards of mercury. Spills were permitted to vaporize. Most cases of intoxication went unrecognized. Goldwater *et al.*[24] documented one such incident in a university laboratory.

Over exposure to mercury in plants manufacturing chlorine is also significantly hazardous.[25]

MERCURY INTOXICATION

Acute mercury poisoning can follow the ingestion of rapidly absorbed mercury preparations, both inorganic and organic. Inhalation of concentrated vapors of elemental mercury can cause acute intoxication. Mercurial ointments applied over a wide surface area can result in acute intoxications as well.

Mercuric chloride has corrosive actions and, when ingested, precipitates the mucous membrane proteins causing an ashen appearance of the mouth, pharynx, and gastric mucosa. The affected areas are intensely painful; there is profuse vomiting and diarrhea, and shock ensues.

Should poisoning occur from alkyl mercury compounds or from inhaling metallic mercury fumes, the characteristic syndrome consists of pneumonitis, lethargy or restlessness, fever, cough, tachypnea, chest pain, and cyanosis, as well as vomiting and diarrhea.

Systemic effects of poisoning begin within a few hours, can last for days, and may terminate fatally. Phenyl mercury and inorganic mercury compounds exert diffuse activity on the capillary walls of the excretory sites (kidney, colon, mouth). Subjects complain of a strong metallic taste, sore gums, and excessive salivation, teeth become loosened, and gingiva subsequently become discolored.

Systemic effects of acute poisoning from ethyl and methyl mercury and from fumes of metallic mercury are referrable to the central nervous system. One observes lethargy, excitement, tremor, and hyperreflexia.

Mercury is an intense kidney poison. The renal lesions produced are confined largely to the tubular epithelium, though the glomeruli are also injured. Disturbed kidney function is manifest within a few minutes after mercury reaches the circulation. The first response is a diuresis if the circulation is adequate. Soon, because of the extensive renal damage, oliguria and finally anuria result.

Occasionally, mercury intoxication can result from rather bizarre accidents. We recall the incident of a family of six, ranging between 4 and 35 years of age, who were exposed to a fairly concentrated metallic mercury mist for hours. It seems a few pounds of mercury kept as a display of sorts by the head of the household, a laboratory technician, was accidentally spilled onto a shag rug about 20×24 feet in size. The unwise attempt to take up the spill with vacuum cleaners succeeded only in dispersing the mercury into the air. Within a few hours all members of the family complained of fever, cough, and restlessness. Those who obviously had the greatest exposure developed a pneumonitis and experienced chest pain. Urinary excretions of mercury by these patients ranged between 500 and 2000 μg l^{-1} within 24 h after exposure. Normal urinary mercury excretions are considered to be less than 10 μg l^{-1}. All recovered; the rug was discarded and the premises treated with flowers of sulfur.

Chronic mercury poisoning can follow exposure in industry, on the farm, or in the home, some rather unexpected. For example, a suburban teenager became chronically poisoned after being permitted to paint her bedroom. She splattered paint lavishly over her person as she covered the walls and then proceeded to sleep in the room. The latex paint used contained a mercurial fungicide.

A middle-aged female who might have been the Mad-Hatter's sibling appeared in the outpatient clinic with miscellaneous complaints. Signs of acrodynia were noted. A history of daily intake of many Carter's Little Liver Pills, which contain calomel, was obtained after much questioning. Her urinary mercury level was 450 μg l^{-1}.

Mercury preparations can produce hypersensitivity reactions which often go unrecognized. Reactions have been caused by cosmetic preparations, gardening materials, etc. 'Pink' disease, occasionally observed in infants and children treated with mercurous chloride (calomel), is thought by some investigators to be a manifestation of sensitization to mercury.[26-28]

TABLE 18.1
Mercury content of tissues following use of mercurochrome for treating omphalocoele in an infant

Tissue	Mercury content (μg g^{-1} wet weight)
Blood	0.83
Brain (cortex)	0.28
Kidney	2.12
Liver	0.77

Even today instances of iatrogenic mercury poisoning, some fatal, do occur. We recall an instance where a newborn infant with an omphalocoele developed bullae during treatment with mercurochrome and succumbed in ten days. The mercury content of tissues obtained at autopsy are summarized in Table 18.1.

Highest concentrations were found in the kidney, followed by blood, liver, and brain in decreasing order. Note that a blood level of 0.83 μg ml^{-1} (83 μg%) was attained. Blood levels of normal subjects are less than 3 μg%, most values ranging between 0.5 and 1.0 μg%. The older literature[29] lists values of 6–7 μg% as normal. These differences can be explained on the basis of the analytical methodologies applied. As with other metal analyses, most values cited in the older literature were obtained with less specific colorimetric procedures.

Little information is available in the literature concerning the tissue mercury content in normal subjects. Brown,[30] circa 1954, suggested that normal

TABLE 18.2
Concentrations of tissue mercury in normal tissue and following exposure to methyl mercury, mercury by inunction, and mercury bichloride (μg g^{-1} fresh tissue)

Tissue	Normal	Following ethyl mercury	Mercury via inunction[a]	Mercury bichloride[a]
Cerebral cortex	0.029–0.038	1.8–4.8	1.5	1.0–1.7
Cerebellar cortex	0.024–0.028	2.7–5.2		
Kidney	1.26–1.31	8.0–9.1	33–63.1	16–70
Liver	0.027–0.037	6.1–6.8	12–24.5	3.2–32.0
Lung	0.013–0.015	2.7–2.9	1.0	0.3–5.7
Skeletal muscle	0.036–0.039	3.8–4.9	—	—
Heart	0.013–0.02	2.9–3.7	—	—
Spleen	0.021–0.03	1.3–2.0	1.5–6.0	1.0–4.3
Testes or ovary	0.011–0.016	1.8–2.5	—	—
Blood	0.009–0.011	15.0–15.5	—	0.15–1.2

[a] Ref. 32–33

tissue contained no more than 0.1 mg% (1 μg g^{-1}). Helmy et al.[31] compared tissue mercury levels of apparently normal subjects with those poisoned by ethyl mercury. Their findings along with other values compiled from the literature are listed in Table 18.2.

ANALYSIS OF MERCURY

Because of its volatility, even at ambient temperatures, the determination of mercury concentrations in biological material presents special challenges. In addition to the usual problems inherent in eliminating contamination from laboratory ware and reagents, there is the additional need for preventing losses of mercury while destroying organic material.

For obvious reasons mercury analyses should not be carried out in a room containing a Van Slyke apparatus. Wearing of cosmetics by technical personnel performing these analyses should be banned as eye shadow, common for daytime wear today, contains mercury salts.

Reagents of acceptable purity do contain traces of mercury. Hydrochloric acid for example, is made from chlorine that has been produced by electrolysis of sodium chloride with amalgam electrodes.[34] We found that the reagent blank of ACS-quality chemicals (sulfuric acid, potassium permanganate, stannous chloride) used in determining mercury by an atomic absorption spectrometric cold vapor technique contained between 12 and 20 μg% of mercury. A urine mercury level of 50 μg l^{-1} (5 μg%) is considered of significance in an industrial worker.

All methods of analysis require the isolation of mercury from its matrix. Classical methods for the destruction of organic material usually employed a mixture of sulfuric and nitric acids. A cold finger condenser, attached to the digestion flask, served to liquify vapors from the digest. Addition of potassium permanganate or peroxide speeded oxidation and ensured more complete destruction of the organic matter.[34] Trichloroacetic acid has been used to precipitate blood and tissue proteins and thus to free mercury.[35] Distillation of mercury in a stream of chlorine has also been used to separate mercury from biological materials.[36,37]

Stock and Cucuel,[38] about fifty years ago, isolated mercury by electrolytic deposition on a gold or silver wire followed by volatilization by heating in a capillary tube.

Mercury can be isolated from acidic solutions by extraction with organic solvents.[39]

A few procedures (circa 1940) describe isolation of mercury as a sulfide.[40,41]

Colorimetric–fluorimetric

Over the years, numerous procedures (and their modifications) involving reaction of mercury ions with dithizone have been described.[42-47] Monovalent and divalent mercury react with dithizone in acidic solutions to yield yellow and orange-coloured complexes, respectively. Measuring the 'mixed color' has been considered the best approach in determining mercury by dithizone.

Dithizone–mercury reactions are problem prone. Other metals, such as copper, silver, gold, palladium, and platinum, react with the chelate in acid solutions. Copper interference can be of significance when biological materials are being analyzed. In addition, the mercuric–dithizone complex is sensitive to light; the orange color begins assuming greenish tones fairly quickly.

Di-β-napthylthiocarbazone in mineral acids forms a red complex with mercury that is stable in light. The reaction has been applied to mercury determinations in biological specimens.[48-50]

Diphenylcarbazide and mercury form a blue–violet complex that can be utilized in the analysis of the metal.[51] However, as regards biological materials, interference by certain essential elements is significant, as copper, iron and zinc also yield colored compounds.

Mercury forms an extractable yellow complex with 2-nitroso-1-naphthol[52] and, over the pH range 2–5, mercury is quantitatively extractable as a cupferrate.[53]

It has been stated that certain dithiocarbamate compounds form quantitatively extractable chelates with mercury over a wide pH range.[54,55] We found ammonium pyrollidine dithiocarbamate and the pH range 2.8–3.8 to be preferrable for the isolation of mercury for atomic absorption spectrometric analysis.[35]

Rhodamine B,[56] quercitin,[57] and thiamine[58] form fluorescent complexes with mercury, but practical procedures employing fluorimetry for the analysis of mercury in biological materials have not yet been described.

Instrumental analysis

The necessity for more definitive and sensitive methods for determining total mercury, as well as the need for distinguishing between the different mercury species, inorganic, aryl and alkyl mercurials, has stimulated the development of gas chromatographic, neutron activation, and atomic absorption procedures. Organic mercurials were identified and estimated by gas chromatography, and total mercury determined by either of the latter two techniques. The difference between the two determinations then represented inorganic mercury present.

Gas chromatography

Westoo in Sweden[59,60] described some of the earliest investigations into the gas chromatographic determination of methyl mercury in foodstuffs. Hydrochloric acid was added to homogenates, the methyl mercury so liberated was reacted with chloride and extracted into benzene; the chloride of methyl mercury was then back extracted into aqueous cysteine to form a methyl mercury cysteine complex. After acidification with hydrochloric acid, the methyl mercury chloride formed was re-extracted into benzene and subjected to analysis by gas chromatography with an electron capture detector. Back extraction of the initial benzene extract into aqueous cysteine and subsequent re-extraction into benzene serves to eliminate many extraneous compounds of little interest which were coextracted with the methyl mercury in the initial benzene extract.

Similar approaches for the gas chromatographic analysis of mercurials have been described by others.[61,62]

Mushak and co-workers[63,64] determined inorganic mercury in waters, blood, urine, and soft-tissue homogenates by gas chromatography with electron capture detection. In essence, inorganic mercury was reacted with various reagents to form aryl and alkyl mercurials. Results obtained compared well with atomic absorption spectrometric analyses but rather poorly with neutron activation analyses. The latter technique yielded higher mercury levels.

More recently Cappon and Smith[65] described a gas chromatographic procedure for the determination of organic and inorganic mercury in various biological matrices, including blood, urine, tissue, feces, hair, water, sediment, fish, milk and grains. Tetramethyl tin served as the methylating agent for inorganic mercury.

Neutron activation analysis

Many approaches to determining mercury by activation analysis have been described, some of which determined mercury directly.[66,67] Methods involving a concentration step exhibited better detection sensitivity.

An early method by Johansen and Steinnes[68] involved irradiation of the sample followed by decomposition of the irradiated sample with nitric and sulfuric acids. Mercury was then isolated by sulfide precipitation at pH 8–9. Though a number of elements in biological materials are coprecipitated with mercury, their radioisotopes apparently did not interfere 'seriously' in the analysis if the mercury concentration was higher than 0.01 ppm.

Adsorption on to charcoal[69] as well as chelation as a dithiocarbamate and extraction[70] have been employed by some investigators to isolate the metal for neutron activation analysis.

Extraction techniques have also been used for isolating mercury for analysis by X-ray fluorescence spectrometry. Marcie[71] extracted the ammonium pyrollidine dithiocarbamate of mercury with methylisobutyl ketone and then evaporated the organic solvent on filter paper.

Boiteau and Robin[72] isolated mercury as a dithizonate prior to X-ray analysis.

Atomic absorption spectrometry

Detection limits for mercury determination by flame atomic absorption spectrometry are 0.2 μg ml^{-1}. Limits are extended about tenfold by flameless techniques. Sodium, potassium, phosphorous, and protein enhance the absorption signal of mercury.

Procedures described for atomic absorption determination of this trace metal are legion. Many current methods utilize some adaption of the cold vapor technique wherein mercury is reduced to the elemental state and swept from a solution by a stream of air into an absorption tube (cell) of an atomic absorption instrument.[73-80] Mercury vapor so measured is essentially free from interference caused by organic matter constituents.

Borohydride generation has been applied occasionally[81] in the atomic absorption spectrometric analysis of the element.

Chelation–extraction techniques described for either flame[35,82,83] or flameless atomization[84] lend themselves readily to the mercury analyses in a clinical laboratory setting.

A preferred method involves precipitation of blood or tissue proteins by 5% trichloroacetic acid for a few hours or overnight. After centrifugation, the supernatant is adjusted to pH 2.5–3.8 before chelation with ammonium pyrollidine dithiocarbamate and extraction into methylisobutyl ketone. Extracts of unknowns and standards, treated similarly, are compared at the 253.7 nm resonance line. Aliquots of 20 μl are introduced into a graphite furnace programmed to dry and char for 40 s each at 75 °C and atomized for 6 s at 2000 °C. Background correction is necessary.

Urine samples can be chelated and extracted directly after adjusting pH.

Blood, tissues and hair can be dissolved in sulfuric–nitric acids at room temperature. Mummified tissues or bone may require one to three days to dissolve. After pH adjustment, mercury is isolated by chelation-extraction.

The necessity for taking a reagent blank throughout the entire procedure cannot be over emphasized.

REFERENCES

1. G. Agricola, *De Re Metallica*, trans. by H. C. Hoover and L. H. Hoover, Dover Publications, New York, 1950.
2. J. Woodall, *The Surgeon's Mate or Military and Domestic Surgery*, London, 1639, cited in L. J. Goldwater, *Mercury: A History of Quicksilver*, York Press, Baltimore, 1972.
3. G. Bornmann, G. Henke, H. Alfes and H. Mollmann, *Arch. Toxicol.*, **26**, 203 (1970).
4. A. Hayes and A. Rothstein, *J. Pharmacol.*, **138**, 1 (1962).
5. J. Teisinger and V. Fiserova-Bergerova, *Ind. Med. Surg.*, **34**, 580 (1965).
6. F. Nielsen-Kudsk, *Acta Pharmacol. Toxicol.*, **23**, 250 (1965).
7. L. Magos, *Brit. J. Ind. Med.*, **25**, 315 (1968).
8. M. Berlin, G. Nordberg and F. Serenius, *Arch. Environ. Health*, **18**, 42 (1969).

9. L. Magos, *Environ. Res.*, **1**, 323 (1967).
10. T. W. Clarkson, J. Gatzy and C. Dalton, Division of Radiation Chemistry and Toxicology, University of Rochester Atomic Energy Project, Rochester, New York UR-582, 1961, cited in *Mercury, Mercurials, and Mercaptans*, (eds M. W. Miller and T. W. Clarkson) Charles C. Thomas, Springfield, 1973.
11. T. W. Clarkson, *Ann. Rev. Pharmacol.*, **12**, 375 (1972).
12. M. Berlin, J. Fazackerly, and G. Nordberg, *Arch. Environ. Health*, **18**, 719 (1969).
13. G. Cassano, P. Viola, B. Ghetti and L. Amaducci, *J. Neuropath. Exptl. Neurol.*, **28**, 308 (1969).
14. G. Nordberg and F. Serenius, *Acta Pharmacol. Toxicol.*, **27**, 269 (1969).
15. F. Nielsen-Kudsk, *Acta Pharmacol. Toxicol.*, **23**, 263 (1965).
16. S. Jensen and A. Jernelow, *Nature (London)*, **223**, 753 (1969).
17. A. Jernelow, in *Conversion of Mercury Compounds in Chemical Fallout*, (eds M. W. Miller and G. G. Berg) Charles C. Thomas, Springfield, 1969.
18. G. N. Edwards, *St. Bartholomew's Hospital Reports*, London, **1**, 141 (1865) cited in D. Hunter, R. C. Bomford and D. J. Russel, *Quart. J. Med.*, **9**, 193 (1940).
19. G. N. Edwards, *St. Bartholomew's Hospital Reports*, London, **2**, 211 (1886) cited in D. Hunter, R. C. Bomford and D. J. Russell, *Quart. J. Med.*, **9**, 193 (1940).
20. T. Okajima, I. Mishima and H. Tokuomi, *Intern. J. Neurol.*, **11**, 62 (1976).
21. H. B. Gerstner and J. E. Huff, *Clin. Toxicol.*, **11**, 131 (1977).
22. F. Baker, S. F. Damliyi, L. Amin-Zaki, M. Murtadha, A. Khalidi, N. Y. Al-Rawi, S. Tikriti, H. I. Dhahir, T. W. Clarkson, J. D. Smith and R. A. Doherty, *Science*, **181**, 230 (1973).
23. P. A. Gronka, R. L. Bobkoskie, G. J. Tomchick, F. Bach and A. B. Rakow, *J. Amer. Dent. Assoc.*, **81**, 923 (1970).
24. L. J. Goldwater, M. Kleinfeld and A. R. Berger, *Arch. Ind. Hyg.*, **13**, 245 (1956).
25. R. Smith, A. Vorwald, L. Patel and T. Mooney, *Amer. Ind. Hyg. Assoc. J.*, **31**, 687 (1970).
26. J. Warkany and D. Hubbard, *Lancet*, **1**, 829 (1948).
27. L. Bivings and G. Lewis, *J. Pediat.*, **32**, 63 (1948).
28. G. Fanconi and A. Botsztejn, *Helv. Paediat. Acta*, **3**, 264 (1948).
29. J. N. Cumings, *Heavy Metals and the Brain*, Charles C. Thomas, Springfield, 1959.
30. I. A. Brown, *Arch. Neurol. Psychiat.*, **72**, 674 (1954).
31. M. I. Helmy, S. A. Rahim and A. H. Abbas, *Toxicology*, **6**, 155 (1976).
32. S. Lomholt, *Arch. Derm. Syph.*, **126**, 1 (1919).
33. T. Sollmann and N. E. Schreiber, *Arch. Intern. Med.*, **57**, 46 (1936).
34. E. B. Sandell, *Colorimetric Determination of Traces of Metals*, Interscience Publishers, Inc., New York, 1959.
35. E. Berman, *Atomic Abs. Newsletter*, **6**, 57 (1967).
36. F. L. Kozelka, *Anal. Chem.*, **19**, 494 (1947).
37. I. M. Weiner and O. H. Müller, *Anal. Chem.*, **27**, 149 (1955).
38. A. Stock and F. Cucuel, *Naturwissenschaften*, **22**, 390 (1934).
39. H. C. Moser and A. F. Voigt, *J. Amer. Chem. Soc.*, **79**, 1837 (1957).
40. J. F. Reith and C. P. Van Dijk, *Chem. Weekblad*, **37**, 186 (1940).
41. A. E. Ballard and C. D. W. Thornton, *Ind. Eng. Chem. (Anal. Ed.)*, **13**, 893 (1941).
42. R. F. Milton and J. L. Hoskins, *Analyst*, **72**, 6 (1947).
43. E. P. Laug and K. W. Nelson, Jr., *J. Assoc. Offic. Agr. Chemists*, **25**, 399 (1942).
44. A. C. Rolfe, F. R. Russell, and N. T. Wilkinson, *Analyst*, **80**, 523 (1955).
45. F. R. Barrett, *Med. J. Austral.*, **2**, 411 (1955).
46. F. R. Barrett, *Analyst*, **81**, 294 (1956).

47. D. C. Abbott and E. I. Johnson, *Analyst*, **82**, 206 (1957).
48. D. M. Hubbard, *Ind. Eng. Chem. (Anal. Ed.)*, **12**, 768 (1940).
49. J. Cholak and D. M. Hubbard, *Ind. Eng. Chem. (Anal. Ed.)*, **18**, 149 (1946).
50. J. Warkany and D. M. Hubbard, *Lancet*, **1**, 849 (1948).
51. F. W. Laird and Sr. A. Smith, *Ind. Eng. Chem. (Anal. Ed.)*, **10**, 576 (1938).
52. G. Gorbach and F. Pohl, *Mikrochem.*, **38**, 258 (1951).
53. J. Stary and J. Smizanska, *Anal. Chim. Acta*, **29**, 546 (1963).
54. H. Bode, *Z. Anal. Chem.*, **144**, 165 (1955).
55. H. Bode and F. Neumann, *Z. Anal. Chem.*, **172**, 1 (1960).
56. C. E. White and R. J. Argauer, *Fluorescence Analysis: A Practical Approach*, Marcel Dekker, New York, 1970.
57. A. Brooker and A. Townsend, *Chem. Commun.*, **24**, 1660 (1968).
58. J. Holzbecher and D. E. Ryan, *Anal. Chim. Acta*, **64**, 333 (1973).
59. G. Westoo, *Acta Chem., Scand.*, **20**, 2131 (1966).
60. G. Westoo, *Acta Chem., Scand.*, **22**, 2277 (1968).
61. I. R. Kamps and B. McMahon, *J. Assoc. Offic. Anal. Chemists*, **55**, 590 (1972).
62. P. Mushak, *Environ. Health Perspec.*, **2**, 55 (1973).
63. P. Mushak, F. E. Tibbets, III, P. Zarnegar and G. B. Fisher, *J. Chromatog.*, **87**, 215 (1973).
64. P. Zarnegar and P. Mushak, *Anal. Chim. Acta*, **69**, 389 (1974).
65. C. J. Cappon and J. C. Smith, *Anal. Chem.*, **49**, 365 (1977).
66. R. H. Filby, A. I. Davis, K. R. Shah and W. A. Haller, *Mikrochim. Acta*, **6**, 1130 (1970).
67. G. F. Clemente and G. G. Mastinu, *J. Radioanal. Chem.*, **20**, 707 (1974).
68. O. Johansen and E. Steinnes, *Int. J. Appl. Radiat. Isotopes*, **20**, 751 (1969).
69. H. A. Van der Sloot and H. A. Das, *Anal. Chim. Acta*, **73**, 235 (1974).
70. T. E. Henzler, R. J. Korda, R. A. Helmke, M. R. Anderson, M. M. Jiminez and L. A. Haskin, *J. Radioanal. Chem.*, **20**, 649 (1974).
71. F. J. Marcie, *Norelco Rep. USA*, **15**, 3 (1968) cited in M. Pinta, *Modern Methods for Trace Element Analysis*, Ann Arbor Science, Ann Arbor, Mich., 1978.
72. H. L. Boiteau and M. Robin, *Coll. Rayons X Matiere*, Siemens Monaco, May 1973, cited in M. Pinta, *Modern Methods for Trace Element Analysis*, Ann Arbor Science, Ann Arbor, Mich., 1978.
73. L. Magos and T. W. Clarkson, *J. Assoc. Offic. Anal. Chemists*, **55**, 966 (1972).
74. W. H. Gutenmann, D. J. Lisk and N. Guer, *Bull. Environ. Contam. Toxicol.*, **8**, 138 (1972).
75. N. P. Kubasik, H. E. Sine and M. T. Volosin, *Clin. Chem.*, **18**, 1326 (1972).
76. A. Taylor and V. Marks, *Brit. J. Ind. Med.*, **30**, 293 (1973).
77. P. J. Nord, M. P. Kadaba and J. R. J. Sorenson, *Arch. Environ. Health*, **27**, 40 (1973).
78. V. T. Liew, A. Cannon and W. E. Huddlestone, *J. Chem. Educ.*, **51**, 752 (1974).
79. A. K. Furr, D. R. Mertens, W. H. Gutenmann, C. A. Bache and D. J. Lisk, *J. Agr. Food Chem.*, **22**, 954 (1974).
80. F. M. Teeny, *J. Agr. Food Chem.*, **23**, 668 (1975).
81. J. Toffaletti and J. Savory, *Anal. Chem.*, **47**, 2091 (1975).
82. B. B. Mesman and B. S. Smith, *Atomic Abs. Newsletter*, **9**, 81 (1970).
83. M. Ceresau, F. C. Wright, J. S. Palmer and J. C. Riner, *J. Agr. Food Chem.*, **21**, 614 (1973).
84. E. Berman, *Appl. Spectros.*, **29**, 1 (1975).

CHAPTER 19

MOLYBDENUM

Molybdenum occurs rather widely in the earth's crust, being found in igneous rocks, shales, limestone and coals. Most important molybdenum-containing ores are molybdenite—a molybdenum sulfide,[1] and wulfenite—a lead molybdenum oxide.

On an average, most soils contain about 2 ppm molybdenum.[2] However, a few soils are sufficiently rich in the element to yield a herbage containing a molybdenum concentration that is toxic to grazing animals. Other soils are so molybdenum deficient that the crop yield is poor.[3] The latter soils are acidic. Striking responses in crop yield, especially with pasture legumes, have been reported following the addition of small quantities of molybdenum ores or salts.[4]

Trace amounts of the metal are found in fresh and sea waters, 0.35 ppb and 0.01 ppm, respectively.[1] Molybdenum is present in all marine animals and plants as well as land animals[5,6] and plants.[7] It is essential to all organisms except perhaps some algae.

The first evidence of a biological role for molybdenum was reported by Bortels[8] in 1930, when he indicated that the element was an essential nutrient for azobacter. Subsequently, irregular occurrence in low concentrations in all plant and animal tissues examined was shown by Ter Meulen.[9] Steinberg[10] found molybdenum necessary for all nitrogen-fixing organisms. Arnon and Stout[11] later demonstrated that it was required by higher plants independent of its role in symbiotic nitrogen fixation.

Two groups, working independently, discovered the essential role of this trace metal in animal nutrition when they demonstrated that the flavoprotein enzyme xanthine oxidase contains molybdenum,[12-14] as well as iron. Aldehyde dehydrogenase contains both metals, while, nitrite reductase contains molybdenum only.[15] Iron and molybdenum are functional in the nitrogenases.[16]

By using highly purified diets, Reid,[17] Higgins,[18] Ellis[19] and their respective co-workers demonstrated the essentiality of molybdenum in diets of lambs and fowl. Deficient molybdenum intake apparently reduced the growth rate in experimental animals and depleted the liver of xanthine oxidase. It is thought that a deficient molybdenum intake may produce xanthine renal calculi in sheep.[20]

Under naturally occurring conditions uncomplicated molybdenum deficiencies have never been reported in either farm animals or man. There seems to be little evidence as yet that this trace metal plays a significant role in any aspect of human health or disease.[21] In truth the metal has been studied very little in clinical situations.

There is some food for thought in the studies of Burrell and co-workers[22,23] concerning a possible role of molybdenum in the occurrence of esophageal carcinoma among an isolated group of Bantu in Transka, South Africa. This group existing upon maize (corn) in a molybdenum deficient area of low fertility has a five-fold increase in carcinoma compared to other populations in the country.

Since Adler and Straub[24] first suggested that molybdenum may be a cariostatic agent, considerable evidence pro and con has been gathered by epidemiological surveys in New Zealand, England, the US, etc.[25-27] and by animal experimentation.[28-30] For the most part the surveys were not too well controlled. Also, analytical technologies employed were rather lacking.

The mechanism of molybdenum action in caries formation is not yet known. It may modify the tooth morphology.[31] In Curzon's survey in high and low molybdenum-containing soil regions in California,[27] a reduced prevalence of caries in children in the molybdenosis region was not observed. However, there was a definitely higher level of molybdenum in the outer enamel of premolars obtained from the high molybdenum region. Furthermore, tooth enamel showed increasing molybdenum content with increasing age of tooth donors.

Human dental enamel has been found to be relatively rich in molybdenum, ranging between 0.7 and 39 μg molybdenum g^{-1}.[32]

TISSUE MOLYBDENUM

Molybdenum concentrations in the tissue of all species of animals is of a low order of magnitude, comparable to manganese. Concentrations in the liver are highest, followed in descending order by the kidney, spleen, lung, brain and muscle.[8,33] Levels were not found to change significantly with age.

Molybdenum concentrations reported for normal man and for the rat are listed in Table 19.1.

Human blood molybdenum levels, among apparent controls reported from various regions, do show variations. Studies circa 1966 indicate that levels range between 0.5 μg ml^{-1} and 1.59 μg ml^{-1}, for the most part.[34,35] Approximately 3% of the population reported by Allaway[34] had blood molybdenum

TABLE 19.1
Molybdenum content in tissues (μg g^{-1} dry weight)

Species	Liver	Kidney	Spleen	Lung	Brain	Muscle	Reference
Man-adult	3.2	1.6	0.2	0.15	0.14	0.14	33
Rat-adult	1.8	1.0	0.52	0.37	0.24	0.06	8

Levels found in fowl are of the same order of magnitude.

concentrations of more than 10 μg ml^{-1}. More recently Baert et al.[36], employing neutron activation analysis found molybdenum levels in serum and packed blood cells to be ten-fold less, about 1.1 ppb and 0.6 ppb, respectively.

Judging from studies on sheep[37-41] and cattle[42,43] blood levels are dependent upon the molybdenum content in the diet. Levels could be raised by adding molybdenum supplements. Changes were reflected also in molybdenum concentrations in tissues and wool.

Foods vary greatly in their molybdenum contents. Legumes, cereal grains, leafy vegetables, liver and kidney are good sources of molybdenum. Fruits, root and stem vegetables, muscle meats, and dairy products are poor. Since molybdenum in grain is mainly concentrated in the outer layers, refined grain products are a poor source. Copper is also lost in the milling process.

Schroeder and co-workers[44] estimated the molybdenum intake in a good standard American diet to be 350 μg per day for adults. Daily intakes ranging between 94 and 162 μg in England[45] and 48–96 μg in New Zealand[46] have been reported. These variations may be due to the differences in the samples themselves, as well as in the analytical techniques used.

Daily molybdenum intakes of 10–15 mg had been reported (circa 1961) to occur in certain regions of Armenia.[47] It would be of interest to study those and other areas again utilizing currently available instrumentation.

At this point minimum molybdenum requirements cannot be estimated.

METABOLISM OF MOLYBDENUM

Molybdenum, copper and sulfate interact. It was first noted that the metabolism of one is dependent upon concentrations of the other more than thirty years ago, when Dick and Bull[48] found increasing the molybdenum intake an effective treatment for copper poisoning among sheep in Australia. Later Dick discovered the inhibiting effect of molybdenum on copper retention to be dependent upon the inorganic sulfate content in the diet.[49] Within certain limits blood molybdenum levels in sheep were found markedly and inversely dependent upon inorganic sulfate intakes.[50]

Studies with sheep[51,52] indicate that sulfate limits molybdenum retention by reducing intestinal absorption, as well as by increasing urinary excretion. Molybdenum is excreted in the urine primarily as molybdate ion. Sulfate, it was postulated,[50] if present in sufficiently high concentration interfered with and prevented the transport of molybdenum across membranes. The mechanism of interference is still unknown.

Apparently, the sulfate effect is quite specific. Citrate, permanganate, selenate, silicate, and tungstate did not show similar activity. Diuretics were found to increase urine volume but not molybdenum excretion.[42]

Smith and Wright[53,54] describe the formation of a stable copper–molybdenum–protein compound in plasma that may explain the lower tissue uptake of copper in the presence of a high copper level in plasma.

Suttle[55] studied the role of organic sulfur in the copper–molybdenum–sulfur interrelationship in ruminant nutrition. It was observed that plasma molybdenum levels were increased by molybdenum supplements; unaffected by molybdenum and sulfur given together; and slightly decreased by sulfur supplements alone. The latter also reduced replenishment of plasma copper pools. Molybdenum and sulfur supplements given together totally inhibited replenishment of plasma copper pools, but molybdenum alone was without apparent effect.

MOLYBDENUM TOXICITY

Manifestations of molybdenum toxicity (molybdenosis) vary among different species. High molybdenum uptakes from diet result in growth retardation and weight loss in different animals observed. However, diarrhea is observed only among cattle.

Molybdenosis in cattle occurs in many parts of the world where the soil molybdenum content is excessive. The condition, first studied in England, is known as 'teart'. New Zealanders refer to the condition as 'peat scours'. Though all cattle are susceptible, young cattle and milking cows are affected more severely, while sheep in those areas, are only slightly affected, and horses not at all. The condition, which can lead to death, has been treated successfully with copper sulfate[56] and potassium sulfate.[57]

Disturbances in phosphorus metabolism resulting in joint abnormalities, osteoporosis, lameness, and spontaneous bone fractures in farm animals, have been reported in areas of high molybdenum soils.[58,59] Decreased reproduction of cattle in such areas has been observed when supplemental copper was not added to their diets.[60] Similar effects of excess molybdenum intake have been observed experimentally in rats.[61,62] Molybdenum in the latter has also been shown to induce thyroid hypofunction.[63]

Toxic manifestations of molybdenum in humans are unknown or unrecognized. The limited data available suggest this trace element and its compounds are of low order of toxicity. Reports that very high molybdenum intakes alter uric acid metabolism in man[47,64] are yet to be substantiated.

USES OF MOLYBDENUM

Molybdenum and compounds have various industrial uses. A ferromolybdenum alloy is employed in manufacturing special steel for tools, rifle barrels, propeller shafts, boiler plate, and screens and grids for radio tubes and X-ray tubes. It is also used in making glass to metal seals.

Molybdenum disulfide finds use as a dry lubricant and lubricant additive. The sesquioxide of molybdenum, combined with ferrous sulfate, has been utilized to a limited degree as a hematinic in treating iron deficiency anemias.

ANALYSIS OF MOLYBDENUM

Colorimetric

Thiocyanate, dithiol, phenylhydrazine, chloranilic acid and mercaptoacetic acid form colored complexes with molybdenum. Iron, bismuth, copper, silver and mercury can interfere unless removed.

Molybdenum(VI) can be separated from as much as 10000 times its weight of ferric iron by running a 0.5 M hydrochloric acid solution through a Dowex resin column with 0.05 M ammonium thiocyanate.[65] Ion exchange has also been used in separating the element from copper, lead, chromium and nickel.[66]

Dithiozone extraction, at approximately pH 3, removes bismuth, copper, silver and mercury, but not molybdenum, from solution.

Molybdenum(VI) in an acidic solution containing a reducing agent, such as stannous chloride, is reduced to molybdenum(V), which forms an orange-colored complex with thiocyanate. The compound is extractable by the oxygen-containing organic solvents (ethers, alcohols, and esters).[67] Solvent extraction increases sensitivity.

Dithiol-(4-methyl-1,2-dimercaptotoluene) forms colored complexes with molybdenum(IV), (V) and (VI) that are extractable by organic solvents, both polar and non-polar.[68] Molybdenum (III) does not react with dithiol.

Molybdenum(VI) is almost totally extractable from mineral acids by pure acetyl acetone, or a mixture of acetyl acetone in chloroform.[69] Molybdenum(VI) is also quantitatively extractable as a cupferrate at a pH up to 1.5,[70] as an 8-hydroxy quinolate at acid pH,[71] and as a dithiocarbamate at acid pH.[72]

Modifications of the thiocyanate procedure have been employed quite extensively in the determination of molybdenum concentrations in biological materials, soils, plants and animals.[73-75] Organic matter was first destroyed by wet or dry ashing.

An early procedure for the analysis of biological matter involved the initial isolation of molybdenum as a cupferrate and subsequent determination with dithiol.[76]

Fluorimetry

The classical fluorimetric method for determining molybdenum involves reaction with carminic acid at around pH 5.[77] The detection sensitivity claimed (0.1 μg ml^{-1}) is less than that attained by colorimetric methods (10–0.5 ng ml^{-1}). Molybdenum also forms a complex with morin.[78,79]

Instrumental methods–miscellaneous

It would appear that instrumental methods for the determination of molybdenum in biological materials are rather slow in developing. The essentially innocuous nature of the element to humans may be a factor in this apparent lack of interest.

Tipton's[80] spectrographic analyses of tissues for trace metal contents, including molybdenum, represents the first departure from colorimetry for determining that element in biological materials. More recently Hambidge[81] employed a DC arc in argon for the spectrographic analysis of molybdenum in serum, urine and hair.

An occasional application of neutron activation analysis[82] or X-ray fluorescence[83] to such analysis has been described. Most practical applications have involved atomic absorption spectrometry.

The sensitivity of a molybdenum analysis by flame atomic absorption spectrometry is said to be about 0.6 μg ml^{-1} at the 313.3 nm resonance line. Greater detection sensitivity can be attained by concentrating the molybdenum with a chelation–extraction technique. Either an air–acetylene flame or a nitrous oxide–acetylene flame can be used. The former is subject to greater interferences, however. Mansell and Emmel[84] analyzed brines for their molybdenum content following chelation with ammonium pyrollidine dithiocarbamate (APDC) or oxine.

A detection sensitivity of 0.2 μg g^{-1} was attained in determining the molybdenum content of dental tissue in rats.[85] After wet digestion in perchloric acid and dilution to approximately pH 1, the sample was chelated with APDC, extracted into methylisobutyl ketone and measured in an air–acetylene flame.

Oxinates of molybdenum extracted from plant ash were analyzed in a nitrous oxide–acetylene flame.[86]

Molybdenum in herbage has been determined by flame atomic absorption[87] and by graphite tube[88] after dry ashing and dissolving the residue in weak hydrochloric acid.

Combining chelation–extraction techniques with graphite furnace analysis should yield more sensitive and practical methods of determining molybdenum in biological materials.

REFERENCES

1. H. J. M. Bowen, *Trace Elements in Biochemistry*, Academic Press, New York, 1966.

2. W. O. Robinson and L. T. Alexander, *Soil Science*, **75**, 287 (1953).
3. K. H. Schutte, *The Biology of the Trace Elements*, Crosby-Lockwood, London, 1964.
4. A. J. Anderson, *Advan. Agron.*, **8**, 163 (1956).
5. E. C. de Renzo, in *Mineral Metabolism*, (eds C. L. Comar and F. Bronner) Academic Press, New York, 1962.
6. E. J. Underwood, *Trace Elements in Human and Animal Nutrition*, Academic Press, New York, 1962.
7. W. Stiles, *Encyclopedia of Plant Physiology*, *Vol. 4*, (ed. W. Rutland) Springer, Berlin, 1958.
8. H. Bortels, *Arch. Mikrobiol.*, **1**, 333 (1930).
9. H. Ter Meulen, *Nature (London)*, **130**, 966 (1932).
10. R. A. Steinberg, *J. Agr. Res.*, **52**, 439 (1936).
11. D. I. Arnon and P. R. Stout, *Plant Physiol.*, **14**, 599 (1939).
12. E. C. deRenzo, E. Kaleita, P. Heytler, J. J. Oleson, B. L. Hutchings and J. H. Williams, *J. Amer. Chem. Soc.*, **75**, 753 (1953).
13. E. C. de Renzo, E. Kaleita, P. Heytler, J. J. Oleson, B. L. Hutchings and J. H. Williams, *Arch. Biochem. Biophys.*, **45**, 247 (1953).
14. D. A. Reichert and W. W. Westerfield, *J. Biol. Chem.*, **203**, 915 (1953).
15. H. R. Mahler in *Mineral Metabolism*, (eds C. L. Comar and F. Bronner) Academic Press, New York, 1961.
16. A. S. Brill, *Transition Metals in Biochemistry*, Springer-Verlag, Berlin, 1977.
17. B. L. Reid, A. A. Kurnick, R. L. Svacha and J. R. Couch, *Proc. Soc. Exptl. Biol. Med.*, **93**, 245 (1956).
18. E. S. Higgins, D. A. Richert and W. W. Westerfeld, *J. Nutr.*, **59**, 539 (1956).
19. W. C. Ellis, W. H. Pflauder, M. E. Muhrer and E. E. Pickett, *J. Animal Sci.*, **17**, 180 (1958).
20. I. J. T. Davies, *The Clinical Significance of the Essential Biological Metals*, Charles C. Thomas, Springfield, 1972.
21. E. J. Underwood, *Trace Elements in Human and Animal Nutrition*, 4th edn., Academic Press, New York, 1977.
22. R. J. Burrell, *J. Nat. Cancer Inst.*, **28**, 495 (1962).
23. R. J. Burrell, W. A. Roach and A. Shadwell, *J. Nat. Cancer Inst.*, **36**, 201 (1966).
24. P. Adler and J. Straub, *Acta Med. Acad. Sci. (Hung.)*, **4**, 221 (1953).
25. T. G. Ludwig, W. B. Healy and F. L. Losee, *Nature (London)*, **186**, 695 (1960).
26. R. J. Anderson, *Caries Res.*, **3**, 75 (1969).
27. M. E. J. Curzon, J. Kubota, B. G. Bibby, *J. Dent. Res.*, **50**, 74 (1971).
28. G. N. Jenkins, *Brit. Dent. J.*, **122**, 435 (1967).
29. G. N. Jenkins, *Brit. Dent. J.*, **122**, 500 (1967).
30. G. N. Jenkins, *Brit. Dent. J.*, **122**, 545 (1967).
31. B. J. Krueger, *J. Dent. Res.*, **45**, 714 (1966).
32. F. Losee, T. W. Cutress and R. Brown, *Trace Substances in Environmental Health*, Vol. 7, *Proceedings*, University of Missouri, Columbia, MO, 1973.
33. I. H. Tipton and M. J. Cook, *Health Phys.*, **9**, 103 (1963).
34. W. H. Allaway, J. Kubota, F. Losee and M. Roth, *Arch. Environ. Health*, **16**, 342 (1967).
35. Y. M. Bala and V. M. Liftshits, *Fed. Proc. Fed. Amer. Soc. Exptl. Biol.*, **25**, T370 (1966).
36. N. Baert, R. Cornelis, J. Hoste, *Clin. Chim. Acta*, **68**, 355 (1976).
37. A. T. Dick, *Austral. Vet. J.*, **28**, 30 (1952).
38. A. T. Dick, *Austral. Vet. J.*, **29**, 18 (1953).
39. A. T. Dick, *Austral. Vet. J.*, **29**, 233 (1953).
40. A. T. Dick, *Austral. Vet. J.*, **30**, 196 (1954).

41. A. T. Dick, in *Inorganic Nitrogen Metabolism*, (eds W. D. McElroy and B. Glass) Johns Hopkins Press, Baltimore, MD, 1956.
42. I. J. Cunningham, in *Symposium on Copper Metabolism*, (eds W. D. McElroy and B. Glass) Johns Hopkins Press, Baltimore, MD, 1950.
43. I. J. Cunningham and K. G. Hogan, *New Zealand J. Agric. Res.*, **1**, 841 (1958).
44. H. A. Schroeder, J. J. Balassa and I. H. Tipton, *J. Chron. Dis.*, **23**, 481 (1970).
45. E. I. Hamilton and M. J. Minski, *Sci. Total Environ.*, **1**, 375 (1972/73).
46. M. F. Robinson, J. M. McKenzie, C. D. Thomson and A. L. Van Rij, *Brit. J. Nutr.*, **30**, 195 (1973).
47. V. V. Kovalskii, C. E. Jarovaja and D. M. Shnavonjan, *Zk. Obshch. Biol.*, **22**, 179 (1961) cited in E. J. Underwood, *Trace Elements in Human and Animal Nutrition*, 4th edn. Academic Press, 1977.
48. A. T. Dick and L. B. Bull, *Austral. Vet. J.*, **21**, 70 (1945).
49. A. T. Dick, *Soil Sci.*, **81**, 229 (1956).
50. A. T. Dick, in *Inorganic Nitrogen Metabolism*, (eds W. D. McElroy and B. Glass) Johns Hopkins Press, Baltimore, MD, 1956.
51. J. F. Scaife, *New Zealand J. Sci. Technol.*, Section A, **38**, 285 (1963).
52. J. F. Scaife, *New Zealand J. Sci. Technol.*, Section A, **38**, 293 (1963).
53. B. S. W. Smith and H. Wright, *J. Comp. Pathol.*, **85**, 299 (1975).
54. B. S. W. Smith and H. Wright, *Clin. Chim. Acta*, **62**, 55 (1975).
55. N. F. Suttle, *Brit. J. Nutr.*, **34**, 411 (1975).
56. W. S. Ferguson, A. H. Lewis and S. J. Watson, *Nature (London)*, **141**, 553 (1938).
57. A. T. Dick, Doctoral thesis, University of Melbourne, Australia, 1954, cited in E. J. Underwood, *Trace Elements in Human and Animal Nutrition*, Academic Press, 1977.
58. G. K. Davis, in *Symposium on Copper Metabolism*, (eds W. D. McElroy and B. Glass) Johns Hopkins Press, Baltimore, 1950.
59. K. G. Hogan, D. F. L. Money, D. A. White and R. Walker, *New Zealand J. Agric. Res.*, **14**, 687 (1971).
60. J. W. Thomas and S. Moss, *J. Dairy Sci.*, **34**, 929 (1951).
61. M. A. Jeter and G. K. Davis, *J. Nutr.*, **54**, 215 (1954).
62. R. Van Reen, *J. Nutr.*, **68**, 243 (1959).
63. M. C. R. Widjajakusuma, P. K. Basrur and G. A. Robinson, *J. Endocrinol.*, **57**, 419 (1973).
64. Y. G. Doesthale and C. Gopalan, *Brit. J. Nutr.*, **31**, 351 (1974).
65. C. E. Crouthamel and C. E. Johnson, *Anal. Chem.*, **26**, 1284 (1954).
66. R. Klement, *Z. Anal. Chem.*, **136**, 17 (1952).
67. E. B. Sandell, *Colorimetric Determination of Traces of Metals*, Interscience Publishers, Inc., New York, 1959.
68. J. H. Hamence, *Analyst*, **65**, 152 (1940).
69. J. P. McKaveney and H. Freiser, *Anal. Chem.*, **29**, 290 (1957).
70. J. Stary and J. Smizanska, *Anal. Chim. Acta*, **29**, 546 (1963).
71. A. R. Eberle and M. W. Lerner, *Anal. Chem.*, **34**, 627 (1962).
72. H. Bode and F. Neumann, *Z. Anal. Chem.*, **172**, 1 (1960).
73. A. T. Dick and J. B. Bingley, *Austral. J. Exptl. Biol. Med. Sci.*, **29**, 459 (1951).
74. F. N. Ward, *Anal. Chem.*, **23**, 788 (1951).
75. C. M. Johnson and I. H. Arkley, *Anal. Chem.*, **26**, 572 (1954).
76. C. S. Piper and R. S. Beckwith, *J. Soc. Chem. Ind.*, **67**, 374 (1940).
77. G. F. Kirkbright, T. S. West and C. Woodward, *Talanta*, **13**, 1637 (1966).
78. G. Almassy and M. Vigvari, *Anal. Abst.*, **4**, 2588 (1957).
79. C. E. White and R. J. Argauer, *Fluorescence Analysis—A Practical Approach*, Marcel Dekker, New York, 1970.

80. I. H. Tipton, in *Metal Binding in Medicine*, (ed. M. J. Seven) J. B. Lippincott Co., Philadelphia, PA, 1960.
81. K. M. Hambidge, *Anal. Chem.*, **43**, 103 (1971).
82. N. V. Bagdavadze and L. M. Mosulishvili, *J. Radional. Chem.*, **24**, 65 (1975).
83. D. D. Runnels, W. R. Chappell and R. Meglen, in *The Molybdenum Project: Geochemical Aspects in Trace Element Geochemistry in Health and Disease*, special paper 155, (ed. J. Freedman) The Geological Society of America, Boulder, Co, 1975.
84. R. E. Mansell and H. W. Emmel, *Atomic Abs. Newsletter*, **4**, 276 (1965).
85. C. A. Helsey, *Talanta*, **20**, 779 (1973).
86. J. Stripar, F. Dolinsek, M. Spenko and J. Furlan, *Landwirtsch Forsch.* **27**, 51 (1974) cited in M. Pinta, *Modern Methods for Trace Element Analysis*, Ann Arbor Science, Ann Arbor, 1978.
87. B. J. Alloway, *J. Agri. Sci.*, **80**, 521 (1973).
88. S. Henning and T. L. Jackson, *Atomic Abs. Newsletter*, **12**, 100 (1973).

CHAPTER 20

NICKEL

Metallic nickel has been known for about 200 years, but interest in its biochemical and toxicological activities is fairly recent. It is found in many ores in combination with sulfur, arsenic or antimony. Nickel occurs free in meteorites. Soils contain 40 ppm nickel on an average compared, for example to 2 ppm molybdenum, 10 ppm lead or 6 ppm arsenic.[1]

Nickel is rather toxic for most plants and fungi. Growth of woody plants is limited or prevented by high nickel concentrations in soils in certain regions.[2,3] Land plant tissue contains about four times more nickel than does animal tissue. Some marine animals, for example oysters and lobster, are richer in the element.

Except for certain industrial exposures, nickel is considered as a relatively non-toxic element. Dietary nickel is poorly absorbed. Contamination of foods during their preparation is not considered a serious hazard.[4]

ESSENTIALITY OF NICKEL

Fairly recently studies have indicated that nickel is an essential element in animal nutrition. Its physiological role has not been delineated as yet. Significant concentrations of nickel have been shown to be present in RNA and DNA.[5,6] Nickel is now considered to be among the newer essential metals.

By feeding diets containing only 40 ppm nickel[7] and later 2–15 ppm nickel,[8] Nielsen and co-workers were able to induce a deficiency in chicks, which was manifested by liver changes, altered shank pigmentation, dermatitis and a lowered hematocrit. Rats exposed to a similar nickel deficient diet throughout their life (fetal to adulthood) exhibited impaired growth and reproduction. Morphological abnormalities in the liver and changes in phospholipid content were also observed in rats.[9-11]

Slow weight gain and higher piglet mortality were observed in pigs fed a nickel-deficient diet.[12]

Schnegg and Kirchgessner,[13] in experiments with rats, were able to show that nickel deprivation causes anemia by impairing the intestinal absorption of iron.

Nickel deficiencies in man have not been demonstrated as yet. Except perhaps in individuals consuming diets exceptionally high in fats, refined carbohydrates, and dairy products, there is little likelihood that inadequate dietary nickel intake may prove a problem in man. Pathological conditions that impede nickel absorption or promote excessive loss should be investigated, however.

Total nickel intake varies with the types of food that are consumed.[14-17] Vegetables are higher in nickel content than animal muscle. Typical Western diets of mixed foods were found to supply 300–500 μg of nickel daily. Nickel requirements are as yet unknown.

METABOLISM OF NICKEL

According to Underwood,[4] only 1–10% of the dietary nickel ingested is absorbed. Animal studies by O'Dell et al.[18] and Schroeder et al.[19] seem to indicate that the animal body neither absorbs, nor retains, nickel readily. A homeostatic control mechanism which only begins to break down when exposed to abnormally high nickel intakes has been postulated.

Of the ingested nickel, 90% is excreted via the feces and 10% via the urine. The Perrys[20] reported that apparently normal, healthy adults excreted 10–70 μg l^{-1} of urine. Sunderman[6] reported slightly lower values.

Under ordinary circumstances, nickel is found in all tissues and body fluids in relatively low concentrations. It does not appear to accumulate with age in any human organ other than the lungs.[16,21] Levels of nickel ranging from 0.02–0.05 μg g^{-1} dry weight and 0.03–1.46 μg g^{-1} have been reported for human liver and lung, respectively.[22] Values reported for calves[18] and rats[19] are somewhat higher. Table 20.1 contains a comparison of nickel levels in these animal tissues.

TABLE 20.1
Nickel contents in animal tissue
(μg g^{-1} dry weight)

Tissue	Calves	Rats
Heart	1.11	3.0
Kidney	2.08	1.7
Liver	0.76	1.3
Lung	2.94	2.4
Spleen	0.00	6.2

Nickel administered parieterally was found to concentrate in the kidneys.[23-26]

Blood nickel concentrations determined on limited populations of most mammals are essentially of the same order of magnitude. Levels in humans ranged between 1.1–4.6 μg l^{-1}; cattle, 1.7–4.4; dogs, 1.8–4.2; horses 1.3–2.5; and in cats, 1.5–6.4. Levels in pigs, rabbits, and guinea pigs were somewhat higher, being 4.2–5.6, 6.5–14.0, and 2.4–7.1 μg l^{-1}, respectively.[22]

Considerable nickel is excreted in sweat. Horak and Sunderman[27] found a mean nickel concentration in sweat of 49 μg l^{-1}, about 20 times that of normal serum. A later study concerning the excretion of trace metals in human sweat yielded similar findings.[28]

TOXICOLOGY OF NICKEL

Acidic foods take up nickel from cooking and storage vessels. However, since it is poorly absorbed, nickel ingested from foods is relatively nontoxic. Exposure to nickel and its compounds in industry or elsewhere is another matter. It is said that there are at least 3000 known nickel alloys.[29] Applications are versatile and widespread: storage batteries, automobile and aircraft parts, spark plugs, electrodes, coins, cooking utensils, etc. Raney nickel, made by fusing equal parts of aluminium and nickel, is employed as a catalyst for the hydrogenation of oils and other organic compounds with gaseous hydrogen.

Certain nickel salts, such as the cyanide, chloride, carbonate, hydroxide, nitrate and sulfate are used in nickel plating processes. Various nickel salts are utilized as pigments. For example, nickel dimethylglyoxime, a red-colored complex, is found in sun-fast paints, lacquers and cosmetics. Nickel phosphate is used as a yellow pigment for water colors and oils. Nickel sulfate serves as a mordant in dyeing and printing fabrics.

Exposure to nickel carbonyl is the most hazardous of all industrial exposures to nickel.[30] This highly volatile compound, formed during a process employed for the production of pure nickel, is toxic even in low concentration. Initial symptoms of exposure, including nausea, dizziness, headaches and chest pain, disappear within a few hours. After 12–36 hours, or as long as 5 days after exposure, severe pulmonary symptoms, tachycardia and profound weakness develop. Death has occurred within 4–13 days. Pathological examinations revealed lesions in the lung, brain, liver, kidney, adrenals and spleen.[1] Subjects who recover from acute nickel carbonyl exposure usually suffer from pulmonary insufficiency for a rather protracted period.[31]

Nickel was recognized fairly early as a potential carcinogen. Stephens (circa 1933–34),[32,33] discussed the high incidence of neoplasms of the respiratory tract and dermatitis among workers in nickel refineries. More recent reviews of the various epidemeological investigations conducted tend to substantiate the view.[30,34-36]

The carcinogenic actions of nickel and its salts have been observed in numerous animal experiments over the years.[37-39] It is interesting that a recent study using rats reported that manganese suppressed the carcinogenesis of nickel sulfide.[40]

Workers exposed to inhalation of aerosols of soluble nickel salts, i.e. those in nickel plating operations and in nickel refineries, are probably at greatest risk.

Monitoring urinary nickel excretion and serum levels of industrial workers is becoming more commonplace. According to Bernack and co-workers,[41] the former is a better index of exposure. Significant increases in urine nickel were not encountered among those engaged in buffing and polishing nickel, or in operations where Raney nickel is used as a hydrogenation catalyst. Some increases were found among arc welders, battery workers and metal sprayers. Workers in nickel plating jobs and in nickel refineries exhibited the greatest increase in urinary nickel excretion.

Spruit and Bongaarts[42,43] observed that the nickel contents in plasma, urine and hair of those occupationally exposed to the metal were ten times those found in unexposed control subjects. Nickel concentrations in the plasma and urines of industrial workers decreased during vacations. These investigators further noted that nickel levels in hypersensitive and nonhypersensitive subjects did not differ.

Hypersensitivity to nickel in industry, i.e. 'nickel itch' is not uncommon.[30] Approximately 9% in the above study were sensitized. Nickel dermatitis caused by wearing nickel-plated watches, other jewelry, spectacle frames, etc. is well known as a dermatological syndrome that is readily cured by removing the source of exposure. A 'split-ear' syndrome caused by nickel-plated earrings has been described.[44]

Stainless steel kitchen appliances and nickel-containing detergents have been implicated as causative agents in contact dermatitis.[45]

Allergic reactions to nickel-containing prostheses,[46,47] surgical devices,[48,49] etc. can be serious at times.

ANALYSIS OF NICKEL

Colorimetric-fluorimetric

Many colorimetric procedures for determination of nickel were based on its reaction with dimethylglyoxime in an organic solvent at a pH around 7.5.[50-54] A detection sensitivity of 17 ng ml^{-1} is claimed for the reaction. In the presence of an oxidizing agent, the sensitivity of the nickel—dimethylglyoxime is about 4 ng ml^{-1}. Copper, palladium, platinum and gold also yield colored complexes with dimethylglyoxime.

Alpha-furildioxime is considered superior to dimethylglyoxime as a nickel-complexing agent.[55]

Nickel is quantitatively extractable as a diethyldithiocarbamate over the pH range 5–11. Though sensitive, the reaction is not specific since copper, bismuth, cobalt and lead are also coextracted. Alexander *et al.*[56] determined nickel contents in acid digestants of food after first separating nickel by extraction with dimethylglyoxime in chloroform. Copper, coextracted, was removed by washing the extract with dilute ammonium hydroxide. Nickel was then reacted with sodium diethyldithiocarbamate; lead interferes.

Nickel and dithiooxalate form a deep red-colored complex in neutral or slightly acidic solutions.[57] Since both iron and cobalt yield strong colors also, the reaction is less than useful for colorimetric analyses of biological materials.

Cluett and Yoe[58] describe a procedure for the determination of nickel in blood by complexing with diethyldithiocarbamate in amyl alcohol. Nickel was separated from most interfering elements by passage through a Dowex-1 anion exchange resin. Lead, which was not removed by the resin, was then absorbed onto calcium carbonate.

The quenching by nickel of the fluorescence of the aluminum—PAN complex (Al-1,2-pyridylazo-2-naphthol) has been proposed as a method for determining nickel.[59] A detection sensitivity of 10^{-9} M is claimed. However, the method can not be considered to be practical for analyses of biological materials. Aluminium in the matrix obviously would interfere by increasing fluorescence emission. Chromium has been shown to interfere considerably with this nickel reaction. Interference from copper, iron and calcium, while minimal when simple solutions are involved, would be considerable indeed when analyzing nickel concentrations in a biological matrix.

Other instrumental methods

Various instrumental techniques have been applied in nickel analyses. Tipton and coworkers determined nickel contents of various tissues spectrographically.[21,60] X-ray fluorescence spectrometry has been employed in analyzing plant materials and waters.[61]

A number of atomic absorption spectrometry procedures both flame[62–64] and flameless[65–69] have been described for a variety of biological matrices. Most methods involved chelation by dithiocarbamate or dimethylglyoxime.

At the 232.0 nm resonance line for nickel, a sensitive atomic absorption spectrometer has a detection limit of 0.01 μg ml^{-1} or 10 μg l^{-1}. Levels of 1.5–4 μg l^{-1} are said to be within the normal range for serum and urine.

A recent procedure worthy of special mention is the furildioxime–electrothermal methods of Mikac-Device *et al.*[70] A 1 ml aliquot of sample was wet ashed. Nickel as a furildioximate was extracted into methylisobutyl ketone and measured in the graphite furnace. Charring and atomization temperatures were 950 °C and 2600 °C, respectively.

REFERENCES

1. H. J. M. Bowen, *Trace Elements in Biochemistry*, Academic Press, New York, 1966.
2. W. Krause, in *Encyclopedia of Plant Physiology*, Vol. 4, (ed. W. Ruhland) Springer, Berlin, 1958.
3. Schutte, K. H., *The Biology of Trace Elements*, Crosby Lockwood, London, 1964.
4. E. J. Underwood, *Trace Elements in Human and Animal Nutrition*, Academic Press, New York, 1977.
5. W. E. C. Wacker and B. L. Vallee, *J. Biol. Chem.*, **234**, 3257 (1959).
6. F. W. Sunderman, Jr., *Amer. J. Clin. Pathol.*, **44**, 182 (1965).
7. F. H. Nielsen, in *Newer Trace Elements in Nutrition*, (eds W. Mertz and W. E. Cornatzer) Marcel Dekker, New York, 1971.
8. F. H. Nielsen, D. R. Myron, S. H. Givand and D. A. Ollerich, *J. Nutr.*, **105**, 1607 (1975).
9. F. H. Nielsen, D. R. Myron, S. H. Givand, T. J. Zimmerman and D. A. Ollerich, *J. Nutr.*, **105**, 1620 (1975).
10. F. H. Nielsen and D. A. Ollerich, *Fed. Proc. Fed. Amer. Soc. Exptl. Biol.*, **33**, 1767 (1974).
11. F. H. Nielsen and H. H. Sandstead, *Amer. J. Clin. Nutr.*, **27**, 515 (1974).
12. M. Anke, M. Grun, G. Dittrich, B. Broppel and A. Henning, *Trace Element Metabolism in Animals*, (ed. W. G. Hoekstra) University Park Press, Baltimore, MD, 1974.
13. A. Schnegg and M. Kirchgessner, *Intern. J. Vitamin Nutr. Res.*, **46**, 96 (1976).
14. N. L. Kent and R. A. McCance, *Biochem. J.*, **35**, 837 (1941).
15. N. L. Kent and R. A. McCance, *Biochem. J.*, **35**, 887 (1941).
16. H. A. Schroeder, J. J. Balassa and I. H. Tipton, *J. Chronic Dis.*, **15**, 51 (1961).
17. E. G. Zook, F. E. Greene and E. R. Morris, *Cereal Chem.*, **47**, 720 (1970).
18. D. G. O'Dell, W. J. Miller, S. L. Moore, W. A. King, J. C. Ellers and H. Juracek, *J. Animal Sci.*, **32**, 769 (1971).
19. H. A. Schroeder, M. Mitchener and A. P. Nason, *J. Nutr.*, **104**, 239 (1974).
20. H. M. Perry and E. F. Perry, *J. Clin. Invest.*, **38**, 1452 (1959).
21. I. H. Tipton and M. J. Cook, *Health Phys.*, **9**, 103 (1963).
22. F. W. Sunderman, Jr., M. J. Decsy and M. D. McNeeley, *Ann. N.Y. Acad. Sci.*, **199**, 300 (1972).
23. A. W. Wase, D. M. Goss and M. J. Boyd, *Arch. Biochem. Biophys.*, **51**, 1 (1954).
24. J. C. Smith and B. Hackley, *J. Nutr.*, **95**, 541 (1968).
25. K. Parker and F. W. Sunderman, Jr., *Res. Commun. Chem. Pathol. Pharmacol.*, **7**, 755 (1974).
26. J. J. Clary, *Appl. Pharmacol.*, **31**, 55 (1975).
27. E. Horak and F. W. Sunderman, Jr., *Clin. Chem.*, **19**, 429 (1973).
28. J. C. Cohn and E. A. Emmett, *Ann. Clin. Lab. Sci.*, **8**, 270 (1978).
29. A. Hamilton and H. L. Hardy, *Industrial Toxicology*, 3rd edn., Publishing Sciences Group, Inc., Acton, Mass., 1974.
30. F. W. Sunderman, Jr., F. Coulston, G. L. Eichorn, J. A. Fellows, E. Mastromatteo, H. T. Reno and M. H. Samitz, *Nickel: A Report of the Committee on Medical and Biologic Effects of Environmental Pollutants*, National Academy of Sciences, USA, Washington, DC, 1975.
31. F. W. Sunderman, *Ann. Clin. Res.*, **3**, 182 (1971).

32. G. A. Stephens, *Med. Press*, **187**, 216 (1933).
33. G. A. Stephens, *Med. Press*, **194**, 287 (1934).
34. L. Fishbein, *J. Toxicol. Environ. Health*, **2**, 77 (1976).
35. F. W. Sunderman, Jr., *Prev. Med.*, **5**, 279 (1976).
36. F. W. Sunderman, Jr., in *Metal Carcinogenesis—Advances in Modern Toxicology*, (eds R. A. Goyer and M. A. Mehlman) Hemisphere Publishing Corp., Washington, DC, 1977.
37. W. W. Payne, *Proc. Amer. Assoc. Cancer Res.*, **50**, 50 (1964).
38. F. W. Sunderman, Jr. and A. J. Donnelly, *Amer. J. Pathol.*, **46**, 1027 (1965).
39. F. W. Sunderman, Jr. and R. M. Maenza, *Res. Commun. Chem. Pathol. Pharmacol.*, **14**, 319 (1976).
40. F. W. Sunderman, Jr., K. S. Kasprzak, T. J. Lau, P. P. Minghetti, R. M. Maenza, N. Becker, C. Onkelinx and P. J. Goldblatt, *Cancer Res.*, **36**, 1790 (1976).
41. E. J. Bernacki, G. E. Parsons, B. R. Roy, M. Mikac-Devic, C. D. Kennedy and F. W. Sunderman, Jr., *Ann. Clin. Lab. Sci.*, **8**, 184 (1978).
42. D. Spruit and P. J. M. Bongaarts, *Dermatologica*, **154**, 291 (1977).
43. D. Spruit and P. J. M. Bongaarts, in *Clinical Chemistry and Chemical Toxicology of Metals*, (ed. S. S. Brown) Elsevier/North Holland, Amsterdam, 1977.
44. D. M. Ashman, A. MacDonald and M. Fewell, *Contact Dermatitis*, **1**, 393 (1975).
45. S. A. Katz and M. H. Samitz, *Acta Derm. Venerol. (Stockh.)*, **55**, 113 (1975).
46. M. H. Samitz and S. A. Katz, *Brit. J. Dermatol.*, **92**, 287 (1975).
47. R. Kubba and R. H. Champion, *Brit. J. Dermatol.*, **93**, (11), 41 (1975).
48. L. F. Tinckler, *Brit. J. Surg.*, **59**, 745 (1972).
49. S. A. Kvorning, *Contact Dermatitis*, **1**, 327 (1975).
50. E. B. Sandell and R. W. Perlich, *Ind. Eng. Chem. (Anal. Ed.)*, **11**, 309 (1939).
51. H. Christopherson and E. B. Sandell, *Anal. Chim. Acta*, **10**, 1 (1954).
52. W. Oelschlager, *Z. Anal. Chem.*, **146**, 339 (1955).
53. W. Nielsch and L. Giefer, *Mikrochim. Acta*, 522 (1956).
54. W. Nielsch, *Z. Anal. Chem.*, **150**, 114 (1956).
55. C. G. Taylor, *Analyst*, **81**, 369 (1956).
56. O. R. Alexander, E. M. Godar and N. J. Linde, *Ind. Eng. Chem. (Anal. Ed.)*, **18**, 206 (1946).
57. J. H. Yoe and F. H. Wirsing, *J. Amer. Chem. Soc.*, **54**, 1866 (1932).
58. M. L. Cluett and J. H. Yoe, *Anal. Chem.*, **29**, 1265 (1957).
59. G. H. Schenk, K. P. Dilloway, J. S. Coulter, *Anal. Chem.*, **41**, 510 (1969).
60. I. H. Tipton, in *Metal Binding in Medicine*, (ed. M. J. Seven) Lippincott, Philadelphia, PA, 1960.
61. J. L. Campbell, B. H. Orr, A. W. Herman, L. A. McNelles, J. A. Thomson and W. B. Coor, *Anal. Chem.*, **47**, 1542 (1975).
62. S. Nomoto and F. W. Sunderman, Jr., *Clin. Chem.*, **16**, 477 (1970).
63. R. A. Duce, J. G. Quinn, C. E. Olney, S. R. Peitrowicz, B. J. Ray and T. L. Wade, *Science*, **176**, 161 (1972).
64. B. F. Willey, C. M. Duke, A. L. Wojcieszak and C. T. Thomas, *J. Amer. Water Works Assoc.*, **64**, 303 (1972).
65. W. M. Edmund, G. R. Giddings and M. Morgan-Jones, *Atomic Abs. Newsletter*, **12**, 45 (1973).
66. C. W. Fuller, *Anal. Chim. Acta*, **62**, 442 (1972).
67. H. Zachariasen, I. Andersen, C. Kostol and R. Barton, *Clin. Chem.*, **21**, 562 (1975).
68. M. Stoeppler, M. Kampel and B. Welz, *Z. Anal. Chem.*, **282** 369 (1976).
69. D. Ader and M. Stoeppler, *J. Anal. Toxicol.*, **1**, 252 (1977).
70. D. Mikac-Devic, F. W. Sunderman, Jr., and S. Nomoto, *Clin. Chem.*, **23**, 948 (1977).

CHAPTER 21

PALLADIUM

Palladium is one of the platinum group of metals. It occurs in nature as a selenide and also alloyed with gold or platinum. The metal is not found widely.

To date there seem to be no reports of ill effects induced in industrial workers exposed to palladium and its compounds. The metal itself alloyed with gold, silver, or copper is used in jewelry manufacture, and in dentistry as a filling material. It is also used in metal plating, in the electrical industry, and as a catalyst.

Palladium chloride is used in photography, in the manufacture of indelible ink, and as a catalyst. The oxide is also employed as a catalyst.

More than 35 years ago Meek and co-workers[1] studied the effects of palladium compounds on man and animals. Only in intravenous administration of palladium chloride in animals were any toxic effects noted.

The fairly recent use of catalytic converters on automobiles in the US has brought palladium and platinum into the position of being future environmental pollutants. Little information concerning concentrations in biological material exists. Koch and Roesmer[2] reported the palladium content of mammalian muscle to be 2 ng g^{-1} dry weight. Recently, Johnson and co-workers[3], in examining two California populations for levels of palladium, platinum and lead in blood, urine, feces and hair, found blood palladium levels to be less than 0.9 μg%; and urine, less than 0.3 μg l^{-1}. Hair contained less than 0.02 μg g^{-1} of palladium, and feces, less than 1 ng g^{-1}. Values in blood, reported by Wolstenholme[4] in 1964, are higher.

During investigations concerning the toxicity and metabolism of palladium, Moore et al.[5] observed that retention of palladium was highest after intravenous administration. More was retained following inhalation than after oral dosing.

Palladium is excreted primarily via the feces following oral ingestion. After intravenous administration equivalent quantities are excreted in both urine and feces. The kidney contains the greatest palladium concentration, followed by the spleen and liver.

After intravenous administration of palladium to pregnant rats, a small amount of the metal was recovered in fetuses.

ANALYSIS OF PALLADIUM

Colorimetry

Colorimetric methods for the determination of palladium do not possess the necessary sensitivity or specificity for the analysis of biological materials. Compounds containing the nitroso-phenylamino groups, for example, p-nitroso-methylaniline and p-nitrosodiphenylamine, have been employed in palladium analyses.[6,7] The colored complexes formed are extractable by organic solvents from a slightly acidic solution.[8]

Certain dioximates, dimethyl glyoxime[9] and α-furildioxime[10] are considered superior reagents for palladium analysis. The latter was the more sensitive. Ferric ion was found to interfere with the extraction of the dimethylglyoxime complex.[11]

Palladium is completely extracted by acetylacetone over the pH range 0–8.[12] The dithiocarbamate complexes are extractable over a wide pH range.[13-15]

The 2-nitroso-1-naphthol complex of palladium is extractable from acidic solutions. Interference from iron and copper can be prevented by addition of ethylenediaminetetraacetate (EDTA).[16]

Palladium can be quantitatively extracted by dithizone from acid solution.[17]

Atomic absorption spectroscopy

Very few applications of atomic absorption spectrometry to palladium analysis in biological material have been reported. The procedure is not technically difficult. However, since palladium is not classed as a hazardous metal, there is little stimulus to develop analytical procedures.

At the 247.6 nm resonance line for palladium, the detection limit by flame atomization is approximately 0.02 μg ml^{-1}. A lean oxidizing flame is employed. The other noble metals do not interfere, apparently.[18] Detection limits are extended by flameless atomization.

Because of the low palladium content in biological materials, samples of greater size than are usually employed in atomic absorption analyses are required. These must be concentrated. For example, Tillery and Johnson[19] used 15 ml blood aliquots and 1 l of urine in their analyses. Palladium was isolated according to a spectrophotometric procedure described by Khattak and Magee[20] and measured in a graphite furnace.

Samples were first ashed in nitric and perchloric acids, evaporated to near dryness, and reconstituted with concentrated hydrochloric acids. A chlorostannous palladium complex was formed and extracted by tri-*n*-octyl-amine.

REFERENCES

1. S. F. Meek, G. C. Harrold and C. P. McCord, *Ind. Med. Surg.*, **12**, 447 (1943).
2. R. C. Koch and J. Roesmer, *J. Food Sci.*, **27**, 309 (1962).
3. D. E. Johnson, J. B. Tillery and R. J. Prevost, *Environ. Health Persp.* **12**, 27 (1975).
4. W. A. Wolstenholme, *Nature (London)*, **203**, 1284 (1964).
5. W. Moore, D. Hysell, L. Hall, K. Campbell and J. Stara, *Environ. Health Persp.*, **10**, 63 (1975).
6. J. H. Yoe and L. G. Overholser, *J. Amer. Chem. Soc.*, **61**, 2058 (1939).
7. L. G. Overholser and J. H. Yoe, *J. Amer. Chem. Soc.*, **63**, 3224 (1941).
8. D. E. Ryan, *Analyst*, **76**, 167 (1951).
9. R. S. Young, *Analyst*, **76**, 49 (1951).
10. O. Menis and T. C. Rains, *Anal. Chem.*, **27**, 1932 (1955).
11. J. G. Fraser, F. E. Beamish and W. A. E. McBryde, *Anal. Chem.*, **26**, 495 (1954).
12. J. Stary and E. Hladky, *Anal. Chim. Acta*, **28**, 227 (1963).
13. H. Bode, *Z. Anal. Chem.*, **143**, 182, (1954).
14. H. Bode and F. Neumann, *Z. Anal. Chem.*, **172**, 1 (1960).
15. J. H. Thompson and M. J. Ravenscroft, *Analyst*, **85**, 735 (1960).
16. K. L. Cheng, *Anal. Chem.*, **26**, 1894 (1954).
17. E. B. Sandell, *Colorimetric Determination of Traces of Metals*, Interscience Publishers, New York, 1959.
18. A. Strasheim and G. J. Wessels, *Appl. Spectros.*, **17**, 65 (1963).
19. J. B. Tillery and D. E. Johnson, *Environ. Health Persp.*, **12**, 19 (1975).
20. M. A. Khattak and R. J. Magee, *Anal. Chim. Acta*, **35**, 17 (1966).

CHAPTER 22

PLATINUM

Toxic effects rarely result from exposure to platinum, the metal, in industry and elsewhere. The metal and its alloys are used in dentistry, jewelry manufacture and in the electrical and chemical industry.

Platinum salts encountered in their industrial applications, namely the manufacture of X-ray fluorescent screens, electroplating, photographic processes, and as catalysts in chemical reactions are another matter. Karasek and Karasek[1] in 1911 found that 8 to 40 employees in photographic studios experienced skin irritation, sneezing and coughing. More severe symptoms among workmen in platinum refineries were described by Hunter et al.,[2] the severity paralleled the degree and length of exposure, and some workmen exhibited skin lesions. A syndrome 'platinum asthma', manifested by profuse rhinorrhea, tightness of the chest, dyspnea, and wheezing, disappeared within an hour after the workplace was left. Additional incidents of platinosis occurring among solderers, electroplaters and jewellers were described subsequently.[3-5]

The use of platinum in catalytic converters in automobiles in the US has initiated investigations regarding potential platinum hazard in the environment. Some are of the opinion that by about 1985 detectable levels of the metal may be present.[6]

METABOLISM OF PLATINUM

Animal studies indicate that platinum and palladium are metabolized similarly.[7] More platinum is retained following inhalation than after oral ingestion. Platinum, also, is detected in fetuses after intravenous administration to pregnant rats.

As with palladium, data in the literature regarding platinum deposition in biological materials is sparse indeed. Koch and Roesmer[8] found the platinum content in mammalian muscle to be 2 ng g^{-1} dry weight. Johnson *et al.*[9] determined platinum levels in blood, urine, feces and hair of inhabitants in two different sections in California. Blood platinum levels averaging 0.049 μg and 0.180 μg% were found in these areas. Levels reported earlier by Wolstenholme are higher.[10]

MEDICAL USES OF PLATINUM COMPOUNDS

In vitro studies, and investigations with animals, demonstrated the anti-neoplastic activity of *cis*-dichlorodiamine platinum (II). The compound was observed to inhibit cell division in sarcoma and leukemia in mice, as well as in gram negative rods.[11] *cis*-Dichlorodiamine platinum inhibits DNA, RNA, and protein synthesis in mammalian cells *in vitro.*[12] When administered to rats, the compound was observed to decrease lymphocytes and reticulocytes. Striking changes in rat thymus and spleen were also observed.[13]

This compound has undergone a number of clinical trials in certain human malignancies, bladder cancer,[14] ovarian cancer,[15] and testicular cancer.[16] It is expected that *cis*-dichlorodiamine platinum (II) will be used more extensively.

ANALYSIS OF PLATINUM

As with palladium, colorimetric methods do not possess the sensitivity necessary for analyses of biological materials but can be combined with atomic absorption spectrometry to advantage. Platinum and palladium are similar in analytical properties and reagents used.[14-16] Coextraction of palladium and platinum is not a disadvantage when metals are determined by atomic absorption spectrometry. Tillery and Johnson[17] determined both palladium and platinum on a single extract in the atomic absorption spectrometric procedure employed in the California study.

Other atomic absorption procedures for platinum analysis have been described.[18-21]

REFERENCES

1. S. R. Karasek and M. Karasek, *Report of Illinois State Commission on Occupational Disease*, 1911, cited in *Industrial Toxicology*, A. Hamilton and H. L. Hardy, Publishing Sciences Group, Acton, Mass, 1974.
2. D. Hunter, R. Milton and K. M. A. Perry, *Brit. J. Ind. Med.*, **2**, 92 (1945).
3. A. E. Roberts, *Arch. Ind. Hyg.*, **4**, 549 (1951).
4. F. A. Patty, *Industrial Hygiene and Toxicology*, Interscience, New York, 1963.
5. E. Browning, *Toxicity of Industrial Metals*, Butterworth, London, 1961.
6. P. E. Brubaker, J. P. Miran, K. Bridbord, F. G. Hueter, *Environ. Health Perspect.*, **10**, 39 (1975).
7. W. Moore, D. Hysell, L. Hall, K. Campbell and J. Stara, *Environ. Health Perspect.* **10**, 63 (1975).
8. R. C. Koch and J. Roesmer, *J. Food Sci.*, **27**, 309 (1962).

9. D. E. Johnson, J. B. Tillery and R. J. Prevost, *Environ. Health Perspect.*, **12**, 27 (1975).
10. W. A. Wolstenholme, *Nature (London)*, **203**, 1284 (1964).
11. B. Rosenberg, L. Van Camp, J. E. Trasko and V. H. Mansour, *Nature (London)*, **222**, 385 (1969).
12. H. C. Harder and B. Rosenberg, *Int. J. Cancer.*, **6**, 207 (1970).
13. H. S. Thompson and G. R. Gale, *Toxicol. Appl. Pharmacol.*, **19**, 602 (1971).
14. R. S. Young, *Analyst*, **76**, 49 (1951).
15. H. Bode and F. Neumann, *Z. Anal. Chem.*, **172**, 1 (1960).
16. M. A. Khattak and R. J. Magee, *Talanta*, **12**, 733 (1965).
17. John B. Tillery and D. E. Johnson, *Environ. Health Perspect.*, **12**, 19 (1975).
18. R. G. Miller and J. U. Doerger, *Atomic Abs. Newsletter*, **14**, 66 (1975).
19. A. H. Jones, *Anal. Chem.*, **48**, 1472 (1976).
20. D. J. Higby, L. Buchholtz, K. Chary, A. Avellanosa, E. S. Henderson, *Proc. Amer. Assoc. Cancer Res.*, **18**, 440 (1977).
21. M. F. Pera and H. C. Harder, *Clin. Chem.*, **23**, 1245 (1977).

CHAPTER 23

SELENIUM

Selenium, like arsenic, is a metalloid. It occurs widely but in low concentrations in nature, usually in the sulfide ores of heavy metals, for example copper, iron, and zinc. Selenium is found in land plants,[1,2] marine plants,[3] land animals,[4] and marine animals.[5,6] Vegetation growing in selenium-rich soils in certain parts of the US, Canada, England, New Zealand and elsewhere has been shown to be toxic to mammals.[2,7-9] There are also selenium-poor areas with their attendant problems.

TOXICOLOGY OF SELENIUM

According to Rosenfeld and Beath[10] selenium toxicity is an ancient disease. The syndrome may have been described by Marco Polo in 1295 when he wrote of a 'poisonous plant growing there (Western China) which if eaten, has the effect of causing the hooves of the animals to drop off'.

Berzelius discovered selenium in 1817. By 1842, Japha[11] proved that the element was definitely toxic to animals. Probably the first authentic record of naturally occurring selenosis is found in a report by a Dr T. C. Madison, an army surgeon in the Nebraska territory, circa 1856. He described a fatal disease among horses grazing in a certain area near the fort. The animals suffered from feet which were so painful that they were unable to move about in search of food. In addition, hair loss from the mane and tail was striking.[12] The condition was not associated, generally, with selenium toxicity until about 1934, when Franke presented his findings.[13]

There are varying degrees of selenosis, ranging from 'alkali disease', a chronic, mild form, to 'blind staggers', an acute state that can result in death.[14]

Chronic intoxication is manifested by dullness, lack of vitality, and emaciation of the animal. Hair loss, soreness of joints and sloughing of hooves are notable. Because of the erosion of joints and long bones, animals are lame and stiff. Other findings include liver cirrhosis and anemia.

In acute selenosis, animals suffer from blindness, abdominal pain, and some degree of paralysis. Salivation, disturbed respiration, and grating of the teeth are observed. Death results from respiratory failure, as well as from starvation and thirst. Animals are unwilling to move about.

The mechanism whereby selenium induces a toxicity has not been established, but inactivation of sulfhydryl-containing enzymes is a possibility. Arsenic can alleviate symptoms of selenosis.[15] Whether selenium toxicity from natural exposure in foods occurs among humans residing in areas where selenosis is endemic among farm animals is equivocal.

INDUSTRIAL USES AND HAZARDS OF SELENIUM

Selenium and its compounds are involved in many and varied industrial applications. Their use in the manufacture of glass and ceramics, ink and paint pigments, plastics, rubber, photoelectric cells, different alloys, etc. involves exposure to hazardous dusts, fumes and vapours. Refinery workers processing ores of copper, gold, nickel and silver are also at risk.

Included among the symptoms of chronic selenium poisoning, as observed in 1925 by Alice Hamilton[16] in copper refinery workers, are strong garlic odor of breath, sore throats, coryza, and gastrointestinal irritation. The occurrence of dermatitis, and a red staining of fingers, teeth and hair have also been reported.[17]

Selenium oxyhalide compounds are strong irritants and vesicants, and can cause pulmonary edema.

SELENIUM DEFICIENCY

The essential role of selenium in animal nutrition was firstly demonstrated by Schwarz and Foltz[18] when they prevented liver necrosis in diet-deficient rats by feeding selenium supplements. Evidence that the element is a dietary essential was subsequently substantiated by other investigators.[19-21]

Shortly after the Schwarz and Foltz report in 1957, two groups, one in Oregon, USA[22] and the other in New Zealand[23] discovered that muscular dystrophy (white muscle disease) in lambs and calves in their respective areas were manifestations of selenium deficiency and could be prevented by feeding selenium supplements. Subsequently it was realized that selenium deficiencies in animals were more widespread than selenosis.

Selenium deficiency in humans has not been clearly established. However, there is some evidence that selenium deficiency may be a complicating factor in certain types of kwashiorkor* in children.[24-26]

---

* Kwashiorkor—a syndrome produced by a severe protein deficiency in the diet. It is characterized by retarded growth, changes in skin and hair pigments, gastrointestinal disorders, anemia, low serum albumin, and mental apathy.

Selenium requirements in man are unknown. Concentrations of selenium found in foods are dependent upon the soil conditions of a specific area, the class of food, and the methods of preparation. Good sources of selenium include kidney, liver, other meat, seafoods, and whole grains. Fruits and vegetables are poor sources, as much of the selenium content in the latter can incidentally be lost in the cooking water[27]. The effect of processing techniques upon the final selenium content in prepared meats and fish has not been too well documented.

SELENIUM DISTRIBUTION IN TISSUES

As expected selenium contents in animal tissues vary with the region of origin, the diet, and the method of analysis. Highest concentrations have been reported for kidney and liver; muscle, blood and bone contain lesser amounts. Interestingly, cardiac muscle was found to contain more selenium than skeletal muscle.[28,29]

Data regarding human tissue levels is rather sparse. Concentrations reported from different parts of the world also vary. For example, liver samples obtained in Canada,[30] England,[31] and New Zealand[32] were reported to contain $0.18–0.66 \mu g\ g^{-1}$, $0.20–0.40 \mu g\ g^{-1}$, and $0.16–0.57 \mu g\ g^{-1}$, respectively.

Dickson and Tomlinson,[30] employing activation analysis, determined selenium levels in both infant and adult tissues. Tissues exhibiting changes with age are listed in Table 23.1.

TABLE 23.1
Average selenium concentration in human tissue ($\mu g\ g^{-1}$ dry weight)

Tissue	Infant	Adult
Adrenal	0.21	0.36
Heart	0.16	0.27
Kidney	0.92	0.63
Liver	0.34	0.39
Lung	0.17	0.21
Lymph node	0.26	0.10
Muscle	0.31	0.40
Pancreas	0.05	0.13
Thyroid	0.64	1.24

An increase in concentration is observed with age in the adrenal, brain, muscle, pancreas and thyroid, but decreased in the heart, kidney and lymph node. The small population examined, and the lack of information regarding their disease status before death, precludes generalizations regarding the significance of these variations with age.

Recently, McConnell and co-workers[33] compared selenium levels in human blood and tissues, in health and disease states, by neutron activation analysis. Apparently normal pancreatic tissue contained from 0.56 to 0.70 μg g^{-1} of selenium while histologically abnormal pancreas contained between 0.93 and 1.03 μg g^{-1}. The highest levels were associated with metastatic disease to the organ and with acute inflammatory processes. Selenium concentrations in normal and abnormal liver tissue ranged from 1.49 to 1.97 μg g^{-1} and 1.34 to 1.46 μg g^{-1}, respectively. Normal synoviae contained from 0.51 to 0.75 μg g^{-1} of selenium and synoviae obtained from patients with rheumatoid arthritis, 0.96–1.58 μg g^{-1}.

Blood concentrations in all species have been investigated more extensively. Allaway et al.[34] reported that 210 specimens obtained from different regions of the US contained from 10 to 34 μg of selenium per 100 ml. Similar values were obtained in the Canadian study.[30] Two surveys in New Zealand[35] revealed blood selenium levels of 5.5–8.1 μg% in whole blood. Values reported from Finland[36] range between 4.2 and 10.9 μg%. Studies from England and Sweden[31,37,38] also reveal variations in blood selenium levels.

Dickson and Tomlinson[30], using radioactivation analysis to determine selenium concentrations in plasma proteins separated electrophoretically, observed the highest concentrations in the α and β globulin fractions. McConnell et al.[33] observed significantly lower selenium levels in sera from cancer patients. Whether this phenomenon is due to the disease process itself or merely reflects the patient's food intake remains to be elucidated.

ANALYSIS OF SELENIUM

Early method for the analysis of selenium in agricultural products, tissues, blood, etc., involved separation of selenium from the biological matrix by distillation with hydrobromic acid and subsequent estimation, either gravimetrically or colorimetrically.[39] Selenium in the hexavalent state can be separated from all other elements except arsenic and germanium by distillation. It is then precipitated in the distillate by reduction with sulfur dioxide and hydroxyl amine hydrochloride. Large-sized samples are required because of the relative insensitivity of the method.

Fluorimetry

Fluorimetric analysis is among the most sensitive of analytical techniques for determining selenium in biological materials, industrial products, etc. Detection sensitivities attained by the diaminobenzidine and diaminonaphthalene fluorescence complexes are 2 ng ml^{-1} and 1 ng ml^{-1}, respectively. A number of procedures employing 3,3-diaminobenzidine have been described;[40-43] 2,3-diaminonaphthalene was preferred as a reagent by subsequent investigators[34,44-48] because of its greater sensitivity. Both the diaminobenzidine and diaminonaphthalene reactions are pH dependent.

Cukor *et al.*[45] combined isotope dilution techniques with fluorimetric methods in determining selenium in plant materials.

Certain forms of selenium in biological materials are volatile, a fact which must be considered in processing and storing samples.

Activation analysis

Bowen and Cawse[48] were among the first to determine selenium by activation analysis. Concentrations in fertilizers, tomato tissue and human blood were measured. Detection sensitivity in the 'ppb' range was attained.

Wester[49] employed neutron activation analysis in determining selenium and other trace metals in human and bovine hearts. In a subsequent investigation, the technique was applied in comparing selenium, arsenic, gold, copper, molybdenum, iron and zinc concentrations in normal and uremic human blood.[50]

Pillay *et al.*[51] determined selenium concentrations in fish by neutron activation analysis, after first isolating the metal by an extraction technique.

Other applications of activation analysis to the determination of selenium levels, in foods[52] and other biological materials,[53] have been described fairly recently.

Miscellaneous instrumental procedures

Strausz *et al.*[54] determined selenium in biological materials by X-ray fluorescence. After sulfuric-nitric acid digestion, selenium was reduced by stannous chloride and coprecipitated with tellurium. The precipitate was filtered and exposed to X-rays. Selenium was detectable over a range of 0.5–25 μg.

A recent publication described the analysis of selenium in blood by proton-induced X-ray emission.[55] A detection sensitivity of less than 10 ppb was claimed.

Thompson[56] determined selenium, arsenic, antimony, and tellurium in animal feeds by flame atomic fluorescence spectrometry. Samples were dissolved in hydrochloric acid, and the individual elements were separated as hydrides.

A procedure for determining selenium in plant, animal, and human materials by electron capture gas liquid chromatography has been presented.[57] Samples are wet ashed in nitric acid and magnesium nitrate. Selenium (IV) in acidic solution is complexed with *o*-phenylenediamine and extracted into toluene. The sensitivity is adequate for the analysis of biological materials.

Atomic absorption spectrometry

Some applications of atomic absorption spectrometry to the analysis of selenium have been described. With an air–acetylene flame, the detection limit at the 196.1 nm resonance line is approximately 0.5 μg ml^{-1} with pure solutions.

Analyses of biological materials by flame atomic absorption at wavelengths in the far ultraviolet region present challenges. Background absorption due to the flame itself reduces the practical detection capability, nevertheless a few applications have been described.[58,59] Mulford[60] was able to attain a detection sensitivity of 0.1 μg ml^{-1} after chelation of selenium with ammonium pyrollidine dithiocarbamate and extraction over the pH range 3–6.

Selenium (IV) is quantitatively extractable at a pH up to 6 in the presence of other dithiocarbamate compounds.[61,62]

Flameless atomization extends the detection limits many fold. Background correction is required.

Separating selenium by borohydride generation[63-67] enables detection at the ppb level. Borohydride generation techniques are not interference free, incidentally.

A procedure for determining selenium in fish and food products by flameless atomization[68] was presented recently. Samples are first solubilized in nitric acid, then atomized in the graphite furnace in the presence of nickel nitrate. A detection limit of 3 ng ml^{-1} and good recoveries are claimed.

Techniques of selenium analysis that combine complexation–extraction procedures and flameless atomization have not yet been fully explored.

REFERENCES

1. H. J. M. Bowen, *Analyst*, **89**, 658 (1964).
2. I. Rosenfeld and O. A. Beath, *Selenium*, Academic Press, New York, 1964.
3. Y. K. Chau and J. P. Riley, *Anal. Chim. Acta*, **33**, 36 (1965).
4. W. B. Dye, E. Bretthauer, H. J. Seim and C. Blincoe, *Anal. Chem.*, **35**, 1687 (1963).
5. H. E. Ganther, C. Gondie, M. L. Sunde, M. J. Kopecky, P. Wagner, S. Hoh and W. G. Hoekstra, *Science*, **175**, 1122 (1975).
6. J. H. Koeman, W. H. M. Peeters, C. H. M. Koudstahl-Hol, P. S. Tjive and J. J. M. Goeij, *Nature (London)* **245**, 385 (1973).
7. S. G. Knott and C. W. R. McCray, *Austral. Vet. J.*, **35**, 161 (1959).
8. G. A. Fleming, *Soil Sci.*, **94**, 28 (1962).
9. K. H. Schutte, *The Biology of Trace Elements*, Crosby Lockwood, London, 1964.
10. I. Rosenfeld and O. A. Beath, *Selenium: Geobotany, Biochemistry, Toxicity and Nutrition*, Academic Press, New York, 1964.
11. A. Japha, *Experimenta nonnulla de vi selenii in organismum animalen*, Dissertation, Halle, 1842, cited in A. L. Moxon and M. Rhian, *Physiol. Rev.*, **23**, 305 (1943).
12. A. L. Moxon and M. Rhian, *Physiol. Rev.*, **23**, 305 (1943).
13. K. W. Franke, *J. Nutr.*, **5**, 597 (1934).
14. E. J. Underwood, *Trace Elements in Human and Animal Nutrition*, Academic Press, New York, 1971.
15. A. L. Moxon, M. Rhian, H. D. Anderson and O. E. Olson, *J. Animal Sci.*, **3**, 299 (1944).
16. A. Hamilton and H. L. Hardy, *Industrial Toxicology*, Publishing Sciences Group, Acton, Mass, 1974.
17. P. Pringle, *Brit. J. Derm.*, **54**, 54 (1942).

18. K. Schwarz and C. M. Foltz, *J. Amer. Chem. Soc.*, **79**, 3293 (1957).
19. E. L. Patterson, R. Milstrey and E. L. R. Stokstad, *Proc. Soc. Exptl. Biol. Med.*, **95**, 621 (1957).
20. J. N. Thompson and M. L. Scott, *J. Nutr.* **97**, 335 (1969).
21. K. E. M. McCoy and P. H. Wesweg, *J. Nutr.* **98**, 383 (1969).
22. O. H. Muth, J. E. Oldfield, L. F. Remmert and J. R. Schubert, *Science*, **28**, 1090 (1958).
23. J. W. McClean, G. G. Thompson, J. H. Claston, *Nature (London)*, **184**, 251 (1959).
24. K. Schwarz, *Fed. Proc. Fed. Amer. Soc. Exptl. Biol*, **20**, 666 (1961).
25. A. S. Majaj and L. L. Hopkins, *Lancet*, **2**, 592 (1966).
26. R. F. Burk, W. N. Pearson, R. J. Wood and F. Viteri, *Amer. J. Clin. Nutr.*, **20**, 723 (1967).
27. K. Aterman, *Brit. J. Nutr.*, **17**, 105 (1963).
28. C. F. Ehlig, D. E. Hogue, W. H. Allaway and D. J. Hamm, *J. Nutr.*, **92**, 121 (1967).
29. K. A. Handreck and K. O. Godwin, *Austral. J. Agr. Res.*, **21**, 71 (1970).
30. R. C. Dickson and R. H. Tomlinson, *Clin. Chim. Acta*, **16**, 311 (1967).
31. E. I. Hamilton, M. J. Minski and J. J. Cleary, *Sci. Total Environ.*, **1**, 341 (1972–73).
32. C. A. Johnson, J. F. Lewin and P. A. Fleming, *Anal. Chim. Acta*, **82**, 79 (1976).
33. K. P. McConnell, W. L. Broghamer, Jr., A. J. Blotcky and A. J. Hurt, *J. Nutr.*, **105**, 1026 (1975).
34. W. H. Allaway, J. Kubota, F. L. Losee and M. Roth, *Arch. Environ. Health*, **16**, 342 (1968).
35. E. J. Underwood, *Trace Elements in Human and Animal Nutrition*, Academic Press, New York, 1977.
36. I. Westernarck, P. Raunu, M. Kirjarenta and L. Lappalainen, *Acta Pharmacol. Toxicol. Suppl.*, **40**, 465 (1977).
37. H. J. M. Bowen and P. A. Cawse, *Analyst*, **88**, 721 (1963).
38. D. Brune, K. Samsahl and P. O. Wester, *Clin. Chim. Acta*, **13**, 285 (1966).
39. W. O. Robinson, H. Dudley, K. T. Williams and H. C. Byers, *Ind. Eng. Chem. (Anal. Ed.)*, **6**, 274 (1934).
40. K. L. Cheng, *Anal. Chem.*, **28**, 1738 (1956).
41. F. B. Cousins, *Austral. J. Exptl. Biol. Med. Sci.*, **38**, 11 (1960).
42. J. H. Watkinson, *Anal. Chem.*, **32**, 381 (1960).
43. W. B. Dye, E. Bretthauer, H. J. Seim and C. Blincoe, *Anal. Chem.*, **35**, 1687 (1963).
44. W. H. Allaway and E. E. Cary, *Anal. Chem.*, **33**, 1359 (1964).
45. P. Cukor, J. Walzcyk and P. F. Lott, *Anal. Chim. Acta*, **30**, 473 (1964).
46. J. H. Watkinson, *Anal Chem.*, **38**, 92 (1966).
47. R. J. Hall and P. L. Gupta, *Analyst*, **94**, 292 (1969).
48. H. J. M. Bowen and P. A. Cawse, *Analyst*, **88**, 721 (1963).
49. P. O. Wester, *Acta Medica Scand. Suppl*, **439**, 7 (1965).
50. D. Brune, K. Samsahl and P. O. Wester, *Clin. Chim. Acta*, **13**, 285 (1966).
51. K. K. S. Pillay, C. C. Thomas, Jr., and C. M. Hyche, *J. Radioanal. Chem.*, **20**, 597 (1974).
52. R. Becker, R. A. Veglia and E. R. Schmid, *Radiochem. Radioanal. Letters*, **19**, 343 (1974).
53. B. Maziere, J. Gros and D. Comar, *J. Radioanal. Chem.*, **24**, 279 (1975).
54. K. L. Strausz, J. T. Purdham and O. P. Strausz, *Anal. Chem.*, **47**, 2032 (1975).
55. M. Berti, G. Buso, P. Colautti, G. Moschini, B. M. Stievano and C. Tregnaghi, *Anal. Chem.*, **49**, 1313 (1977).
56. K. C. Thompson, *Analyst*, **100**, 307 (1975).

57. C. F. Poole, N. J. Evans and D. G. Webberly, *J. Chromatog.*, **136**, 73 (1977).
58. C. S. Rann and A. N. Hambly, *Anal. Chim. Acta*, **32**, 346 (1965).
59. W. Slavin, *Atomic Absorption Spectroscopy*, Interscience Publishers, New York, 1968.
60. C. E. Mulford, *Atomic Abs. Newsletter*, **5**, 88 (1966).
61. H. Bode, *Z. Anal. Chem.*, **144**, 165 (1955).
62. H. Bode and F. Neumann, *Z. Anal. Chem.*, **172**, 1 (1960).
63. F. J. Schmidt and J. L. Royer, *Anal. Letters*, **6**, 17 (1973).
64. S. Ng and W. McSharry, *J. Assoc. Offic. Anal. Chemists*, **58**, 987 (1975).
65. M. R. Church and W. H. Robeson, *Int. J. Environ. Anal. Chem.* **3**, 323 (1974).
66. J. A. Fiorino, J. W. Jones and S. G. Capar, *Anal. Chem.*, **48**, 121 (1976).
67. O. E. Clinton, *Analyst*, **102**, 187 (1977).
68. G. I. C. Shum, H. C. Freeman and J. F. Uthe, *J. Assoc. Offic. Anal. Chemists*, **60**, 1010 (1977).

CHAPTER 24

SILICON

Next to oxygen, silicon is the most abundant element in nature. Almost 28% of the earth's crust is composed of silicon. The element is present in all animal and plant matter, both marine and land. Silicon occurs as silica (silicon dioxide) and as silicates (metallic salts of silicic acid).

More than 40 years ago King and Belt[1] in a review of the physiological and pathological aspects of silica state: 'The almost universal distribution of silicon makes it probable that silicon will at least be a likely contaminant if not an essential ingredient of all protoplasm'.

H. Schulz[2-5] in the early 1900s made the first thorough investigation of silica in human material. He noted some differences between normal and diseased tissue. Subsequently, with the development of more precise and sensitive methods of analysis, tissue deposition and metabolism were studied by others.

Belt and co-workers[1,6] determined silicon contents in fetal and adult tissues in Canada. Concentrations found in most fetal tissues ranged between 4 and 22 mg% (40–220 μg g^{-1}) dry weight while values for most human adult tissues varied between 10 and 200 mg% (100–2000 μg g^{-1}). Silica concentrations in lung and lymph nodes were found to increase considerably with age. Between infancy and adulthood lung concentrations varied from 14 to 2000 mg% (140–20 000 μg g^{-1}) and peribronchial lymph nodes, 27–5000 mg% (270–50 000 μg g^{-1}). Silicon levels in blood[7] averaged less than 1 mg% (10 μg ml^{-1}).

Values reported by recent investigators seem to be of the same order of magnitude. Hamilton et al.,[8] who reported their findings in terms of wet weight, found the following concentrations in adult human tissues: brain, 18.6–27.4 μg g^{-1}; kidney, 29–51 μg g^{-1}; liver, 19.8–47.4 μg g^{-1}; lung, 46.7–68.1 μg g^{-1}; lymph node, 274–704 μg g^{-1}; muscle, 40.1–41.9 μg g^{-1}. Silicon levels in whole blood are said to range about 1 μg ml^{-1}.[9]

ESSENTIALITY OF SILICON

The electron microprobe studies of Carlisle,[10] showing the localization of silicon in active growth areas in bones of young mice and rats, gave the first indications of a physiological role for the element. In a subsequent study where weanling rats were fed either silicon-poor or silicon-supplemented diets, silicon was found to hasten bone mineralization.[11] Somewhat later it was demonstrated that silicon is essential for growth and skeletal development in rats[12] and chicks.[13] Silicon has been shown, also, to be an integral part of mucopolysaccharides, which are essential structural components in cartilage and connective tissue.[14–17]

The minimum dietary silicon requirements for satisfactory growth and for the maintenance of health are unknown. Because the element is ubiquitous and so plentiful, naturally occurring deficiency states are beyond imagination. Foods of plant origin are richer in silicon than are foods of animal origin; refined grain products contain substantially less than whole grain.

METABOLISM OF SILICON

Silicon, as solid silica, monosilicic acid, mucopoly saccharides and other organically bound forms, is ingested in foods. The mechanism of absorption from these sources is not entirely known. Increased urinary silica excretion following increased oral intake had been observed by investigators in the early part of this century.[18–20] Increased silicon levels in blood were not observed, however.

In later investigations, King et al.[7] observed that herbivorous animals excreted 10–20 times as much silica in urine as did carnivores. They also demonstrated that intravenous administration of soluble silica could increase urinary excretion up to 150-fold without causing much of an increase in blood levels.[21]

Subsequent studies by Jones and Handreck demonstrated that the body retains only approximately 1% of the silica ingested,[22] most of that ingested is excreted via the feces. Observations on man[23] and other animals[24–29] indicate that the urinary excretion of silicon is equivalent to 4% or less of the daily intake.

SILICON TOXICITY

Silicosis is a disease of the lung which has long afflicted miners and others engaged in occupations where various forms of silica, such as quartz, sandstone, flint, or diatomaceous earths, are employed. Exposure to siliceous dusts can be encountered in the manufacture of and use of abrasives, i.e. grinding and polishing, ceramic manufacture, construction, etc.

Pathological manifestations of exposure to silica have long been of interest. Peacock[30] and Greenhow[31] originally demonstrated 'sand' in human lungs.

Much later Policard and Morel[32] studied 'siliceous material' in lungs by spectrographic means. X-ray diffraction methods were applied by others.[33-35]

Particles of silica and of asbestos, which are fibrous silicates of complex composition, have long been known to stimulate severe fibrogenic reactions in the lung and elsewhere in the body. The reaction arises initially from phagocytosis of silica particles by alveolar macrophages. Collagen synthesis by neighboring fibrobasts is stimulated by the death of these macrophages.[36,37] It has been postulated that perhaps these macrophages liberate a fibrogenic agent.[38]

The action of silica on tissues can be explained further perhaps by certain *in vitro* studies. For example, it was demonstrated about 40 years ago that silicic acid can react with lipoprotein and protein films.[39] Also, the fact that colloidal silica was shown to be hemolytic indicates that it can induce changes in the permeability of biological membrane systems.[40]

ANALYSIS OF SILICA

At present, colorimetric methods form the basis of most microanalytical procedures for the analysis of silicon. Most procedures are based on the formation of a yellow silicomolybdic acid complex, its reduction to yield a blue color and subsequent measurement spectrophotometrically.[41-46] Phosphate and iron can interfere.

Fluorescence analysis by formation of a silicon–benzoic complex has been suggested.[47] A detection limit of 0.08 μg ml^{-1} has been attained.

Applications of spectrography and X-ray diffraction to analyses of silicon in biological material were mentioned previously.[32-35]

Recently Olmez and co-workers[48] reported on the analysis of silica in atmospheric dust samples.

Few practical applications of atomic absorption spectrometry to analyses of biological materials for silicon have been developed, primarily because existing colorimetric methods fulfill most analytical needs. Detection limits of 0.1 μg ml^{-1} are attained at the 251.6 nm resonance line when using a nitrous oxide–acetylene flame. These limits can be extended by flameless atomization.

REFERENCES

1. E. J. King and T. H. Belt, *Physiol. Rev.*, **18**, 329 (1938).
2. H. Schulz, *Pfluger's Arch.*, **84**, 67 (1901).
3. H. Schulz, *Pfluger's Arch.*, **144**, 346 (1912).
4. H. Schulz, *Biochem. Ztschr.*, **46**, 376 (1912).
5. H. Schulz, *Biochem. Ztschr.*, **70**, 464 (1915).
6. T. H. Belt, D. Irwin and E. J. King, *Canad. Med. Assoc. J.*, **34**, 125 (1936).
7. E. J. King, H. Stantial and M. Dolan, *Biochem. J.*, **27**, 1002 (1933).
8. E. I. Hamilton, M. J. Minski and J. J. Cleary, *Sci. Total Environ.*, **1**, 341 (1972–73).

9. E. J. Underwood, *Trace Elements in Human and Animal Nutrition*, Academic Press, New York, 1977.
10. E. Carlisle, *Science*, **167**, 279 (1970).
11. E. Carlisle, *Fed. Proc. Fed. Amer. Soc. Exptl. Biol.*, **29**, 565 (1970).
12. K. Schwarz and D. B. Milne, *Nature (London)*, **239**, 333 (1972).
13. E. Carlisle, *Fed. Proc. Fed. Amer. Soc. Exptl. Biol.*, **31**, 700 (1972).
14. K. Schwarz, *Proc. Natl. Acad. Sci., USA*, **70**, 1608 (1973).
15. K. Schwarz and S. C. Chen, *Fed. Proc. Fed. Amer. Soc. Exptl. Biol.* (Abst. 2795), **33**, 704 (1974).
16. E. Carlisle, *Fed. Proc. Fed. Amer. Soc. Exptl. Biol.*, **33**, 1758 (1974).
17. E. Carlisle, *Fed. Proc. Fed. Amer. Soc. Exptl. Biol.*, **32**, 930 (1973).
18. H. Schulz, *Pflugers Arch.*, **144**, 350 (1912).
19. E. Salkowski, *Ztschr. Physiol. Chem.*, **83**, 143 (1913).
20. M. Gonnermann, *Ztschr. Physiol. Chem.*, **94**, 163 (1919).
21. E. J. King, H. Stantial and M. Dolan, *Biochem. J.*, **27**, 1007 (1933).
22. L. H. P. Jones and K. A. Handreck, *J. Agr. Sci.*, **65**, 129 (1969).
23. P. F. Holt, *Brit. J. Ind. Med.*, **7**, 12 (1950).
24. F. Sauer, D. H. Laughland and W. M. Davidson, *Canad. J. Biochem. Physiol.*, **37**, 183 (1959).
25. F. Sauer, D. H. Laughland and W. M. Davidson, *Canad. J. Biochem. Physiol.*, **37**, 1173 (1959).
26. R. F. Keeler and S. A. Lovelace, *J. Exptl. Med.*, **109**, 601 (1959).
27. R. J. Emerick, L. B. Embay and O. E. Olson, *J. Animal Sci.*, **18**, 1025 (1959).
28. M. C. Nottle, *Austral. J. Agr. Res.*, **17**, 175 (1966).
29. C. H. Bailey, *Amer. J. Vet. Res.*, **28**, 1743 (1967).
30. T. B. Peacock, *Trans. Pathol. Soc. (London)*, **12**, 36 (1861), cited in E. J. King and T. H. Belt, *Physiol. Rev.*, **18**, 329 (1938).
31. E. H. Greenhow, *Trans. Pathol. Soc. (London)*, **16**, 59 (1865) cited in E. J. King and T. H. Belt, *Physiol. Rev.*, **18**, 329 (1938).
32. A. Policard and A. Morel, *Med. du Travail*, **4**, 56 (1932).
33. G. L. Clark and D. H. Reynolds, *Univ. of Toronto Studies (Geol. Ser.)*, **38**, 13 (1935) cited in E. J. King and T. H. Belt, *Physiol. Rev.*, **18**, 329 (1938).
34. O. M. Faber, *Staub. Haele*, **1**, 26 (1936) cited in E. J. King and T. H. Belt, *Physiol. Rev.*, **18**, 329 (1938).
35. V. Hicks, O. McElroy and M. E. Warga, *J. Ind. Hyg. Toxicol.*, **19**, 177 (1937).
36. A. C. Allison, J. S. Harington and M. Birbeck, *J. Exptl. Med.*, **124**, 141 (1966).
37. T. Nash, A. C. Allison and J. S. Harington, *Nature (London)*, **210**, 259 (1966).
38. A. G. Heppleston and J. A. Styles, *Nature (London)*, **214**, 521 (1967).
39. J. H. Schulman and E. K. Rideal, *Proc. Roy. Soc.*, **122B**, 46 (1937).
40. J. D. Hurley and J. Margolis, *Nature (London)*, **189**, 1010 (1969).
41. E. J. King, *J. Biol. Chem.*, **18**, 25 (1928).
42. J. T. Wood and M. G. Mellon, *Ind. Eng. Chem. (Anal. Ed.)*, **13**, 760 (1941).
43. A. O. Gettler and C. J. Umberger, *Amer. J. Clin. Pathol.*, *(Tech. Suppl.)*, **9**, 1 (1941).
44. F. G. Straub and J. A. Grabowski, *Ind. Eng. Chem. (Anal. Ed.)*, **16**, 574 (1944).
45. E. J. King, B. D. Stacy, P. F. Holt, D. M. Yates and D. Pickles, *Analyst*, **80**, 441 (1955).
46. M. E. Jankowiak and R. R. LeVier, *Anal. Biochem.*, **44**, 462 (1971).
47. G. Elliot and J. A. Radley, *Anal. Chem.*, **33**, 1623 (1961).
48. J. Olmez, N. K. Aras, G. E. Gordon and W. H. Zoller, *Anal. Chem.*, **46**, 935 (1974).

CHAPTER 25

SILVER

Silver is an ancient and precious metal known to people in the Mediterranean region as early as 4000 BC, copper and gold being the two metals preceding it. The Greeks mined silver extensively. It is found in the native state or associated with copper, gold and lead. Though a toxic metal, it cannot be ranked with arsenic, antimony, lead, etc. There is no evidence, to date, that silver has an essential function in plant and animal metabolism.

INDUSTRIAL USES AND HAZARDS

In addition to its use in the manufacture of tableware and jewelry, silver is alloyed with metals, such as aluminum, antimony, cadmium, copper and lead, to increase strength, hardness and corrosion resistance. It is employed as a coating for steel used in aviation bearings, and in telegraph and telephone instruments. Silver is used in the manufacture of solder and also in dentistry.

The toxic activity of silver is primarily local. The metal inactivates sulfhydryl enzymes and also combines with amino, imidazole, carboxyl and phosphate groups. Its systemic action is not extensive because silver is absorbed slowly. Silver ions are bound to protein and are also precipitated as silver chloride at the sites of its application.

Minute amounts of silver do gain access to the body from the mucous membranes and from burns with silver nitrate. Absorbed silver is widely distributed in the body, expecially in the subepithelial portions of the skin. Appreciable concentrations impart a characteristic gray to bluish pigmentation known as argyria. The pigment consists of silver sulfide and metallic silver. Any injury sustained is cosmetic only. It remains for life.

Argyria, which develops slowly, within two to twenty-five years, as a result of industrial exposure, is localized in those areas coming in contact with the metal, i.e. the skin and the respiratory tract. Generalized argyria may follow ingestion or inhalation.[1,2]

Occupational argyria, though apparently not recorded, must have been familiar to the ancient Greeks, who mined and processed silver ores or made silver-coated thread.

SILVER IN THERAPEUTICS

Salts of silver were used therapeutically by the Mohammedan school of physicians in the 8th Century.[1] Avicenna c. 980 AD used silver in silvering pills and prescribed silver filings as a blood purifier, for palpitation of the heart, and for offensive breath. He described what is probably the first case of localized argyria, a bluish discoloration of the eyes resulting from the ingestion of silver.[3]

In the alchemists' order of things, silver corresponded to the moon and the brain. Since the moon was thought to be related to brain disorders, silver came to be used in treating insanity and epilepsy. Silver nitrate was known as 'lunar caustic'.

Paracelsus advised the internal use of silver dissolved in distillate of nitre and vitriol (nitric and sulfuric acids!). Angelo Sala, an early seventeenth century physician, administered silver nitrate internally for various uses: as a purge, and in the treatment of brain 'afflictions'. He described the occurrence of generalized argyria from such use.[1]

Silver and its compounds fell into disuse as a medicinal agent during the early part of the eighteenth century. However, the use of silver nitrate, revived later for treating epilepsy and other nervous conditions, continued into the late nineteenth century.

Silver was combined with arsphenamine, employed in treating syphillis, during the early part of this century.

Today silver compounds are employed for their local action only. The nitrate and silver protein complexes are utilized to some extent for their antiseptic and germicidal properties. Silver nitrate is also used for its local caustic or corrosive action.

TISSUE DISTRIBUTION OF SILVER

Recent investigations concerning silver deposition in the tissues and blood, of a population not occupationally exposed to silver, indicate that the metal is present in less than 'ng g^{-1}' quantities.[4] Kehoe and co-workers[5] had reported higher levels in their studies c. 1940. Differences may be due to the selection of the populations examined, as well as to analytical technology.

Pathological manifestations in argyria, arising from therapeutic use or possible industrial exposure, were studied extensively as the rare opportunity presented itself to investigators. Gettler et al.[6] discuss the case of a 70-year-old man who had been the 'blue man in a circus', and gave his history of previous work in a silver mine. Their results of the chemical analysis of his tissues are of historical interest only.

The Gerlachs[7] employed spectrography in determining gold and silver contents in tissues of patients being treated with metal compounds for various reasons. Gaul and Staud[8,9] analyzed biopsy material from subjects receiving silver arsphenamine and other organic and colloidal silver medications.

Blumberg and Nelson[10] discuss a patient who was being investigated for possible lead poisoning because of a 'very peculiar ashen pallor'. Though he was edentulous, a very definite, gray-blue pigmentation with punctate distribution was noted on the gums. Spectrographic analysis of blood revealed a silver level estimated at 0.05 mg% (0.5 μg ml^{-1}). 'Normal' blood was found to contain 0.5 μg% (5 ng ml^{-1}) or less. A history obtained subsequently revealed that the patient had been taking capsules containing 16 mg of silver nitrate, three times daily, for his gastrointestinal symptoms.

ANALYSIS OF SILVER

Colorimetric–fluorimetric

Colorimetric and fluorimetric procedures for the determination of silver do not possess the required sensitivity for analyses of biological materials. However, extraction of the various silver complexes formed can be applied to advantage in isolating silver for spectroscopic analysis.

Silver forms quantitatively extractable chelates with dithizone at acid pH.[11] The dithiocarbamate complexes of the metal are quantitatively extractable from mineral acids up to pH 11.[12,13] 2-Nitroso-1-napthol forms an extractable brown substance with silver,[14] and the metal complexes with rhodanine compounds.[15]

Silver forms fluorescing complexes with eosin plus 1,10-phenanthroline,[16] and 8-hydroxyquinoline-5-sulfonic acid.[17,18]

Instrumental methods

A number of instrumental techniques for the analysis of silver contents in biological materials have appeared during the present decade. Nakarni and Ehmann[19] and Cornelis et al.[20] described methods involving neutron activation analysis. Marich et al.[21] reportedly were able to detect silver at the nanogram level by employing a laser microprobe. Atomic absorption spectrometry, using both flame[22-25] and flameless atomization,[25] have been applied in determining silver contents in various biological materials.

The detection sensitivity for silver at the 328.1 nm resonance line is 0.1 μg ml^{-1} by flame atomic absorption spectrometry, and the limits are extended by flameless atomization. Chelation–extraction techniques can be utilized to both isolate silver from matrix constituents and to concentrate the amount present.

Urine samples, after pH adjustment, can be chelated and extracted directly before analysis.[26] Tissue samples and blood specimens must be wet ashed initially.

REFERENCES

1. W. R. Hill and D. M. Pillsbury, *Argyria: The Pharmacology of Silver*, The Williams and Wilkens Company, Baltimore, MD, 1939.
2. J. E. M. Wigley and P. M. Deville, *Proc. Roy. Soc. Med.*, **37**, 648 (1944).
3. S. Ayres, *Arch. Derm. Syph.*, **38**, 645 (1938).
4. E. I. Hamilton, M. J. Minski and J. J. Cleary, *Sci. Total Environ.*, **1**, 341 (1972–73).
5. R. A. Kehoe, J. Cholak and R. V. Storey, *J. Nutr.*, **19**, 579 (1940).
6. A. O. Gettler, C. P. Rhoads and S. Weiss, *Amer. J. Pathol.*, **3**, 631 (1927).
7. Walther Gerlach and Werner Gerlach, *Virchow's Arch. f. Pathol. Anat.*, **28**, 209 (1931).
8. L. E. Gaul and A. H. Staud, *Arch. Derm. Syph.*, **30**, 433 (1934).
9. L. E. Gaul and A. H. Staud, *J. Amer. Med. Assoc.*, **104**, 1387 (1935).
10. H. Blumberg and C. T. Nelson, *J. Amer. Med. Assoc.*, **103**, 1521 (1934).
11. H. Fischer, G. Leopoldi and H. Van Uslar, *Z. Anal. Chem.*, **101**, 1 (1935).
12. H. Bode, *Z. Anal. Chem.*, **143**, 182 (1954).
13. H. Bode and F. Neuman, *Z. Anal. Chem.*, **172**, 1 (1960).
14. G. Gorbach and F. Pohl, *Mikrochem.*, **38**, 258 (1951).
15. E. B. Sandell and J. J. Newmayer, *Anal. Chim. Acta*, **5**, 445 (1951).
16. M. T. El-Ghamry, W. Frei and G. W. Higgs, *Anal. Chim. Acta*, **47**, 41 (1969).
17. D. E. Ryan and B. K. Pal, *Anal. Chim. Acta*, **44**, 385 (1969).
18. R. L. Wilson and J. D. Ingle, Jr., *Anal. Chem.*, **49**, 1066 (1977).
19. R. A. Nakarni and W. D. Ehmann, in *Trace Substances in Environmental Health*, IV, Proc. 4th Annual Conference Trace Substances in Environmental Health, Columbia, MO, 1971.
20. R. Cornelis, A. Speecke and J. Hoste, *Anal. Chim. Acta*, **68**, 1 (1973).
21. K. W. Marich, P. W. Carr, W. J. Treytl and D. Glick, *Anal. Chem.*, **42**, 1775 (1970).
22. B. F. Grabowski and W. G. Haney, *J. Pharm. Sci.*, **61**, 1488 (1972).
23. R. C. Rooney, *Analyst*, **100**, 471 (1975).
24. R. A. Greig, *Anal. Chem.*, **47**, 1682 (1975).
25. C. J. Pickford and G. Rossi, *Atomic Abs. Newsletter*, **14**, 78 (1975).
26. T. Takeuchi, M. Suzuki and M. Yanagisawa, *Anal. Chim. Acta*, **36**, 258 (1966).

CHAPTER 26

TELLURIUM

Tellurium, also known as *metallum problematum* or *aurum paradoxum*, is a toxic metal resembling arsenic in its physiological action, though related to selenium and sulfur chemically. The chief source of the metal is as a by-product in copper refining. It is also found associated with ores containing lead, silver and gold.

Its principal uses in industry include the coloring of glass and ceramics, rubber compounding to increase resistance to heat and aging, and the manufacture of alloys. It hardens lead and increases the machinability of bronze and stainless steel.

Radioactive iodine is produced through use of tellurium.

The most prominent symptoms following the inhalation of tellurium dusts and fumes are dry mouth, metallic taste, inhibition of sweat, languor, somnolence, loss of appetite, nausea and vomiting. The garlic breath odor is troublesome.[1]

Hydrogen telluride, like arsine and hydrogen selenide is a hemolytic poison.[2]

ANALYSIS OF TELLURIUM

Prior to the development of atomic absorption spectrometric instruments, tellurium contents in materials were primarily measured by gravimetric and titrimetric methods.[3] Practical procedures utilizing atomic absorption[4-8] and atomic fluorescence[9] have been described.

The detection limit for tellurium by flame atomic absorption spectrometry at the 214.3 nm resonance line is about 0.1 μg ml^{-1}. A lean, oxidizing flame is used. Limits are extended markedly with flame atomization. Borohydride generation techniques and chelation extraction methods can be employed to both isolate and further concentrate the metal.

Tellurium (IV) is quantitatively extractable by dithiocarbamates over the pH range of about 3–8.[10-11]

The dithizonate of tellurium is extractable from 0.1 to 1 M mineral acid solution.[12]

REFERENCES

1. A. Hamilton and H. L. Hardy, *Industrial Toxicology*, Publishing Sciences Group, Inc., Acton, Mass., 1974.
2. S. H. Webster, *J. Ind. Hyg. Toxicol.*, **28**, 179 (1946).
3. M. B. Jacobs, *The Analytical Chemistry of Industrial Poisons, Hazards, and Solvents*, Interscience Publishers, New York, 1949.
4. R. E. Kinser, *J. Amer. Ind. Hyg. Assoc.*, **27**, 260 (1966).
5. S. R. Koirtyohann and J. W. Wen, *Anal. Chem.*, **45**, 1986 (1973).
6. J. T. Cheng and W. F. Agnew, *Atomic Abs. Newsletter*, **13**, 123 (1974).
7. T. H. Lockwood and L. P. Limitiaco, *J. Amer. Ind. Hyg. Assoc.*, **36**, 57 (1975).
8. J. A. Fiorino, J. W. Jones and S. G. Capar, *Anal. Chem.*, **48**, 121 (1976).
9. K. C. Thompson, *Analyst*, **100**, 307 (1975).
10. H. Bode, *Z. Anal. Chem.*, **143**, 182 (1954).
11. H. Bode and F. Neumann, *Z. Anal. Chem.*, **172**, 1 (1960).
12. H. Mabuchi, *Bull. Chem. Soc. (Japan)*, **29**, 842 (1956).

CHAPTER 27

THALLIUM

The fact that thallium is one of the most toxic of metals has been appreciated since its discovery in 1861 by Crookes and Lamy working independently. The latter was the first investigator to report upon the toxicity of thallium salts.[1] A contemporary, Paulet, subsequently concluded from experiments with dogs that thallium was more toxic than lead.[2]

Thallium salts have been used to a limited extent clinically despite their toxic action. Combermale, while treating night sweats in tuberculosis with thallium, noted that patients suffered a complete hair loss.[3] Later the metal salts were administered internally to cause hair loss while treating cases of favus and ringworm of the scalp. Subsequently, thallium salts were incorporated into depilatory creams. Teleky[4] described further toxic effects from therapeutic use.

Thallium and its salts have no place in modern therapeutics.

INDUSTRIAL USES OF THALLIUM

The metal and its salts have found various industrial applications. Thallium readily amalgamates with mercury and forms alloys with other metals as well. It is used, for example, in the manufacture of switches and closures meant to operate at subzero temperatures. Thallium salts are also used in the manufacture of pigments, dyes, luminous paints, artificial gems and window glass. The oxide and bromoiodide salts of thallium, because of their high refractive index, are employed in manufacturing lenses and prisms for various optical systems. Thallous carbonate is used in making artificial diamonds. Thallium sulfate is employed extensively as a rat poison and ant bait.

Thallium intoxications from industrial exposures are very rare. Non-industrial exposures are definitely another matter. Poisonings due to thallium occur much more frequently than are reported in the world literature. Since symptoms at different stages can resemble numerous other syndromes including lead, arsenic, mercury, or gold intoxications, diabetic polyneuritis, Guillan-Barre syndrome, periarteritis nodosa, porphyria, carbon monoxide poisoning etc., it is a rare physician indeed who is sufficiently astute to include the possibility of thallium poisoning in his differential diagnosis. Incidentally, prior to the ready availability of good spectroscopic instrumentation, analytical technology was less than supportive in establishing the correct diagnosis. The challenge of diagnosing thallotoxicosis is graphically described by Prick et al.[5]

TOXICOLOGY OF THALLIUM

Thallous salts are readily absorbed from the mucous membranes of the mouth, the gastrointestinal tract, and from the skin. Since the metal is excreted slowly, it acts as a cumulative poison and as such can be found in all tissues.

Symptoms of acute thallotoxicosis, referable primarily to the nervous system and the gastrointestinal tract, become quite evident within 8 to 24 hours following ingestion of a toxic amount. The subject suffers severe paroxysmal abdominal pain, vomiting and diarrhea; hemorrhage follows along with albuminuria and oliguria; circulatory collapse ensues. Delerium, convulsions, often coma and high fever can appear rapidly. The subject's extremities may become so painful that even pressure from a sheet is intolerable. Cardiac symptoms of sufficient severity to suggest myocardial infarction are often observed. Death from respiratory failure can occur within 20 to 40 hours after the appearance of the initial symptoms.

In sub-acute and chronic thallotoxicosis symptoms are directed mainly toward the nervous system. A paresthesia of the hands and feet, noted initially, progresses to a definite peripheral polyneuritis. Legs are involved more often than arms. The subject becomes ataxic. Aptosis, strabismus, facial palsy, and mydriasis can also occur. Psychotic signs have been noted; skin changes, including eruptions of sorts and a hyperkeratosis of the palmar surface of the hands and soles of the feet, develop. A stomatitis, often ulcerative in nature, may occur. Gastrointestinal symptoms characterized by loss of appetite, vomiting and some pain are not severe in chronic poisoning. However, the picture is complicated further by the development of liver and kidney damage.

Alopecia, the most characteristic sign of thallotoxicosis does not appear for at least 20 days following the initial dose of thallium. The hair strip across the forehead and the distal third of the eyebrows often remains fairly intact, however.

The mechanism by which thallium acts is still unknown. However, potassium and thallium are known to be antagonists, competitors at receptor sites on all membranes or enzyme active centers. It has been shown that potassium counteracts many of the toxic actions in thallium poisoning and promotes renal excretion of the metal. While potassium chloride was found to have a protective effect upon rats poisoned with thallium sulfate, sodium chloride did not exhibit such activity.[6]

DISTRIBUTION OF THALLIUM

Thallium, unlike lead and cadmium, is not a ubiquitous trace metal, an ordinary environmental contaminant. It is not a normal constituent of blood, tissues and biological fluids. Its presence indicates an exposure of some sort, industrial perhaps, or from the use of thallium-containing rodenticides or ant baits. The latter practice was very common in the inner city areas in Chicago a few years ago.

Thallium sulfate was incorporated into jelly-like preparations and spread on bread to be placed at rat holes. Occasionally young ghetto children consumed these baits first. Ant or rat baits resembling macaroon cookies were also available. Legislation banning their use has not completely eliminated these thallium preparations since they are still obtainable in adjoining states.

Long after the ban was imposed, three siblings, 2–4 years old, were admitted to a sister hospital. All exhibited ataxia and hair loss. However, the youngest was the most affected. Routine laboratory data was non contributory.

A history revealed that a rodenticide spray obtained out of state had been used around the home. The mother did the spraying. Blood and urine thallium levels were determined on the siblings and parents, and the findings are summarized in Table 27.1.

TABLE 27.1
Blood and urine thallium levels in an exposed family

Subject	Blood (μg%)	Urine (μg l^{-1})
2 year old	30	1240, 984
3 year old	33	298, 270
4 year old	10	115, 110
mother	0	10
father	0	0

An enterprising investigator sent to the home to locate the specific source of exposure learned that the siblings spent considerable time playing under a certain bed, 'their hideout'. Intrigued by the scale-like deposit on the bed springs, he decided to submit a piece of the bed frame and a spring for analysis. Material scraped from the springs contained approximately 10 mg of thallium per gram of scrapings. It seems the children liked to sit in their 'hideout' and chew on the bedsprings. All eventually recovered. Thallium is detectable in urine for months following an exposure.

Minimal amounts of thallium have been found in some autopsy tissues obtained from subjects of different ages originating in the Chicago inner city. Tissue distributions are listed in Table 27.2.

TABLE 27.2
Tissue distribution of
thallium in Chicago inner city

Tissue	Thallium (μg g^{-1} wet weight)
Brain	0–0.02
Kidney	0–0.08
Liver	0–0.19
Spleen	0–0.002
Stomach	0–0.01

As a means of comparison, thallium contents found in the tissues of a 42-year-old black male who died 4 hours after hospital admission because of a possible myocardial infarction are listed in Table 27.3. Autopsy revealed

TABLE 27.3
Tissue thallium content in subject
initially diagnosed as a myocardial
infarct

Tissue	Thallium content (μg g^{-1} wet weight)
Kidney	2.92
Liver	3.13
Stomach	0.11

that the subject did not die from an infarction. All tissues were markedly congested. The stomach wall from the lesser to greater curvature varied from pink to bright red in color. There was no evidence of desquamation.

Unfortunately the stomach contents, having been washed down the drain by an astute pathology fellow, were not available for analysis.

A complete toxicological evaluation of tissues for other toxic metals, drugs, and poisons yielded nothing additional. The source of exposure to thallium was not revealed.

A rather more dramatic instance of thallotoxicosis is that of a 77-year-old white female who lived alone on a farm in southern Illinois. The subject, who had been hospitalized 'because of suspected peptic ulcer and possible myocardial ischemia', apparently improved considerably during the course of her hospital stay of approximately 18 days. Complaints were few 'except for some pain in the back, abdomen, and feet'. Appetite was poor. Suddenly, early in the morning of her eighteenth hospital day, the subject developed 'a severe spell of vomiting, experienced difficulty in breathing, after which she slipped into a coma and expired two hours later. Pathological diagnosis included myocardial infarction and congestive heart failure as causes of death. The gastrointestinal bleeding was attributed to 'stress' since ulcerations were not present. Marked congestion of lungs, liver, kidney and brain was noted.

A series of circumstances subsequent to her death prompted exhumation of the body six years later. Results of analyses performed on digests of tissue samples referred from the State of Illinois Public Health Laboratories[7] are listed in Table 27.4.

TABLE 27.4
Thallium content in
suspected homicide

Tissue	Thallium ($\mu g\ g^{-1}$)
Bone	57.20
Brain	55.20
Kidney	37.00
Liver	53.60
Muscle	13.00
Tongue	31.60
Hair-proximal	26.20
Hair-distal	14.69

Soil samples and vault materials contained less than 0.01 $\mu g\ g^{-1}$ thallium.

On the basis of the above findings and the pathological description of tissues, one can conclude that the subject was being poisoned chronically for a considerable period of time. Death may be due in part to a myocardial infarction. However, the possibility of an additional toxic dose of thallium being administered during visiting hours the day or night before death cannot be excluded.

Thallium being odorless, tasteless and colorless can easily be administered in food or a beverage as was the situation in recent cases in England.[8] The favorite vehicles of the perpetrator of the above and approximately five other homicides were doughnuts and other pastries.

ANALYSIS OF THALLIUM

Colorimetry–fluorimetry

Colorimetric methods for measuring the thallium contents in biological materials lack selectivity. Thallium (I) can be quantitatively extracted from cyanide[9] or cyanide-citrate media[10] over the pH range 9–12. Lead, cadmium, bismuth, and stannous tin are coextracted. Zinc, mercury and nickel can also interfere.[11]

Thallium can be separated from lead by extraction as chlorothallic acid from a hydrochloric acid solution. However, iron, antimony (V), mercury, tin, and gold are coextracted. The use of anion[12] and cation[13] exchange resins for separating thallium have been described but for pure solutions only.

One need not dwell on the exquisite challenges of determining thallium in biological matrices by colorimetry. Small wonder, then, that levels reported in the older literature are so high[14,15] compared to reports of similar instances in more recent publications. The values reported were undoubtedly due in part to incomplete separation from the lead, cadmium, etc. present in the respective matrices.

Thallium forms extractable chelates with numerous reagents. The metal is quantitatively extractable by acetylacetone over the pH range 2–10.[16] It is extractable as a dithiocarbamate over a wide range,[17–19] and as an oxinate from pH 3.5 to 11.5.[20] At a pH of approximately 7, thallium is extractable as a thenoyltrifluoracetonate in benzene.[21]

Fluorimetric methods of thallium analysis also lack specificity. Carminic acid and Rhodamine B react with other metals.[22,23] Furthermore, the colorimetric methods employing dithizone can exhibit great sensitivity of detection.

Instrumental applications

Instrumental applications to practical thallium analyses have primarily involved anodic stripping voltammetry and atomic absorption spectrometry. A rare paper describes other instrumental techniques, for example, mass spectrometry in blood thallium determinations.[24]

Emission spectrometry (arc-spark, flame) is too insensitive for useful applications to analyses of biological materials.

Anodic stripping voltammetry (ASV)

Ariel and Bach[25] described the determination of thallium in urine by ASV in 1963. Other applications followed. [26,27] Franke and co-workers[28] employed differential pulse anodic stripping voltammetry to identify thallium and other heavy metals in urine.

ASV could be applied to analysis of other biological matrices. Samples must be ashed. The technique is not free from interferences of other trace metals.

Atomic absorption spectrometry

With the advent of atomic absorption spectrometry, thallium analyses in the clinical laboratory ceased being a game of chance and became specific and decidedly more exact. The detection limit at the 276.8 nm resonance line is approximately 0.05 μg ml^{-1} by flame atomic absorption spectrometry. A lean oxidizing flame is used. The detection sensitivity is of course extended by flameless atomization. Deuterium background correction is a necessity with the latter even when thallium contents in extracts of biological matrices are being measured.

Organic matter must be removed prior to instrumental analysis. Blood proteins can be either wet ashed or precipitated with 5% trichloroacetic acid prior to chelation–extraction. Tissues are wet ashed. Urine samples can be chelated and extracted directly after adjusting the pH. A series of standards must be carried through the identical procedures.

An early method for thallium analysis by flame atomic absorption involved chelation with sodium diethyldithiocarbamate at pH 6.0–7.5 and extraction by methylisobutyl ketone.[29] Fifteen millilitres of blood or 30 ml of urine were employed. The procedure, using smaller-sized samples, was subsequently adapted to flameless atomization.[30]

The fact that lead and cadmium are coextracted is an advantage when employing atomic absorption spectrometric analysis. One extract serves for the three determinations. Only the hollow—cathode lamp need be changed. EDTA does not interfere with the extraction of thallium or cadmium.

Other atomic absorption procedures have been described.[31-34] Some applied chelation–extraction methods in combination with the graphite rod[32] and Delves cup.[34]

Conditions for analysis in the graphite furnace are the same as those for lead determination.

REFERENCES

1. A. Lamy, *Compt. Rend. Acad. Sci.*, **57**, 442 (1863) cited in A. O. Gettler and L. Weiss, *Amer. J. Clin. Pathol.*, **13**, 422 (1943).
2. Paulet, *Compt. Rend. Acad. Sci.*, **57**, 494 (1863) cited in A. O. Gettler and L. Weiss, *Amer. J. Clin. Pathol.*, **13**, 422 (1943).
3. Combermale, *Bull. Acad. Med.* (*Paris*), **39**, 572 (1898).
4. L. Teleky, *Weiner Med. Wochschr.*, **78**, 506 (1928).
5. J. J. G. Prick, W. G. S. Smitt and L. Muller, *Thallium Poisoning*, Elsevier, Amsterdam, 1955
6. A. Lund, *Acta Pharmacol. Toxicol.*, **12**, 260 (1956).
7. J. Spikes, J. Pirl and staff, Toxicology Laboratory, Illinois Department of Public Health, Chicago, Ill., (personal communication).
8. J. B. Cavanaugh, N. H. Fuller, H. R. M. Johnson and P. Rudge, *Quart. J. Med.*, **43**, 293 (1974).
9. E. B. Sandell, *Colorimetric Determination of Traces of Metals*, Interscience Publishers, New York, 1959.
10. C. W. Sill and H. E. Peterson, *Anal. Chem.*, **21**, 1268 (1949).
11. L. A. Haddock, *Analyst*, **60**, 394 (1935).
12. K. A. Kraus, F. Nelson and G. W. Smith, *J. Phys. Chem.*, **58**, 11 (1954).
13. T. Nozaki, *J. Chem. Soc.*, (*Japan*), **77**, 493 (1956).

14. A. O. Gettler and L. Weiss, *Amer. J. Clin. Pathol.*, **13**, 422 (1943).
15. A. Heyndrickx, *Acta Pharmacol. Toxicol.*, **14**, 20 (1957).
16. J. Stary and E. Hladky, *Anal. Chim. Acta*, **28**, 227 (1963).
17. H. Bode, *Z. Anal. Chem.*, **143**, 182 (1954).
18. H. Bode, *Z. Anal. Chem.*, **144**, 165 (1955).
19. H. Bode and F. Neumann, *Z. Anal. Chem.*, **172**, 1 (1960).
20. T. Moeller and A. J. Cohen, *Anal. Chem.*, **22**, 686 (1950).
21. F. Hagemann, *J. Amer. Chem. Soc.*, **72**, 768 (1950).
22. C. E. White and R. J. Argauer, *Fluorescence Analysis: A Practical Approach*, Marcel Dekker, New York, 1970.
23. T. S. West, in *Trace Characterization: Chemical and Physical*, (eds W. W. Meinke and B. F. Scribner) NBS Monograph, No. 100, US Govt Printing Office, Washington, DC, 1969.
24. E. Weinig and P. Zink, *Arch. Toxicol.*, **22**, 255 (1967).
25. M. Ariel and D. Bach, *Analyst*, **88**, 30 (1963).
26. U. Eisner and M. Ariel, *J. Electroanal. Chem.*, **11**, 26 (1966).
27. A. R. Curtis, *J. Assoc. Offic. Anal. Chemists*, **57**, 1366 (1974).
28. J. P. Franke, P. M. J. Coenegracht and R. A. de Zeeuw, *Arch. Toxicol.*, **34**, 137 (1975).
29. E. Berman, *Atomic Abs. Newsletter*, **6**, 57 (1967).
30. E. Berman, in *Applied Atomic Spectroscopy*, Vol. 2, (ed. E. L. Grove) Plenum Press, New York, 1978.
31. A. S. Curry, J. F. Read and A. R. Knott, *Analyst*, **94**, 744 (1969).
32. N. P. Kubasik and H. T. Volosin, *Clin. Chem.*, **19**, 954 (1973).
33. G. M. Shkolnik and R. F. Bevell, *Atomic Abs. Newsletter*, **12**, 112 (1973).
34. N. P. Singh and M. M. Joselow, *Atomic Abs. Newsletter*, **14**, 42 (1975).

CHAPTER 28

TIN

Tin, another of the ancient metals, was known before 3500 BC. Bronze is an alloy of tin and copper. The general importance of the metal need not be stressed.

Though used widely, lasting harmful effects from exposure to inorganic tin in industrial applications or via the food chain have not been reported. Inorganic tin is poorly absorbed from mucous membranes and the gastrointestinal tract. Containers for food and liquid are plated with the metal because of its corrosion-resistant qualities and the ease with which it can be soldered. Electrical, radio, and automobile parts are coated with tin. Modern pewter is made with the metal, and its colored salts are used in decorating china and porcelain. Stannic chloride is utilized in polishing glass and metals, as a mordant in printing and dyeing fabrics, and in the manufacture of ruby and of alabaster-colored glass.

A benign respiratory disease, stannosis, resulting from exposure to dust or fumes of tin or tin oxide has been described.[1] Stannic chloride (tin tetrachloride) used in weighting silks and as a stabilizer for colors in soaps and fabrics among others, can be highly irritating to the mucous membranes and eyes. Stannous acetate and stannous chloride also may be irritating.

Alkyl tin compounds employed in the plastics and paint industries and used as a fungicide in agriculture are of a greater order of toxicity.[2-4] Experimental animals exposed to these compounds have developed liver damage and exhibited signs of central nervous system involvement. Other than the acute burns of the skin and eyes, resulting from brief contacts with these compounds,[5] human toxicities due to industrial exposure have not been reported.

THERAPEUTIC USE OF TIN COMPOUNDS

There was a traditional belief in parts of France that tin and its compounds were of value in controlling furunculosis (boils). The basis of the use of tin powder probably stemmed from the observation that the tin workers in Beauve never seemed to suffer from boils. Ingestion of stannous oxide did little good, but neither did it do any harm. However, the dispensing of alkyl tin compounds led to a tragedy.[4]

A proprietary preparation, 'Stalinon', containing diethyl tin, traces of triethyl tin and linoleic acid was sold throughout France prior to 1960 for treating furuncles, other skin infections due to staphylococci, osteomyelitis, acne and anthrax! During the regimen of 90 mg% of diethyl tin daily for eight days, 217 people, 100 of whom died, are known to have been poisoned.

Alajouanine et al.[6] presented data on 201 cases, including 98 deaths. Symptoms of poisoning appeared after four days approximately. Severe, persistent headache was the most constant symptom. Other common symptoms included vomiting, urinary retention, vertigo, abdominal pain, weakness, visual disturbances (especially photophobia). Weight loss was rapid. Pyrexia was absent. Paresis and psychic disturbances were observed in severe cases. Death occurred in coma following respiratory or cardiac failure.

Only ten of the affected population observed by Alajouanine and coworkers recovered completely. Convalescence was very slow. Neither inorganic nor organic tin compounds have any rational place in human therapeutics.

DISTRIBUTION OF TIN

Tin is widely distributed in nature, being found in marine and land plants and animals. Soils contain between 2 to 200 ppm of tin.[7]

Boyd and De[8] in their spectrographic studies of animal and human tissues c. 1933 reported tin to be present in most tissue samples examined. Concentrations found ranged between 0.5 and 4 μg g^{-1} of tin (dry tissue). Kehoe and co-workers[9] detected tin in all tissues examined except the brain. Levels reported in the fresh tissues examined varied between 0.1 μg g^{-1} in muscle to 0.5 μg g^{-1} in bone. Mean concentrations in other tissues ranged around 0.2 μg g^{-1}.

Dundon and Hughes[10] in 1950 described the findings in the chance discovery of a benign pneumoconiosis in an elderly man hospitalized for a totally unrelated problem. The subject had, for many years, been exposed to fumes and dusts of stannic oxide while he tended a detinning furnace in which stannic oxide was recovered from tin scrap by a calcining process. However, he had not been exposed to these dusts and fumes for 18 years prior to his hospital admission. Tissue concentrations found in this subject are compared, in Table 28.1, with those reported by Kehoe et al.[9] in normal, apparently unexposed adults. Both groups of investigators employed the spectrographic techniques of Cholak and Story[11] for their analyses.

The concentration in the lungs is rather striking. Dundon and Hughes[10] concluded that systemic absorption of tin from lungs following heavy exposure

TABLE 28.1
Tin concentrations in tissues
(μg g^{-1} wet weight)

Tissue	Detinner	Normal[9]
Lung	1100	0.45
Liver	1.8	0.6
Spleen	6.4	0.2
Bone	0.1	0.5

to stannic oxide by inhallation occurs to an insignificant extent, if at all.

Somewhat later, Schroeder et al.[12] demonstrated that the tin content in tissues was related to geographical location. Age is also a factor, as older subjects have consumed more tin. Tipton's earlier studies[13] also indicated a wide range of variation in tin concentrations of tissues.

In a recent study, Hamilton and co-workers,[14] using a low temperature ashing technique to destroy organic matter prior to spark source mass spectrometric analysis, found adult brain tissue to contain 0.26–0.34 μg g^{-1} of tin (wet weight). Mean concentrations found in lung, liver and kidney were 0.8, 0.4 and 0.2 μg g^{-1}, respectively. Lymph nodes contained between 0.9 and 2.1 μg g^{-1}. Bone ash samples from soft water areas yielded somewhat lower tin contents than samples from hard-water areas; mean concentrations being 3.7 μg g^{-1} compared to 4.1 μg g^{-1}. The tin content in blood varied from 0.007 μg g^{-1} to 0.010 μg g^{-1}.

Undoubtedly tin concentrations in tissues are diet dependent. Kehoe et al.,[9] Schroeder et al.,[12] and Tipton et al.[15] report that the mean daily intakes of dietary tin, by adults in the US, are 17 mg, 3.5 mg, and 1.5 mg, respectively. Schroeder's group observed that a diet containing considerable quantities of canned fruit juices, fish and vegetables might supply approximately 38 mg of dietary tin daily, compared to 1mg supplied in diets that included mainly fresh meats, vegetables and cereals. Hamilton and Minski[16] in a more recent study, reported that an average English diet supplied from 145 to 229 μg of tin daily.

The difference in tin contents in diet are due primarily to the composition and geographical location of the populations studied. The analytical techniques employed are also an important factor.

Foods, standing in contact with the tin plating of a can, accumulate large quantities of tin. As with other metals higher levels are leached by acidic foods such as tomato soups and ketchup. Fresh tomatoes were found to contain 0.02 μg g^{-1} of tin, while tomato soup contained about 1 μg g^{-1}. Coating the tins with a lacquer or resin will inhibit leaching of tin. A recent study comparing tin contents of foods packed in resin-coated and non-coated tin-plated cans indicated that while the resin coating remained intact, the quantity of tin leached out by foods in the former was reduced by a factor of 50.[17] As expected, tin contents of foods packed in non-tin containers is considerably less.

According to the investigations of Kehoe,[9] the Perrys,[18] and Tipton and Cook,[19] almost all tin ingested with the diet is excreted via the feces. The respective investigators observed urinary tin excretions of only 16, 8–11, and 23 μg per day. The poor absorption of tin undoubtedly accounts for its low order of toxicity with normal exposures. In sub-acute experiments with rats fed 250–5300 mg of tin per kg of diet, de Groot[20] observed an inhibition in growth and a reduction in hemoglobin levels. The latter could be corrected by iron and copper supplements.

Biochemistry of tin

Schwarz et al.[21] in 1970 demonstrated that tin was an essential nutrient for the growth of rats. The mechanism of biochemical activity of tin is unknown. It may act as an oxidation–reduction catalyst.[22] Specific biochemical lesions associated with tin deficiency have not been described. Furthermore growth inhibition, or other anomalies induced in different experimental animals by tin deprivation, has not been reported as yet.

ANALYSIS OF TIN

Colorimetry–fluorimetry

As is the case with other metals, colorimetric methods for the determination of tin suffer from lack of specificity and sensitivity. Dithiol, i.e. toluene-3,4-dithiol(4-methyl-1,2-dimercaptobenzene) is considered to be about the most useful reagent for measuring traces of the metal.[23] Stannous tin in acid solution forms a magenta–red color upon warming. Since stannic tin reacts slowly, it is best reduced to the stannous state by addition of mercaptoacetic (thioglycollic) acid before complexation. Other heavy metals reacting with dithiol, interfere with the tin analysis. Silver, mercury, lead, arsenic, and cadmium form yellow complexes; bismuth yields a brick-red chelate; cobalt, copper, nickel and ferric iron form black precipitates. Nevertheless, a number of procedures employing dithiol have been described.[24-29] Stannous dithiolate assumes a yellow hue when extracted by organic solvents.

Stannous, but not stannic, tin is extracted from mineral acids by an organic solvent containing diethyl ammonium diethyldithiocarbamate. Copper and bismuth are coextracted. As a consequence, one can remove these interferences with the dithiol determination of tin by first converting tin to the stannic form, extracting with diethyl ammonium diethyldithiocarbamate, then chelating with dithiol. Dickinson and Holt[30] used this approach in determining tin concentrations in food ash.

The cupferrates of both stannous and stannic tin are extractable from a mineral acid medium.[31] Iron, bismuth, antimony (III) are coextracted. However, the cupferrates of arsenic, antimony (V), lead, silver, zinc, cadmium, and selenium remain behind in the aqueous phase.

Tin(IV) as a diethyl dithiocarbamate is quantitatively extracted over the pH range 4–6.2.[32] The 8-hydroxyquinolate complex, extractable at pH 2.5–5.5, has been utilized in determining tin.[33-35]

Sandell[23] considers distillation of tin halides to be one of the better means of separating the metal from its matrix constituents. Distillation is usually carried out at a temperature of 140 °C from a sulfuric–hydrochloric–hydrobromic acid solution. A mixture of the chloride and bromide of tin is more volatile than either halide alone. Volatilization of tin bromide has been applied in isolating tin from biological materials prior to formation of a dithiol complex.[36,37]

Colorimetric determination of some of the organic tin compounds has been described. Diethyl tin can be extracted from a 10% trichloroacetic acid medium by dithizone in chloroform.[38] A red-colored complex is formed. Triethyl tin is apparently not extracted. Both organo-tin compounds can be extracted by dithizone from a borate buffer at pH 8.4. Interference from copper, zinc, lead, and cobalt is prevented by the addition of EDTA.

Fluorimetry

Fluorimetry has been applied to a limited degree in tin analysis. Stannous tin in an ammoniacal solution yields a blue fluorescence with 6-nitro-2-naphthyl-amine-8-sulfonic acid.[39,40] Other metals which produce a fluorescence include iron, titanium, and vanadium.

Flavanol[41] in weak sulfuric acid (0.5–1 N) and nitroaminonaphthalene monosulfonic acids[42] form fluorescent compounds with stannous tin also.

Miscellaneous instrumental applications

Activation analysis has been utilized in measuring tin concentrations in a variety of materials. One of the more recent applications is that of Mahler and co-workers,[43] who applied neutron activation analysis in determining the tin, manganese, copper, gold, silver and sodium levels in fingernails and hair in normals, uremics and patients on dialysis.

A procedure combining activation analysis with selective extraction techniques has been employed in determining the tin contents in rocks.[44]

Chelation–extraction methods have been applied with other instrumental analyses, such as X-ray fluorescence, in determining tin and other trace metal concentrations in various waters.[45] Metals are chelated with ammonium pyrollidine dithiocarbamate, then concentrated by extracting the water aliquots with a lesser volume of methylisobutyl ketone.

Noble et al.[46], using proton induced X-ray fluorescence spectrometry, determined nine elements simultaneously in wines. Tin, aluminum, copper, iron, zinc and nickel have deleterious effects on the color, aroma and taste of wines.

Emission spectrometry

Emission spectrometry was the first instrumental technique applied to analyses of tin and other trace metal contents in biological materials. The earlier applications were mentioned previously.[8-13,15]

Theuer and co-workers[47] c. 1971 investigated the placental transfer of iron and tin in rats given various ferric and tin salts. Fetal tin concentrations were found to be elevated when maternal diets contained tin salts, but bore no apparent relationship to dietary tin levels. Little relationship between placental tin levels and dietary tin concentrations was discernible.

Analyses were performed by emission spectrometry using the Stallwood jet method of excitation.

Atomic absorption spectrometry

Few applications of atomic absorption spectrometry to determine tin levels in biological material have been described. There were technical difficulties with applications of flame atomization.

At the 224.6 nm line, the detection limit in a hydrogen–air flame is about 0.2 μg ml^{-1}. There are other analytical lines. Interference from copper, lead, nickel and zinc was observed. While interference was not encountered in an air–acetylene flame, the detection sensitivity was approximately 1/3 that attained with an air–hydrogen flame. A detection limit of 0.02 μg ml^{-1} is claimed for an argon–hydrogen flame. However, this flame is unsuitable for organic solvents.[48,49] Hence, the impracticality of determinations of tin contents in biological materials by flame atomic absorption spectrometry.

The detection sensitivity, extended many fold by flameless atomization, can be enhanced further by employing chelation–extraction methods of isolating and concentrating the tin present.

Beck et al.[50] described the use of the graphite tube in measuring tin and manganese levels in serum.

The potential of flameless atomic absorption spectrometry applied to tin analysis is yet to be realized.

REFERENCES

1. A. Hamilton and H. L. Hardy, *Industrial Toxicology*, Publishing Sciences Group, Acton, Mass., 1974.
2. J. M. Barnes and P. N. Magee, *J. Pathol. Bact.*, **75**, 267 (1958).
3. J. M. Barnes and H. B. Stoner, *Brit. J. Ind. Med.*, **15**, 15 (1958).

4. J. M. Barnes and H. B. Stoner, *Pharmac. Rev.*, **11**, 211 (1959).
5. W. H. Lyle, *Brit. J. Ind. Med.*, **15**, 193 (1958).
6. T. Alajouanine, L. Derobert and S. Thieffrey, *Rev. Neurol.*, **98**, 85 (1958).
7. H. J. M. Bowen, *Trace Elements in Biochemistry*, Academic Press, London, 1966.
8. T. C. Boyd and N. K. De, *Indian J. Med. Res.*, **20**, 789 (1933).
9. R. A. Kehoe, J. Cholak and R. V. Story, *J. Nutr.*, **19**, 579 (1940).
10. C. C. Dundon and J. P. Hughes, *Amer. J. Roentgenol. Radium Therapy*, **63**, 797 (1950).
11. J. Cholak and R. V. Story, *Ind. Eng. Chem. (Anal. Ed.)*, **10**, 619 (1938).
12. H. A. Schroeder, J. J. Balassa and I. H. Tipton, *J. Chronic Dis.*, **17**, 483 (1964).
13. I. H. Tipton, in *Metal Binding in Medicine*, (eds M. J. Seven and L. A. Johnson) J. B. Lippincott, Philadelphia, Pa., 1960.
14. E. I. Hamilton, M. J. Minski and J. J. Cleary, *Sci. Total Environ.*, **1**, 341 (1972/73).
15. I. H. Tipton, P. L. Stewart and P. G. Martin, *Health Physics*, **12**, 1683 (1966).
16. E. I. Hamilton and M. J. Minski, *Sci. Total Environ.*, **1**, 375 (1972/73).
17. J. J. M. de Goeij and J. J. Kroon, IAEA/FAO/WHO Symposium on Nuclear Techniques in Comparative Studies of Food and Environmental Contamination, Otaniemi, Finland, 1973, cited in E. J. Underwood, *Trace Metals in Human and Animal Nutrition*, Academic Press, New York, 1977.
18. H. M. Perry, Jr. and E. F. Perry, *J. Clin. Invest.*, **38**, 1452 (1959).
19. I. H. Tipton and M. J. Cook, *Health Physics*, **9**, 103 (1963).
20. A. P. de Groot, *Food Cosmet. Toxicol.*, **11**, 955 (1973).
21. K. Schwarz, C. B. Milne and E. Vinyard, *Biochem. Biophys. Res. Commun.*, **40**, 22 (1970).
22. K. Schwarz, *Fed. Proc.*, **33**, 1748 (1974).
23. E. B. Sandell, *Colorimetric Determination of Traces of Metals*, Interscience Publishers, New York, 1959.
24. R. E. D. Clark, *Analyst*, **61**, 242 (1936).
25. R. E. D. Clark, *Analyst*, **62**, 661 (1937).
26. G. Jantsch, A. Hummer-Kroupa and I. Gansinger, *Z. Anal. Chem.*, **128**, 451 (1948).
27. M. Fransworth and J. Pekola, *Anal. Chem.*, **26**, 735 (1954).
28. T. C. J. Ovenston and D. Kenyon, *Analyst*, **80**, 566 (1955).
29. H. Onishi and E. B. Sandell, *Anal. Chim. Acta*, **14**, 153 (1956).
30. D. Dickinson and R. Holt. *Analyst*, **79**, 104 (1954).
31. N. H. Furman, W. B. Mason and J. S. Pecola, *Anal. Chem.*, **21**, 1325 (1949).
32. H. Bode, *Z. Anal. Chem.*, **143**, 182 (1954).
33. C. H. R. Gentry and L. G. Sherrington, *Analyst*, **75**, 17 (1950).
34. P. W. Wyatt, *Analyst*, **80**, 374 (1955).
35. A. R. Eberle and M. W. Lerner, *Anal. Chem.*, **34**, 627 (1962).
36. J. Schwaibold, W. Borchers and G. Nagel, *Biochem. Z.*, **306**, 113 (1940).
37. N. H. Low, *Analyst*, **67**, 283 (1942).
38. W. N. Aldridge and J. E. Cremer, *Analyst*, **82**, 37 (1957).
39. J. R. A. Anderson and J. L. Garnett, *Anal. Chim. Acta*, **8**, 393 (1953).
40. J. R. A. Anderson and S. L. Lowry, *Anal. Chim. Acta*, **15**, 246 (1956).
41. C. F. Coyle and C. E. White, *Anal. Chem.*, **29**, 1486 (1957).
42. J. R. A. Anderson and J. L. Garnett, *Anal. Chim. Acta*, **17**, 452 (1957).
43. D. J. Mahler, A. F. Scott, J. R. Walsh and G. Haynie, *J. Nucl. Med.*, **11**, 739 (1970).
44. A. Alian and W. Sanad, *Radiochem. Radioanal. Letters*, **17**, 155 (1974).
45. M. Pinta, *Modern Methods for Trace Element Analysis*, Ann Arbor Science, Ann Arbor, Mich., 1978.

46. A. C. Noble, B. H. Orr, W. B. Cook and J. L. Campbell, *J. Agr. Food Chem.*, **24**, 532 (1976).
47. R. C. Theuer, H. W. Mahoney and H. P. Sarett, *J. Nutr.*, **101**, 525 (1971).
48. L. Capacho-Delgado and D. C. Manning, *Atomic Abs. Newsletter*, **4**, 317 (1965).
49. L. Capacho-Delgado and D. C. Manning, *Spectrochim. Acta*, **22**, 1505 (1966).
50. K. C. Beck, J. H. Reuter and E. M. Perdue, *Geochim. Cosmochim. Acta*, **38**, 341 (1974).

CHAPTER 29

TITANIUM

Titanium has been recognized since the 1800s. Like aluminum it is abundant in soils and the lithosphere. For example, rocks contain 400 to 5700 ppm, and soils, 5000 ppm.[1] Since the metal is so poorly absorbed, concentrations in herbage are about 10^{-5} that in soils.[2] A wide range of titanium concentrations in human and animal tissues have been reported.[3,4] Tipton and Cook[3] found most soft tissues of human adults to contain from 0.1 to 0.2 μg g^{-1} (wet weight). Lungs, however, averaged over 5 μg g^{-1} of titanium.

Information concerning the intake of titanium with the diet is sparse also. The 30-day study of two individuals reported by Tipton et al.[5] revealed a daily intake of 0.37–0.41 mg.

To date there is no evidence that titanium performs any vital function in animals or is a dietary essential for any living organism.[6]

USES OF TITANIUM

Titanium is a useful metal. It imparts tensile strength to steel and renders aluminum resistant to attacks by salt solutions and organic acids. Titanium forms alloys with lead, nickel, tin, copper, iron, chromium and cobalt. The metal is used in medicine for the fixation of fractures.

Titanium dioxide is used in paint and enamel, in ink manufacture, in ceramics and plastics, and in leather finishing. It is employed medically in creams and powders used as a protectant against external irritation and sunlight.

The trichloride is a powerful reducing agent and is so used in chemistry. It is also utilized as a stain remover in laundering. Titanium tetrachloride has been used in the manufacture of iridescent glass and artificial pearls. The military use titanium chloride in generating smoke screens.

Industrial exposures to titanium and its compounds are considered essentially innocuous. However, high concentrations of titanium oxide dust may cause irritation of the respiratory tract. The tri- and tetrachloride salts also irritate the eyes and respiratory tract.

Powders of titanium are pyrophoric. Careless handling has been known to cause explosions.

ANALYSIS OF TITANIUM

Colorimetry

Certain reagents used in the determination of titanium yield rather sensitive detection limits. All suffer from interference from vanadium, molybdenum, phosphorous, iron and other common metals. Separations from other elements have been achieved more or less by precipitation from alkali media,[7-9] electrolysis with a mercury cathode, ion exchange,[10,11] and extraction.

When applying electrolysis with a mercury cathode to a dilute sulfuric acid solution of a matrix containing different metals, copper, chromium, cobalt, zinc cadmium, bismuth, iron, thallium, silver, gold, etc. are plated out on the cathode while titanium, zirconium, vanadium, and phosphorus remain in solution.

Titanium as a cupferrate can be isolated from dilute mineral acid solutions by extraction with organic solvents.[12,13] Iron vanadium, molybdenum, and zirconium are coextracted. Iron as chloroferric acid can be removed from the aqueous phase, for the most part, by prior extraction with ethylisopropyl ether or β,β'-dichlorodiethyl ether.[7,12] The latter solvent also removes vanadium.

Hydrogen peroxide added to an acidic solution containing titanium yields a yellow color which formed the basis of the classical method for titanium determination.[14-16] Solvent extraction of peroxidized titanium at pH 3–5 in the presence of 8-hydroxyquinoline enhances the detection sensitivity considerably.[17]

Compounds containing a phenolic ring have been employed in the colorimetric analysis of titanium. Tiron (disodium-1,2-dihydroxybenzene-3,5-disulfonate) yielding a strong yellow color over the pH range 4.3 to 9.6 has been utilized in different applications.[18,19]

Common metals, such as calcium, magnesium, manganese and aluminum do not interfere with the sulfosalicyclic acid reaction with titanium at pH 3–5. Since ferrous iron also does not interfere, the iron effect can be eliminated by the addition of mercaptoacetic acid.[20-22]

Sensitive detection limits are attained with chromotropic acid over the pH range of 3–6.2.[23] Both the reagent and the titanium complex are light sensitive.[3] A procedure utilizing the reagent after removal of iron by electrolysis has been described.[24]

Other compounds, including salicylic acid[25] and ascorbic acid,[26] have been proposed as reagents for the analysis of titanium.

Activation analysis

The feasibility of proton activation analysis of titanium and other metals was indicated by Schweikert and Albert[27,28] in 1966 and again by Schweikert and Swindle[29] later. Delmas and coworkers[30] described the non-destructive analysis of titanium and other elements in rocks by this technique. Application to the determination of titanium contents in other matrices has been limited.

X-ray fluorescence analysis

X-ray fluorescence analysis has been employed for more than two decades as an analytical tool by geologists. Hooper[31] in 1964 determined titanium, aluminum, calcium, iron, manganese, potassium and silicon in rocks. Hammerle and Pierson[32] more recently described the monitoring of titanium, iron, nickel, manganese and zinc concentrations in the air by X-ray fluorescence techniques.

Emission spectrometry

Some applications of emission spectrometry to determinations of titanium in biological material were mentioned previously.[3-5] The instrumentation has found recent use in geological analysis[33] and in the analysis of industrial products[34] and petroleum products[35] as well.

Atomic absorption spectrometry

The detection limit for titanium at the 364.3 nm resonance line, employing a nitrous oxide-acetylene reducing flame, is said to be about 0.1 μg ml^{-1}. An acetylene flame is not suitable because the element forms stable oxides.

Procedures for flame atomic absorption determination of titanium in alloys,[36,37] cement,[38] bauxite,[39] and propylene[40] have been described.

Flameless atomization extends the detection limits for titanium many fold. Combined with a chelation–extraction technique (cupferron, 8-hydroxyquinoline, etc.) to concentrate the trace metal further, practical analyses of titanium concentration in biological materials by flameless atomic absorption spectrometry would be feasible.

REFERENCES

1. H. J. M. Bowen, *Trace Elements in Biochemistry*, Academic Press, New York, 1966.
2. D. J. Swaine, *Commonw. Bur. Soils Tech. Commun.*, No. 48, 1955, cited in E. J. Underwood, *Trace Metals in Human and Animal Nutrition*, Academic Press, New York, 1962.
3. I. H. Tipton and M. J. Cook, *Health Physics*, **9**, 103 (1963).
4. E. I. Hamilton, M. J. Minski and J. J. Cleary, *Sci. Total Environ.*, **1**, 341 (1972/73).
5. I. H. Tipton, P. L. Stewart and P. G. Martin, *Health Phys.*, **12**, 1683 (1966).

6. E. J. Underwood, *Trace Elements in Human and Animal Nutrition*, Academic Press, New York, 1977.
7. E. B. Sandell, *Colorimetric Determination of Traces of Metals*, Interscience Publishers, New York, 1959.
8. W. F. Pickering, *Anal. Chim. Acta*, 9, 324 (1953).
9. W. F. Hillebrand, G. E. F. Lundell, H. A. Bright and J. I. Hoffman, *Applied Inorganic Analysis*, Wiley, New York, 1953.
10. W. E. Brown and W. Rieman, *J. Amer. Chem. Soc.*, 74, 1278 (1952).
11. K. A. Kraus, F. Nelson and G. W. Smith, *J. Phys. Chem.*, 58, 11 (1954).
12. T. C. J. Ovenston, C. A. Parker and C. G. Hatchard, *Anal. Chim. Acta*, 6, 7 (1952).
13. J. Stary and J. Smizanska, *Anal. Chim. Acta*, 29, 546 (1963).
14. H. E. Merwin, *Amer. J. Sci.*, 28, 119 (1909).
15. G. H. Ayres and E. M. Vienneau, *Ind. Eng. Chem. (Anal. Ed.)*, 12, 96 (1940).
16. M. Bendig and H. Hirschmüller, *Z. Anal. Chem.*, 120, 385 (1940).
17. K. Gardner, *Analyst*, 76, 485 (1951).
18. J. H. Yoe and A. R. Armstrong, *Anal. Chem.*, 19, 100 (1947).
19. C. V. Potter and C. E. Armstrong, *Anal. Chem.*, 20, 1208 (1948).
20. M. Ziegler and O. Glemser, *Z. Anal. Chem.*, 139, 92 (1952).
21. K. Saarni and S. Suikkanen, *Z. Anal. Chem.*, 143, 112 (1954).
22. A. Okac and L. Sommer, *Anal. Chim. Acta*, 15, 345 (1956).
23. W. W. Brandt and A. E. Preiser, *Anal. Chem.* 25, 567 (1953).
24. R. Rosotte and E. Jaudon, *Anal. Chim. Acta*, 6, 149 (1952).
25. J. H. Muller, *J. Amer. Chem. Soc.*, 33, 1506 (1911).
26. J. Korkisch and A. Farag, *Mikrochim. Acta*, 659 (1958).
27. E. Schweikert and P. Albert, *Compt. Rend. Acad. Sci.*, 262, 87 (1966).
28. E. Schweikert and P. Albert, *Compt. Rend. Acad. Sci.*, 262 342 (1966).
29. D. L. Swindle and E. A. Schweikert, *Anal. Chem.*, 45, 2111 (1973).
30. R. Delmas, J. N. Barrandon and J. L. Debrun, *Analusis*, 4, 339 (1976).
31. P. R. Hooper, *Anal. Chem.*, 36, 1271 (1964).
32. R. H. Hammerle and W. R. Pierson, *Environ. Sci. Technol.*, 9, 1058 (1975).
33. Y. Besnus and R. Rouault, *Analusis*, 2, 111 (1973).
34. Y. Osumi, A. Kato and Y. Miyake, *Z. Anal. Chem.*, 255, 103 (1971).
35. A. P. Kuznecova, Z. I. Otmahova and G. A. Kataev, *Zav. Lab.*, 39, 957 (1973) cited in M. Pinta, *Modern Methods for Trace Element Analysis*, Ann Arbor Science, Ann Arbor, Mich., 1978.
36. J. B. Headridge and D. P. Hubbard, *Anal. Chim. Acta*, 37, 151 (1967).
37. R. A. Mostyn and A. F. Cunningham, *Atomic Abs. Newsletter*, 6, 86 (1967).
38. L. Capacho-Delgado and D. C. Manning, *Analyst*, 92, 553 (1967).
39. J. A. Bowman and J. B. Willis, *Anal. Chem.*, 39, 1210 (1967).
40. D. Druckman, *Atomic Abs. Newsletter*, 6, 113 (1967).

CHAPTER 30

VANADIUM

Vanadium, an element discovered in the nineteenth century, is of considerable industrial importance. The metal alloys with iron, manganese, chromium, etc. Approximately 90% of the world's supply is diverted into steel manufacture. Vanadium steels are rust resistant and of great tensile strength.

Vanadium pentoxide is employed as a catalyst in the chemical industry, for reactions such as the oxidation of alcohol to acetaldehyde. It is also used as a depolarizer, as a developer in photography, and in the manufacture of yellow glass. The compound, because it inhibits ultra violet light transmission, is also incorporated in glass for windows and eyeglasses.

Ammonium vanadate is used as a mordant in dyeing and printing cottons and silks, in dyeing leather and furs, in the manufacture of dyes like aniline black, and in synthesizing polymers of propylene and ethylene, etc.

Vanadyl dichloride is used as a mordant in dyeing and printing fabrics also.

Vanadium compounds, such as zirconium vanadium blue, are used to a limited extent in the ceramic industry, primarily as glazes for wall tiles.

TOXICITY OF VANADIUM

Vanadium exposures are considered to be a bit of an occupational health problem in industrialized areas. Concern regarding vanadium as an environmental pollutant has also been expressed.[1] Vanadium, possibly in an organic form such as porphyrin, is present in all coals and petroleum oils in low concentrations.

Dutton[2] first described vanadium poisoning as early as 1911. Detailed reviews on the toxicity of vanadium and its salts for man and animals have appeared subsequently.[3-7] The respiratory tract is the most common route of entry in industrial exposure. There may be some exposure via the gastrointestinal system as part of the overall airborne exposure, however.

Symptoms observed include conjunctivitis with smarting eyes, rhinitis with watery discharge, sore throat, and persistent cough with pressure over the chest. Severe chronic bronchitis with bronchiectasis may also occur. Gastrointestinal symptoms, depression of blood-forming organs, injury to the kidneys, and injury to the central nervous system following industrial exposure to vanadium have not been reported.

'Green tongue', a harmless though colorful sign of vanadium exposure resulting from the deposition of vanadium salts of reduced valence form was first described by Wyers[8] in 1946.

Exposures to the metal occur in various industrial operations such as, cleaning oil and gas fired boilers; refinery operations employing vanadium as a catalyst; the manufacture of vanadium pentoxide from vanadium concentrated slag; glass and ceramic manufacture; manufacturing of ferrovanadium in pelletizing plants; and all operations involving the handling, grinding and bagging of vanadium-containing materials. The toxicity of vanadium compounds increases with valence, the pentoxide being the most toxic.

Symptoms consisting of the respiratory signs described, as well as 'green tongue', have been reported to occur in cleaners coming in contact with the petroleum coat in boilers.[9]

Signs of vanadium poisoning in men exposed to exhausts from gas turbines using residual fuels were found to occur between 1 and 14 days following exposure. In addition to the respiratory symptoms listed previously, men also complained of nose bleeds and skin irritations.[10]

Bronchiospasms, which persist for 2 or 3 days following exposures, have been reported.[11]

Tebrock and Machle[12] observed men in a vanadium-bearing phosphor plant over a seven-year period and did not note the appearance of any chronic or systemic effects resulting from exposure to the metal. Significant X-ray changes did not occur during this interval.

TOXICITY TO ANIMALS

Vanadium added to diets of chicks and rats in concentrations of 25 ppm ($2.5 mg g^{-1}$) was found to depress growth and increase mortality.[13] The ingested metal concentrated in the bone and kidney of chicks and the kidneys of adult rats. EDTA prevents absorption of vanadium and hence occurrence of toxicity.

More recently, Johnson and coworkers[14] in studying vanadium toxicity in rats obsered that toxic effects are related to the extent of accumulation of the metal in the liver and kidneys.

Other investigators[15] noted that the L-ascorbic acid content of liver tissue in rats intoxicated with vanadium is lowered. The normal histological picture of the liver and kidney was also disturbed. Supplementation with L-ascorbic acid did not reverse the effects of vanadium intoxication.

In the experiments of Dimond et al.[16] orally administered vanadium salts apparently exerted little toxic activity in adult man other than inducing abdominal cramps and diarrhea.

VANADIUM METABOLISM

The work of Hopkins and Mohr,[17] Schwarz and Milne[18] in 1971 and later researches of Nielsen and Ollerich[19] indicate that vanadium has some essential physiological role. Chicks[17,19] and rats[18] fed a diet deficient in the element exhibited impaired growth and reproduction, a disturbance in lipid metabolism, altered erythrocyte levels and iron metabolism and impairment of hard-tissue metabolism.

Lewis[20] found lowered cholesterol levels among workmen exposed to the metal.

Vanadium, by some mechanism, affects cystine metabolism. Mountain and co-workers,[21,22] who observed that the cystine content was lowered in the hair of rats fed vanadium and in the fingernails of men exposed to the metal industrially, concluded that the synthesis of cystine and its reduction product, cysteine, was decreased as a result of exposure to excess concentrations. Bergel et al.[23] demonstrated later that destruction of cystine and cysteine increased in the presence of vanadium.

The reduction of the coenzyme A content in the livers of rats fed sodium vanadate in the diet or administered intraperitoneally is probably due to interference with cystine metabolism.[24] Thioethanolamine, one of the compounds involved in the synthesis of coenzyme A, is derived from the decarboxylation of cysteine.

The Bernheims[25] forty years ago demonstrated that vanadium was catalytically active in various oxidations in certain tissues, for example, the oxidation of phospholipids in liver, kidney and brain tissue, and postulated a possible role in the body for the element.

Subsequently, findings reported regarding vanadium actions on cholesterol, triglycerides, and phospholipids have been contradictory. Experimental design, and dosage levels were contributory factors. A decrease in cholesterol following vanadium feeding has been reported by Curran[26] and Azarnoff et al[27,28] in experiments with rats. However, changes in serum cholesterol or lipoproteins were not observed by Somerville and Davies[29] in human subjects with hypercholesteremia and ischemic heart disease. Schroeder et al.[30] showed a similar lack of effect on serum cholesterol levels in subjects over 60 years of age. Curran et al.[31] in their experiments with man found serum triglycerides and phospholipids increased during vanadium feedings.

Data presented concerning the metal's effect upon monamine oxidase activity and hematoporesis are also conflicting.

Hathcock et al.[32] reported that vanadium may uncouple oxidative phosphorylation.

Ascorbic acid metabolism may be affected by the metal. Watanabe *et al.*[33] observed that the urinary excretions of ascorbic acid by workers in a vanadium industry were less than that of unexposed controls. Urinary vanadium excretion, on the other hand, ranged between 21 and 259 μg l^{-1} among the exposed group compared to 3.5–11 μg l^{-1} among the control subjects.

Mitchell and Floyd.[34] found ascorbic acid to be an effective antidote in experimental vanadium poisoning. However, in the later studies of Chakraborty *et al.*,[15] toxic effects were not reversed by *L*-ascorbic acid supplementation.

Though shown to be essential for laboratory animals, there does not appear to be any firm evidence regarding the essentiality of vanadium for man. Information regarding vanadium activity at physiological levels is sparse.

DISTRIBUTION OF VANADIUM

Vanadium has been found in all plant and animal species which have been examined. Generally, concentrations determined in land plants and animals are less than those found in marine forms. Variations among species have been observed. For example, the mushroom *amanita muscaria* contains 100 times as much of the metal as do other mushrooms.

Investigators noted that among the higher plants legumes contain greater amounts of the metal. Vanadium appears to be more concentrated in the root nodules which carry the nitrogen-fixing bacteria than in the leafy portions.[35] One can but wonder about possible interactions with molybdenum.

The ascidian, a sea squirt, is a very efficient accumulator of vanadium[36] and a rather fascinating creature. Levels in its blood are about 10 000 times more concentrated than the sea water in which it resides.[37] In some ascidian species vanadium is contained in green blood cells called vanadocytes. The metal in the trivalent state is complexed to pyrrole rings and is associated with unusually high concentrations of sulfuric acid.[38] It does not seem to be involved in oxygen transport. In other species this vanadium-pyrrole complex of unknown function is found in the blood plasma rather than in specific cells.

Vanadium was not detected in all tissue samples examined by Tipton *et al.*[39,40] Lung samples contained between 0 and 18 μg g^{-1} (dry weight). Over one half the lung samples analyzed contained the element. Increased concentrations were noted with increasing age. Significantly more vanadium was present in lungs obtained from Middle Eastern populations than in samples from people in the Far East, Africa and the US. Furthermore, the metal was found in 12% of the aortas and 13% of the kidney specimens from the Middle East. Kidney, heart, brain and muscle from adults in the US did not contain the element.[41] Concentrations between 0.02 and 0.03 μg g^{-1} were found in adult liver, pancreas, spleen and prostate gland.

Values obtained earlier by Bertrand[42] were higher as were the more recent levels reported by Hamilton et al.[43] The latter reported concentrations of 0.022–0.038 μg g^{-1} (wet weight) for brain; 0.03–0.04 μg g^{-1}, liver; 0.12–0.28 μg g^{-1}, testes; 7–13 μg g^{-1}, muscle; 0.080–0.12 μg g^{-1} for lung; and 0.2–0.6 μg g^{-1} in lymph nodes.

Allaway et al.[44] in the study of blood donors in 19 cities in the US found 90% of the samples to contain less than 1 μg vanadium per 100 ml; 2 μg per 100 ml was the highest level. Higher values have been reported by others.[45,46]

Söremark[47] in his investigations of miscellaneous biological materials obtained considerably lower values for most of the tissues listed above. These differences may be due to the populations investigated, the numbers sampled, their diets, and their general environment. The analytical techniques employed also influence the data presented.

Information regarding the vanadium content of foods is not too plentiful and is quite variable. Schroeder et al.[46] had estimated that an institutional diet provided 1.2 mg of the element daily. Calculations based on Söremark's data[47] would indicate that the figure could be rather high. Myron et al.[48] in their more recent investigations estimated institutional diets to contain between 18.1 and 22.7 μg of vanadium daily.

ANALYSIS OF VANADIUM

Colorimetry

A number of reagents form, with vanadium, colored complexes, that can be applied to its determination. Since the metal is present in rather low levels in biological materials and matrix constituents represent potential interferences, it must be isolated and concentrated before determination.

Formation of a colored complex with cupferron has been utilized in determining vanadium.[49,50] The cupferrate is quantitatively extractable from dilute mineral acid. Iron, copper, antimony, bismuth, molybdenum are co-extracted.

A pH of 3.5–4.5 is optimal for the extraction of the magenta-colored complex of vanadium as 8-hydroxyquinolate.[51,52] Ferric iron, copper, molybdenum, bismuth, etc. are coextracted. Interferences can be minimized by extracting the solvent layer with an aqueous solution of pH 9.4. The vanadium complex, which is water soluble, is isolated and can be reextracted relatively interference free into an organic solvent for measurement. Procedures using 8-hydroxyquinoline have been applied quite successfully in analysis of biological materials.[53,54]

A method based on the formation of phosphotungstovanadic acid by adding sodium tungstate and phosphoric acid to an acid vanadate solution is approximately as sensitive as the 8-hydroxyquinoline procedure.[55] The yellow color is stable for days. Phosphotungstovanadic acid can be extracted with butyl alcohol to improve the sensitivity somewhat.[56] Copper, cobalt and chromium will interfere.

Vanadium is quantitatively extractable as a diethyldithiocarbamate at pH 3–6.[57] Iron is the major interference. Chromium and molybdenum are also coextracted.

Vanadium forms a blue-colored complex with PAN, 1-(2-pyridylazo)-2-naphthol, which is extractable over the pH range 3.5–4.5. This chelate has been applied in the analysis of organic materials.[58]

Formaldoxime in basic solution or catechol at pH 4.5–6 are the most sensitive of the reagents proposed for the colorimetric determination of vanadium.[53] Metals also reacting with the former include cobalt, copper, iron, manganese, and nickel. A procedure incorporating the separation of vanadium by extraction as an 8–hydroxyquinolate prior to reaction with formaldoxime has been described.[59] Molybdenum is the major interference with the vanadium-catechol reaction.

Chromatographic techniques, for example, a Dowex-1 anion exchange resin column[60] or paper chromatography,[61] have been applied in separating vanadium from other metals before colorimetric determination. Electrolysis with a mercury cathode has also been utilized.[54]

Fluorimetry

Relatively few applications of fluorimetry to the analysis of vanadium have been described.

Fluorescence detection with resorcinol, as described by Rao and Rao[62] has little practical value in analyses of biological material since the limit of sensitivity is 2.5 ppm (2.5 μg ml^{-1}).

Bognar and Jellinek[63] were able to attain a sensitivity of 0.02 μg ml^{-1} with a titrimetric procedure involving fluorescence quenching. Vanadium which was present catalyzed the bromate-bromide-ascorbic acid reaction to release bromine, which quenched the fluorescence of indicators, such as Rhodamine B, cresyl violet, trypaflavine, acridine red, or a mixture of Rhodamine B and trypan red.

A more recent method of Koh and Ryan,[64] claiming a detection capability of 1 ng ml^{-1}, employs benzoic acid and zinc amalgam as reagents. Iron forms fluorescent complexes with these compounds also. Titanium, while it gives no fluorescence itself, will increase vanadium fluorescence. Sodium, potassium, magnesium, calcium, cadmium, copper, cobalt, lead, nickel, manganese, chromium, aluminum and molybdenum apparently do not interfere.

Electrometric methods

Electrometric methods have been utilized to a limited extent in vanadium analysis. Jerman and Jettmar[65] describe the determination of the metal in industrial air by polarography. Techniques applying coulometry,[66] ampero-metry,[67] and potentiometry[68] have also been proposed.

Because of the lack of suitable reversible reactions vanadium is one of the more difficult metals to detect by anodic stripping voltammetry (ASV). Applications of ASV in the determination of vanadium concentrations in biological materials have not been reported as yet.

Gas chromatography

Vanadium (IV) as an acetylacetonate has been successfully detected by gas chromatography.[69,70] The complex elutes within a reasonable time at column temperatures between 150 °C and 180 °C. Peaks are sharp and easily separated from acetylacetonates of beryllium, aluminum, chromium (III) and other metals.

The trifluoroacetylacetonate chelates being more volatile can be chromato-graphed at lower temperatures (125 °C).

Separation of 43 metal acetylacetonates, including the vanadium chelates, has been achieved by hyperpressure gas chromatography.[71] Chromatography was carried out at 800–1000 psi and 115 °C.

Neutron activation analysis

Neutron activation analysis has been applied in determining the vanadium contents of various matrices. Valberg and Holt[72] detected vanadium in blood by this technique as did Livingston and Smith[73] in their studies of the vanadium contents of biological materials. Investigations of Söre-mark[47], and later Hopkins and Mohr[17], alluded to previously utilized neutron activation analysis.

The complexity of sample preparation required is determined by the nature of the matrix. For example vanadium, along with other elements in air particulates collected on a filter, has been detected by direct irradiation,[74] an approach which would yield erroneous results with more complex matrices.

Detection limits are dependent upon the nature of the sample. For example, detection limits of 0.001, 0.015, 0.013, and 0.56 μg g^{-1} vanadium have been attained for filtered air samples, water, plastics, and fish, respectively.[1] The high sodium level in biological materials represents the major complica-tion.[75] It and the analyte in question must be separated prior to irradiation.

Lambert et al.[76] by first removing the analyte by solvent extraction were able to determine vanadium contents of diets at the nanogram level. Chela-tion–extraction prior to irradiation also serves to concentrate the element.

Linstedt and Krueger[77] estimated vanadium levels in lake waters after chloroform extraction of an oxine complex.

Irradiation after extraction has been applied to analyses of clays and feldspars.[78]

X-ray fluorescence spectrometry

The X-ray fluorescence analysis of vanadium also suffers from interferences. Morris[79] determined vanadium in waters by extracting the ammonium pyrollidine dithiocarbamate chelate with methylisobutyl ketone, evaporating the extract, and subjecting the residue to X-ray fluorescence.

Alexander[80] investigated vanadium in biological materials using X-ray fluorescence. The technique has been applied to analyzing other matrices, for example fuel oils.[81-83]

Emission spectroscopy

Various means of sample preparation and quantification of the analyte have been employed in the different procedures described in the emission spectroscopy literature. Complexity of preparation is dependent upon the matrices investigated and the desire of the analyst for accuracy and precision.

Most procedures presented for determination of metals in biological materials, air particulates, etc. involve a variation of the 'universal method' of sample preparation. In essence the ashed sample is diluted with graphite, and lithium carbonate or lithium fluoride before analysis.[84]

Tipton and co-workers[85] employed graphite as the diluent. Quantification was achieved by adding metallic compounds as required to a synthetic matrix simulating biological ash. Bedrosian et al.[86] freeze-dried biological material, diluted with graphite before analysis, and quantified in a like manner.

Landers et al.[87], working with a simpler matrix (air filters), merely rolled a one-inch square of filter paper and placed it in the analytical gap. Calibration standards consisted of solutions applied and dried on similar filter paper.

In order to minimize inaccuracies caused by differences between chemical forms of standards and metallic constituents in unknown matrices, Morrow and Brief[88] preferred to ash samples, fuse the ash with lithium tetraborate and pelletize with graphite before subjecting to spark discharge.

The use of the laser microprobe in detecting trace metals in small, discrete areas such as cells has been described.[89,90] However, despite the very sensitive detection limit of 10^{-15} g, detection of vanadium in body fluid or tissues by laser microprobe has not been reported as yet.

Spark source mass spectrometry has been applied in determining vanadium and other elements in biological materials, etc.[43,91,92] Evans and Morrison[91] employed the technique as a 'survey method' on a small population of samples. Inter-sample comparison was poor compared to that obtained by emission spectroscopic analysis performed in their laboratories. For example, the vanadium contents of two sheep lungs were estimated to be 0.33 and 12 μg g^{-1}, respectively, compared to levels of less than 0.3 and 0.8 μg g^{-1} attained by the latter method.

Hamilton et al.[43] subsequently obtained values of 0.08 to 0.12 μg g^{-1} in human adult lung tissue.

Atomic absorption spectrometry

The detection sensitivity claimed for vanadium by flame atomic absorption spectrometry is about 1.7 μg ml^{-1} at the 318.3 nm resonance line. A fuel-rich reducing flame is used. Since the element forms stable oxides that are only partially dissociated in the flame, a nitrous oxide–acetylene flame is necessary to attain lower detection limits. Detection limits are extended many fold by flameless atomization and chelation–extraction techniques.

Atomic absorption spectrometry has been utilized extensively in determining vanadium in fuel oils, steels, minerals and waters. Few applications have been made to date in analyses of biological materials, however, primarily because the metal is not as yet of great interest in routine clinical investigations. Myron et al.[48] employed this technique in their recent investigations of institutional diets.

Capacho-Delgado and Manning[93] described one of the earlier flame applications to the analysis of vanadium in steels and gas oils.

Mansell and Emmel[94] estimated the element in brines after chelation with ammonium pyrollidine dithiocarbamate (APDC) and extracting into methylisobutyl ketone (MIBK). Sachdev et al.[95] later measured vanadium in waters after extracting the cupferrate with MIBK. Chau et al.[96] used 5,7-dichloro-8-hydroxyquinoline as the chelating agent.

Vanadium levels in natural waters[97,98] and mineral oils[99] have been estimated successfully by flameless atomization methods.

Analysis of geological materials require application of separation techniques. Korkisch and Gross[100] filtered an acid solution onto a Dowex 1–8 column, eluted the metal with a methanol/hydrochloric acid solution and after evaporating the eluate to a desired volume determined vanadium by flame analysis.

Terashima[101] prepared samples for flame analysis by first evaporating the solution of rocks, clay or ores that had been decomposed in a mixture of hydrofluoric, perchloric and nitric acids, then dissolving the residue in hydrochloric acid. After pH adjustment, the metal of interest is isolated by extraction as an acetyl acetonate. Standards were prepared in a stimulated matrix.

Techniques used in determining vanadium in beach asphalts[102] and silicates[103] by atomization in the graphite furnace were described recently.

Vanadium contents of glasses and carbonates have been determined in the graphite furnace following decomposition in hydrofluoric and perchloric acids.[104] The residue remaining after evaporation was redissolved in perchloric acid, buffered to pH 6 and extracted with a solvent containing sodium diethyl dithiocarbamate.

Suggested charring and atomizing temperatures are 1100–1600 °C and 2500 °C, respectively. The use of background correction is advisable.

REFERENCES

1. Committee on Biologic Effects of Atmospheric Pollutants, Division of Medical Sciences of National Research Council, *Vanadium*, National Academy of Sciences, Washington, DC, 1974.
2. L. F. Dutton, *J. Amer. Med. Assoc.*, **56**, 1648 (1911).
3. J. Symanski, *Arch. Gewerbepath. Gewerbehyg.*, **9**, 295 (1939).
4. S. G. Sjoberg, *Acta Med. Scand. Suppl.*, **138**, 238 (1950).
5. H. E. Stokinger, *AMA Arch. Ind. Health*, **12**, 675 (1955).
6. H. E. Stokinger, in *Industrial Hygiene and Toxicology*, (ed. F. A. Patty) Interscience, New York, 1963.
7. H. T. G. Faulkner, *Vanadium Toxicology and Biological Significance*, Elsevier, New York, 1964.
8. H. Wyers, *Brit. J. Ind. Med.*, **3**, 177 (1946).
9. N. Williams, *Brit. J. Ind. Med.* **9**, 50 (1952).
10. R. C. Browne, *Brit. J. Ind. Med.*, **12**, 57 (1955).
11. C. E. Lewis, *AMA Arch. Ind. Health*, **19**, 497 (1959).
12. H. E. Tebrock and W. Machle, *J. Occup. Med.*, **10**, 692 (1968).
13. J. N. Hathcock, C. H. Hill and G. Matrone, *J. Nutr.*, **82**, 106 (1964).
14. J. L. Johnson, H. J. Cohen and K. V. Kajagopalan, *Biochem. Biophys. Res. Commun.*, **56**, 940 (1974).
15. D. Chakraborty, A. Bhattacharijya, K. Majumdar and G. C. Chatterjee, *Int. J. Vitamin Nutr. Res.*, **47**, 81 (1977).
16. E. G. Dimond, J. Caravaca and A. Benchimol, *Amer. J. Clin. Nutr.*, **12**, 49 (1963).
17. L. L. Hopkins, Jr. and H. E. Mohr, in *Newer Trace Elements in Nutrition*, W. Mertz and W. E. Cornatzer (eds.), Marcel Dekker, New York, 1971.
18. K. Schwarz and D. B. Milne, *Science*, **174**, 426 (1971).
19. F. H. Nielson and D. A. Ollerich, *Fed. Proc. Fed. Amer. Soc. Exptl. Biol.*, **32** 329 (1973).
20. C. E. Lewis, *AMA Arch. Ind. Health*, **19**, 419 (1959).
21. J. T. Mountain, L. L. Dekker and H. E. Stokinger, *AMA Arch. Ind. Hyg. Occup. Med.*, **8**, 406 (1953).
22. J. T. Mountain, F. R. Stockwell, Jr., and H. E. Stokinger, *AMA Arch. Ind. Hyg. Occup. Med.*, **12**, 494 (1955).
23. F. Bergel, R. C. Bray and K. R. Harrop, *Nature (London)*, **181**, 1654 (1958).
24. E. Mascitelli-Coriandoli and C. Citterio, *Nature (London)*, **183**, 1527 (1959).
25. F. Bernheim and M. L. C. Bernheim, *J. Biol. Chem.*, **127**, 353 (1939).
26. G. L. Curran, *J. Biol. Chem.*, **210**, 765 (1954).
27. D. L. Azarnoff and G. L. Curran, *J. Amer. Chem. Soc.*, **79**, 2968 (1957).
28. D. L. Azarnoff, F. E. Brock and G. L. Curran, *Biochem. Biophys. Acta*, **51**, 397 (1961).
29. J. Somerville and B. Davies, *Amer. Heart J.*, **54**, 54 (1962).
30. H. A. Schroeder, J. J. Balassa and I. H. Tipton, *J. Chron. Dis.*, **16**, 1047 (1963).
31. G. L. Curran, D. L. Azarnoff and R. E. Bolinger, *J. Clin. Invest.*, **38**, 1251 (1959).
32. J. W. Hathcock, C. H. Hill and S. B. Tove, *Canad. J. Biochem.*, **44**, 983 (1966).
33. H. Watanabe, H. Murayama and S. Yamooka, *Japan. J. Ind. Health*, **8**, 23 (1966).
34. W. B. Mitchell and E. P. Floyd, *Proc. Soc. Exptl. Biol. Med.*, **85**, 206 (1954).
35. D. Bertrand, *Bull. Am. Museum Natl. Hist.*, **94**, 403 (1950).
36. P. J. Peterson, *Sci. Progress*, **59**, 505 (1971).
37. M. Henze, *Hoppe-Seylers Z. Physiol. Chem.*, **72**, 494 (1911).

38. E. P. Levine, *Science*, **133**, 1352 (1961).
39. I. H. Tipton and M. J. Cook, *Health Phys.*, **9**, 103 (1963).
40. I. H. Tipton and J. J. Shafer, *Arch. Environ. Health*, **8**, 58 (1964).
41. I. H. Tipton, H. A. Schroeder, H. M. Perry, Jr. and M. J. Cook, *Health Phys.*, **11**, 403 (1965).
42. D. Bertrand, *Bull. Soc. Chim. Biol.*, **25**, 36 (1943).
43. E. J. Hamilton, M. J. Minski and J. J. Cleary, *Sci. Total Environ.*, **1**, 341 (1972–73).
44. W. H. Allaway, J. Kubota, F. Losee and M. Roth, *Arch. Environ. Health*, **16**, 342 (1968).
45. E. M. Butt, R. E. Nusbaum, T. C. Gilmour, S. L. DiDio and S. Mariano, *Arch. Environ. Health*, **8**, 52 (1964).
46. H. A. Schroeder, J. J. Balassa and I. H. Tipton, *J. Chronic Dis.*, **16**, 1047 (1963).
47. R. Söremark, *J. Nutr.*, **92**, 183 (1967).
48. D. R. Myron, T. J. Zimmerman, T. R. Shuler, L. M. Klevay, D. E. Lee and F. H. Nielsen, *Amer. J. Clin. Nutr.*, **37**, 527 (1978).
49. D. Bertrand, *Bull. Soc. Chim. (France)*, **9**, 121 (1942).
50. R. Bock and S. Gorbach, *Mikrochim. Acta*, 593 (1958).
51. E. B. Sandell, *Ind. Eng. Chem. (Anal. Ed.)* **8**, 336 (1936).
52. N. A. Talvitie, *Anal. Chem.*, **25**, 604 (1953).
53. D. Bertrand, *Bull. Soc. Chim. Biol.*, **23**, 391 (1941).
54. E. B. Sandell, *Colorimetric Determination of Traces of Metals*, Interscience, New York, 1959.
55. E. R. Wright and M. G. Millon, *Ind. Eng. Chem. (Anal. Ed.)*, **9**, 251 (1937).
56. R. M. Sherwood and F. W. Chapman, Jr., *Anal. Chem.*, **27**, 88 (1955).
57. H. Bode, *Z. Anal. Chem.*, **144**, 165 (1955).
58. F. W. Staten and E. W. D. Huffman, *Anal. Chem.*, **31**, 2003 (1959).
59. M. Tanaka, *Mikrochim. Acta*, 701 (1954).
60. K. A. Kraus, *J. Phys. Chem.*, **58**, 11 (1954).
61. F. H. Pollard, G. Nickless and A. J. Banister, *Analyst*, **81**, 577 (1956).
62. V. P. Rao and G. G. Rao, *Z. Anal. Chem.*, **161**, 406 (1958).
63. J. Bognar and O. Jellinek, *Mikrochim. Acta*, 1013 (1968).
64. K. J. Koh and D. E. Ryan, *Anal. Chim. Acta*, **57**, 295 (1971).
65. L. Jerman and V. Jettmar, *Z. Ges. Hyg.*, **14**, 12 (1968).
66. L. P. Rigdon and J. E. Harrar, *Anal. Chem.*, **41**, 1673 (1969).
67. F. Sierra, C. Sanchez-Pedrino, T. Perez-Ruiz and C. Martinez-Lozano, *An. Quim.*, **66**, 479 (1970) cited in *Vanadium*, National Acad. Sci., Washington, DC, 1974.
68. F. Cassani, *Chem. Ind. (Milan)*, **51**, 1248 (1968) cited in *Vanadium*, National Acad. Sci., Washington, DC, 1974.
69. R. E. Sievers, B. W. Ponder, M. L. Morris and R. W. Moshier, *Inorg. Chem.*, **2**, 693 (1963).
70. R. W. Moshier and R. E. Sievers, *Gas Chromatography of Metal Chelates*, Pergamon Press, Oxford, 1965.
71. N. M. Karayannis and A. H. Corwin, *J. Chromatog. Sci.*, **8**, 251 (1970).
72. L. S. Valberg and J. M. Holt, *Life Sci.*, **3**, 1263 (1964).
73. H. D. Livingston and H. Smith, *Anal. Chem.*, **37**, 1285 (1965).
74. K. Dams, J. A. Robbins, K. A. Rahn and J. W. Winchester, *Anal. Chem.*, **42**, 861 (1970).
75. H. R. Ralston and E. S. Sato, *Anal Chem.*, **43**, 129 (1971).
76. J. P. F. Lambert, R. E. Simpson, H. E. Mohr and L. L. Hopkins, Jr., *J. Assoc. Offic. Anal. Chemists*, **53**, 1145 (1970).
77. K. D. Lindstedt and P. Krueger, *Anal. Chem.*, **42**, 113 (1970).
78. H. Zaffrezic, A. Decarreau, J. P. Carbonnel and N. Deschamps, *J. Radioanal. Chem.*, **18**, 49 (1973).

79. A. W. Morris, *Anal. Chim. Acta*, **42**, 397 (1968).
80. G. V. Alexander, *Applied Spectros.*, **18**, 1 (1964).
81. E. N. Davis and B. C. Hoeck, *Anal. Chem.*, **27**, 1880 (1955).
82. C. C. C. Kang, E. W. Keel and E. Solomon, *Anal. Chem.*, **32**, 221 (1960).
83. C. W. Dwiggins, Jr. and H. N. Dunning, *Anal. Chem.*, **32**, 1137 (1960).
84. P. W. J. M. Boumans, *Theory of Spectrochemical Excitation*, Plenum Press, New York, 1966.
85. I. H. Tipton, M. J. Cook, R. L. Steiner, C. A. Boyle, H. M. Perry, Jr. and H. A. Schroeder, *Health Phys.*, **9**, 89 (1963).
86. A. J. Bedrosian, R. K. Skogerboe and G. H. Morrison, *Anal. Chem.*, **40**, 854 (1968).
87. D. W. Lander, R. L. Steiner, D. H. Anderson and R. L. Dehm. *Applied Spectros.*, **25**, 270 (1971).
88. N. L. Morrow and R. S. Brief, *Environ. Sci. Technol.*, **5**, 786 (1971).
89. D. Glick, *Ann. N.Y. Acad. Sci.*, **157**, 265 (1969).
90. K. W. Marich, P. W. Carr, W. J. Tietzl and D. Glick, *Anal. Chem.*, **42**, 1775 (1970).
91. C. A. Evans, Jr. and G. H. Morrison, *Anal. Chem.*, **40**, 869 (1968).
92. R. Brown and P. G. T. Vossen, *Anal. Chem.*, **42**, 1820 (1970).
93. L. Capacho-Delgado and D. C. Manning, *Atomic Abs. Newsletter*, **5**, 1 (1966).
94. R. E. Mansell and H. Emmel, *Atomic Abs. Newsletter*, **4**, 365 (1965).
95. S. L. Sachdev, J. W. Robinson and P. W. West, *Anal. Chim. Acta*, **37**, 12 (1967).
96. Y. K. Chau and K. Lum-Shue-Chan, *Anal. Chim. Acta*, **50**, 201 (1970).
97. D. A. Segar and J. G. Gonzalez, *Anal. Chim. Acta*, **58**, 15 (1972).
98. C. J. Pickford and G. Rossi, *Analyst*, **97**, 647 (1972).
99. S. H. Omang, *Anal. Chim. Acta*, **56**, 470 (1971).
100. J. Korkisch and H. Gross, *Talanta*, **20**, 1153 (1973).
101. S. Terashima, *Japan Analyst*, **22**, 1317 (1973).
102. L. A. May and B. J. Presley, *Atomic Absorp. Newsletter*, **13**, 144 (1974).
103. C. Riandez, P. Linhares and M. Pinta, *Analusis*, **3**, 303 (1975).
104. C. W. Fuller, *Atomic Absorp. Newsletter*, **12**, 40 (1973).

CHAPTER 31

ZINC

Zinc ores mixed with copper were employed by the ancients in making brass. Oxides of zinc, obtained from furnace fumes by constructing dust chambers to collect the sublimate, were used for medicinal purposes about 2000 years before zinc metal itself became generally known.[1]

Erasmus Ebener of Nurnberg is said to have been the first European to recognize metallic zinc and make brass directly with zinc globules recovered from the reduction of ores.[2] The metal may have been known in the Far East (China and India) much earlier.

Zinc ores are widely distributed. Calamine and zincite consist primarily of zinc oxide. Franklinite is composed of oxides of zinc and iron, while gahnite consists of zinc and aluminum oxides. Willemite is a zinc silicate ore; sphalerite is essentially zinc sulfide.

USES OF ZINC AND ITS SALTS

The metal has various industrial uses, the most ancient being the manufacture of bronze and brass. It is employed as a coating on iron and steel. Zinc sheets are utilized for building purposes.

Zinc salts, being astringent, mildly antiseptic and corrosive in action, have limited application in medicine. Zinc chloride, a soluble salt, is more irritating than the oxide. The oxide, stearate, oleate, and sulfate are incorporated into lotions for the relief of pruritic conditions. Zinc acetate is employed as a styptic; zinc sulfate has been utilized as an emetic.[3]

THE ROLE OF ZINC

Zinc has been considered an essential element since Raulin[4] in 1869 demonstrated that it was necessary in the nutrition of the mould *Aspergillus niger*.

The metal's essentiality for higher plant forms was not recognized until more than half a century later.[5] Zinc was subsequently shown to be an essential constituent in animal nutrition.[6-10] Its role as an essential nutrient for man became established following investigations by Prasad et al.[11,12] in Iran and Egypt.

A number of enzymes including alkaline phosphatase, aldolase, alcohol dehydrogenase, carboxy peptidase, carbonic anhydrase, and lactic acid dehydrogenase are zinc dependent.[13]

ZINC DISTRIBUTION

A normal human body is said to contain between 1.4 and 2.3 g of zinc.[14] The element is present within all body cells. Highest concentrations are found in the eye, prostate, kidney, liver, muscle, heart, skin, and pancreas. Examples of typical tissue concentrations reported in the literature[15] are listed in Table 31.1.

TABLE 31.1
Zinc levels in normal tissues

Tissue	Concentration (μg g^{-1} fresh)
Adrenal	12
Brain	14
Eye	281 (dry wt)
Heart	33
Kidney	55
Liver	55
Lung	15
Muscle	54
Pancreas	29
Prostate	102
Skin (as nails)	151
Spleen	21
Testis	17

Zinc is present in plasma and in all the cellular elements of blood. According to Prasad and co-workers,[16] normal plasma zinc levels range from 100 to 124 μg per 100 ml. Serum levels are said to be approximately 16% higher, a finding attributed to the liberation of zinc from platelets during the clotting process.[17] Perhaps Prasad's population was better fed than the usual population seen at Cook County Hospital, but we find that serum zinc levels of apparently normal adults are within the range of 70 to 110 μg per 100 ml.

About one-third of the plasma zinc is loosely associated with the albumin fraction and two-thirds are more firmly bound to the globulins.[18] A specific zinc transport protein has not been identified as yet.

Zinc contained in formed elements of blood is found chiefly as a component of the metalloenzyme carbonic anhydrase.[19] A small fraction is contained in other enzymes, such as alkaline phosphatase.[20]

The highest concentrations of zinc in normal tissue are encountered in the eye. Differences between species and among individuals are great. Concentrations are greater in the eyes of carnivorous mammals than those of herbivorous species. For example, levels of 14.6, 69.0 and 9.10 mg g^{-1} on a dry weight basis have been reported for dog, fox, and marten, respectively.[21] Mean concentrations of 0.356 and 0.192 mg g^{-1} have been found in eyes of sheep and cattle.[22,23] The human choroid contains about 0.281 mg g^{-1}.[24] The function of zinc in the eye is, as yet, unknown.

Skin and its appendages, hair and nails, have been known to be rich in zinc since the studies described by Lutz[25] in 1926. Levels are affected by nutrition. Hambidge et al.[26] in their recent survey in Denver, found that hair concentrations in apparently normal infants aged 3–12 months averaged only 74 µg g^{-1}, compared to levels of 180 µg g^{-1} among adults and 174 µg g^{-1} among neonates of the same population.

Mean zinc concentrations in the nails and hair of normal subjects investigated by Smith[27] were 151 µg g^{-1} and 173 µg g^{-1}, respectively. Strain et al.[28] found mean levels of 119.6 µg g^{-1} in the hair of North American males while a mean value of 103 µg g^{-1} was obtained for samples from normal Egyptians. However, zinc concentrations in the hair of zinc-deficient dwarfs averaged 54.1 µg g^{-1}.

As a matter of interest, we determined zinc and other trace metal concentrations in the hair, nails, skin fragments, bone, and wrappings of a 2000-year-old mummy, apparently the body of a 6- to 12-month-old child with a malformation of the ear structure encountered in 85% of children with leukemia or lymphoma. Zinc and copper levels are listed in Table 31.2.

The findings concerning concentrations of the toxic metals listed in Table 31.3 provide some points of conjecture concerning the general environment and possible therapeutic agents employed in treating a child who was not thriving.

TABLE 31.2
Zinc and copper concentrations in a Peruvian mummy (6–12 month-old child)

Tissue	Zinc (µg g^{-1} dry weight)	Copper (µg g^{-1} dry weight)
Hair	40.6	23.8
Nails	44.6	21.5
Scalp skin	60.7	10.9
Leg skin	12.0	10.8
Bone (thoracic centra)	53.8	10.4
Food bowl scrapings	19.6	10.9
Wrappings	18.8	38.5

TABLE 31.3
Toxic metal concentrations in a Peruvian mummy[a] (µg g⁻¹)

Tissue	Arsenic	Cadmium	Lead	Mercury	Thallium
Hair	0.0	1.4	20.3	21.5	1.4
Nails	0.0	2.4	4.1	16.2	4.0
Scalp skin	0.0	2.8	12.1	1.7	0.0
Leg skin	0.0	2.8	0.0	1.8	0.0
Thoracic centra	0.0	4.1	6.1	4.3	0.0
Bowl scrapings	21.5	2.3	1.1	1.0	1.8
Wrappings	0.0	2.4	11.7	12.4	0.0

[a] Specimens were supplied by Ms Cathy Elbaum, Committee on Evolutionary Biology, University of Chicago, May 1975.

Note that arsenic was detected only in scrapings from the child's food bowl. Perhaps arsenical therapy, quite common to ancient cultures, had been initiated less than two weeks before death. Mercury, lead, and cadmium concentrations are also of interest.

METABOLISM OF ZINC

Zinc influences growth rate and bone development, the integrity of the skin, and the development and function of the reproductive organs. The element influences wound healing.[29] The rate of wound closure has been shown to be directly proportional to zinc hair concentrations. Wound healing in healthy young men given zinc sulfate supplements following surgery was accelerated compared to the healing rate in non-medicated control subjects.[30]

Zinc requirements are governed by state of health, activities and age. It is estimated that under normal conditions about 1–2 mg of zinc are required daily. A usual Western diet supplies between 10 and 15 mg.

Foods vary greatly in zinc content. Wheat germ, bran and oysters contain the greatest concentrations, while sugar and citrus fruits contain the least. Legumes, nuts, eggs, fish, meat, vegetables, milk, etc. lie in between.

From 5 to 10% of the dietary zinc intake is absorbed.[31] Absorption is dependent upon dietary composition. For example, phytates and fiber decrease zinc absorption.[32,33] Zinc-deficient states can be induced by diets high in these components. However, amino acids in proteins bind zinc tenaciously. Histidine and cysteine compete with phytates most effectively and so prevent zinc deficiencies.[34]

Zinc absorption occurs mainly in the duodenum. The intestines also serve as the major excretory organ of ingested zinc.[31] A comparatively small amount is excreted in the urine. Normal urinary excretion in adults varies between 0.1 and 0.7 mg per day.[35] Significant quantities of zinc are lost in sweat.[29,36]

Urinary zinc excretions in certain pathological states, such as nephrosis, post-alcoholic hepatic cirrhosis, certain types of porphyria, are increased. A concurrent decrease in serum zinc levels is observed.

Various therapeutic agents can influence serum and urinary zinc levels. Other trace elements are also affected. High urinary zinc levels have been noted in patients receiving isoniazid, an anti-tuberculous drug, and tetracycline, a broad-spectrum antibiotic.[37] Serum zinc levels decrease and urinary zinc levels increase during chelation therapy with such chelating agents, such as penicillamine[38] and EDTA.[39] One of the hazards of chelation therapy of lead poisoning, for example, is the side effects produced by EDTA, penicillamine, BAL, etc. Usually the populations involved are poorly nourished. Further depletion of zinc and certain other trace elements from the body stores that may result, may possibly not be too beneficial.

One can but question the long-term indiscriminate administration of penicillamine to a population of young children with blood lead levels below 50 μg per 100 ml. It would be advisable to improve their nutrition and monitor their trace metal status with greater care.

Planas-Bohne and Olinger[40] postulated that the toxicity of Ca—DTPA (sodium calcium diethylenetriaminepenta-acetate) was due to zinc and manganese depletion of tissues which followed administration of the agent. For example, zinc concentrations in fresh liver, kidney, and small intestine of control rats were 56.7, 26.2, and 22.4 μg g^{-1}, respectively, compared to levels at 26.5, 24.8, and 15.5 μg g^{-1} in similar tissues from Ca—DTPA-treated animals.

Serum zinc levels decrease in stress situations wherever tissue injury occurs; for example, infective processes, burns, surgical trauma, and myocardial infarction.[41] Serum zinc levels found to occur in various pathological conditions are listed in Table 31.4 below.

TABLE 31.4
Serum zinc levels in various pathological states

Conditions	Serum zinc (μg per 100 ml)	Reference
Burns	47	42
Surgical trauma	55	43
Suppurative infection	61	44
Acute infection	68	45
Myocardial infarction	56–68	46–48
Acute viral hepatitis	25	49

We have found levels of 45 μg per 100 ml and less in patients in shock-like states following surgery.

According to Low and Ikram[50] plasma zinc concentrations in acute myocardial infarction have diagnostic and prognostic implications. Plasma levels decrease within the first 3 days and return to normal in about 14 days. Good correlation between minimum plasma zinc value and peak plasma values of cardiac enzymes has been observed. The mean minimum plasma zinc level of all patients with myocardial infarct studied was 68 μg per 100 ml. Those who did not suffer a cardiac arrest showed a minimum level of 72 μg per 100 ml while patients who suffered a cardiac arrest or death had minimum plasma zinc levels of 55 μg per 100 ml. Mean minimum zinc levels of the control group or subjects with ischaemia and non-cardiac chest pain were 84 μg per 100 ml.

Elevated serum zinc levels are encountered rarely. Levels between 150 and 200 μg per 100 ml have been observed in subjects receiving zinc supplements. The high plasma zinc levels seen following ingestion of toxic quantities of zinc salts are transitory.

Smith[51] recently reported on a heritable hyperzincemia in a black family. The propositus had a serum zinc level of 275 to 280 μg per 100 ml. Levels ranging up to 435 μg per 100 ml were found among his siblings and children. There was not a concurrent increase in hair, bone or erythrocyte zinc.

ZINC DEFICIENCIES

Zinc deficiencies have been demonstrated in birds, experimental and domestic animals, and man. The syndrome manifests itself by growth retardation, anorexia, lesions of the skin and appendages, impaired development and function of the reproductive organs. Birds exhibit gross bone disorders. Poor wound healing and susceptibility to infections in man are also indicative. The severity of involvement is dependent upon the degree of zinc deficiency.

Included among the causes of zinc-deficient states in man are excessive intake of cereal foods with a high fiber and phytate content, which render the zinc ingested unavailable for absorption. Malabsorption syndrome, and renal and liver disease, alcoholism with the accompanying inadequate diet also induce deficiencies. Failure to include zinc in fluids of total parenteral nutrition has resulted in conditioned zinc deficient states.[52]

Klingberg et al.[38] recently reported the development of zinc deficiency in a patient with Wilson's disease being treated with penicillamine. The syndrome was characterized by a parakeratosis, alopecia, keratitis and lowered zinc levels in hair, plasma and erythrocytes. Some improvement was observed after zinc supplementation.

Acrodermatitis enteropathica, an inherited disease seen primarily in infants and children, can prove fatal if untreated. Severe erythematous and vesiculo-bullous skin lesions, alopecia and diarrhea are characteristic symptoms. Lesions are most prominent at the extremities and adjacent to body orifices. Moynahan[53] recognized the manifestations as due to zinc deficiency and found zinc therapy to be curative for the condition. Hambidge et al.[54] found that zinc supplementation, the treatment of choice in *acrodermatitis enteropathica*, is without toxicity at the effective dosage level. Such therapy leads to a rapid, consistent, and sustained cure.

Sever[55] is of the opinion that zinc deficiency in man is potentially teratogenic.

ZINC TOXICITY

True zinc intoxication is an infrequent occurrence. Many of the toxic effects attributed to zinc may have been due to other metals present, namely, cadmium, lead, antimony and arsenic.

Metal fume fever, also called brass founders' ague, brass chills, zinc shakes, galvo, spelter shakes, has been attributed to zinc oxide fumes and dusts.[13,56,57] Inhalation of zinc fumes causes fever, malaise, depression, coughing, which might induce vomiting, salivation and headache.

Zinc, unlike lead, cadmium, arsenic, antimony, is not cumulative. Batchelor, *et al.*[58] in their observations of workmen exposed for 2–35 years to zinc oxide dusts poor in lead and cadmium did not uncover any acute or chronic illness attributable to zinc. According to Hamilton and Hardy[59] chronic industrial poisoning due to zinc is probably nonexistent.

Acute zinc intoxication manifest by nausea, vomiting, and severe anemia is reported to have occurred in a patient in renal failure who had been dialyzed with water stored in a galvanized tank.[60] Undoubtedly other toxic metals were also involved.

Toxic quantities of zinc can be brought into solution when acidic foods (vegetables, meats, fruit juices, salads, cocoa, etc.) are prepared or stored in galvanized utensils. Gastrointestinal symptoms and fever have been reported to occur within 20 min or to be delayed for as long as 12 hours following ingestion. Symptoms may be due to other metals present (lead, cadmium, iron, arsenic). The emetic action of zinc is a protective mechanism.[61,62] However, dehydration, electrolyte imbalance, dizziness, lethargy, and muscular uncoordination may follow.

Occurrence of acute renal failure following ingestion of zinc chloride has been reported.[63] The purity of the zinc salt is not known.

ANALYSIS OF ZINC

Zinc is a ubiquitous metal. Maintaining an essentially zinc-free environment from sample collection through to completion of the analysis is a challenge greater than that encountered in lead determinations. Laboratory ware, both glass and plastic, contains zinc contaminants. Blood samples cannot be collected in the usual evacuated tubes because the rubber stoppers contain zinc in highly significant quantity. Certain polypropylene tubes are quite zinc free.

Acids and other reagents contain zinc; therefore, irrespective of the analytical technique employed, standard concentrations and blanks must be carried through the entire analytical procedure concurrently with the unknown samples.

Colorimetric methods

Diphenylcarbazone (dithizone) and di-β-naphtholthiocarbazone are the primary reagents that have been employed in the colorimetric determination of zinc in biological materials. Though quite sensitive, neither is an ideal reagent since numerous trace metals form colored complexes under similar conditions. Zinc is quantitatively extracted by dithizone[64-68] in an organic solvent over the pH range 6–9.5 and by di-β-naphtholthiocarbazone[69,70] at pH 8–10. Lead, cadmium, and thallium can be major interferences with the former, but only lead and cadmium may interfere with the latter.

Copper, nickel, cobalt, iron, mercury and most of the bismuth present can be separated from zinc in the initial solvent–chelate extract by extraction with 0.2 N hydrochloric acid. The aqueous phase contains the zinc, cadmium, thallium and lead while the other metals remain in the solvent phase. The acid solution is then made alkaline and reextracted with the chelate.

Zinc as a benzoylacetonate[71] is quantitatively extracted over the pH range of 7–9. The oxinate[72] is extractable at pH 4–5.

The diethyldithiocarbamate of zinc is quantitatively extracted at pH 4–11.[73,74] Interferences are legion. The reagent has been employed in determining zinc in biological materials.[75]

Xylenol orange,[76,77] Rhodamine B,[78] 2-naphthol-4-sulphonic acid,[79] and Eriochrome blue black R[80] are the miscellaneous reagents utilized in colorimetric zinc analyses.

Fluorimetry

Fluorescent complexes of zinc have been employed in measuring concentrations in biological materials.[81-83] One of the first procedures described used benzoin as a reagent.[84] Most of the techniques presented subsequently involved formation of fluorescent complexes with 8-hydroxyquinoline[85,86] or 8-hydroxyquinoline-5-sulfonic acid.[87]

Chromatography

Chromatographic techniques have been employed alone or in conjunction with other techniques in determining zinc in various materials. As an example, the metal was isolated by anion exchange resin chromatography and determined colorimetrically.[88-91] Hunter and Miller[92] isolated zinc from other elements by anion exchange resins and measured the metal by titration with disodium ethylenediaminetetracetate.

Kahle and Reif[93] employed cation exchange for separating constituents of blood plasma and determined zinc by polarography.

Paper chromatography has been utilized as both a separation and identification technique.[94-96]

Gas chromatography

Zinc as an acetylacetonate has been separated by gas chromatography[97,98] with varying degrees of success. The complexes are not thermally stable. Better separations and more reproducible retention times can be achieved with the more volatile trifluoroacetylacetonate.[99]

Karayannis and Corwin[100] were able to volatize and elute 43 metal acetylacetonate complexes, including that of zinc, by employing high pressure gas chromatography at 115 °C and 800–1000 psi.

Polarography

Various applications of polarography to the analysis of zinc concentrations in plant and animal materials have been described. Cholak *et al.*[69] determined zinc by polarography, as well as colorimetry, after extraction with di-β-naphthylthiocarbazone.

Ginzburg and Elpener[101] estimated the copper and zinc content in healthy and carious teeth by polarographic methods.

The technique has been applied to analysis of zinc contents in soils,[102] plants,[103] blood,[104] and pharmaceuticals.[105]

Franke and de Zeeuw[106] described the use of differential pulse anodic stripping voltammetry (DPASV) as a screening procedure to identify zinc, cadmium, lead and other heavy metals in urine.

Neutron activation analysis

Many of the earlier investigations concerning zinc concentrations in different matrices, biological and other, employed neutron activation analysis.[107-110] The technique was also applied subsequently to estimating the metal in waters,[111-114] rocks,[115] fingernails and hair,[116] human brain,[117] fish,[118] foods,[119] and miscellaneous biological materials.[120,121]

Some investigators separated zinc by ion exchange before irradiation.[115] Others preferred extraction procedures[118,120] as a means for isolating zinc from the respective matrices.

X-ray spectroscopy

X-ray spectroscopy, a highly useful tool for the direct, nondestructive measurements of elements in biological materials, has been applied to a limited degree for the estimation of zinc and other metals in biological materials.[122-126] Grebe and Esser,[126] as early as 1936, reported on their investigations of zinc, cadmium, lead, thallium, copper, magnesium, cobalt, etc. concentrations in tissues. Natelson, however, is considered the pioneer in the applications of X-ray fluorescence spectrometry to the analyses of zinc and other trace metals in the clinical laboratory.[127-129]

Though an elegant instrumental technique for determining elements in biological materials in submicrogram quantities, the use of X-ray spectrometry did not gain wide acceptance in the clinically oriented laboratory. Matrix effects challenge the accuracy and precision of analysis. For example, iron can interfere since it has an absorption effect upon the zinc signal.[130] Atomic absorption instrumentation introduced in the early 1960s, proved more adaptable to the specific needs of the clinical laboratory requiring the capability for determining zinc and other trace metals with acceptable specificity, precision, accuracy, and in volume as well. Detection limits for most elements are also better with atomic absorption.

However, X-ray fluorescence spectrometry continued to be employed extensively in other specialty areas where large sized samples are available for routine analyses. Numerous applications to geological samples (rocks and soils) have been described.[131-134] Precision and accuracy were improved by employing synthetic matrices for standard preparations.

Methods for determining zinc concentrations in plant and grains have been described.[134-135]

The technique has been utilized in water analysis in conjunction with different separation methods including cation exchange resins and membranes[136,137] and various chelation–extraction procedures.[135,138] It has also been employed in estimating zinc and other trace metal levels in atmospheric particulates.[139-140]

Emission spectroscopy

Zinc was among the elements detected in tissues and biological fluids by the early investigators utilizing flame and arc spectrographic methods.[141-147] Emission spectrography was employed in the studies of Tipton et al.[15,24] and Hambidge and co-workers[26,54] mentioned previously.

Atomic absorption spectrometry

Because of the favorable detection limits, relative freedom from spectral interferences and comparative simplicity in acceptable analytical methods, atomic absorption spectrometry is the preferred technique for the estimation of zinc concentrations in blood, other body fluids, tissues, etc. Detection limits for zinc at the 213.8 nm resonance line are stated as 2 ng ml^{-1} by flame atomization. A normal blood serum, for example, contains between 0.7 and 1.1 μg ml^{-1} of zinc. Analytical sensitivity can be augmented by employing chelation–extraction procedures and/or by the use of flameless atomization.

As stated previously, eliminating the zinc contaminants in the analytical environment is the major challenge in zinc analysis.

Applications of the technique to analyses of zinc contents in different matrices have been legion. We shall restrict our comments to the numerous applications of animal, human[148-167] and plant and agricultural materials.[168-172]

All samples require varying degrees of pretreatment. Blood proteins can be removed by ashing procedures (wet or dry) or by simple precipitation with 5% trichloroacetic acid (TCA), employing 4 parts of TCA and 1 part of sample, as an example.

Tissues of plant and animal origin must be ashed initially. After dilution to a suitable volume and pH adjustment, an aliquot is chelated with the reagent of choice and extracted prior to analysis. Determining the zinc content of hair and nails in the graphite furnace directly without any chemical pretreatment can only be criticized. The non-specific molecular absorption generated by matrix constituents markedly enhances the absorption signals from zinc.

Urine samples can be chelated and extracted directly after pH adjustment. Merely diluting urine samples is inadvisable. The nonspecific molecular absorption generated by sodium and other components can be overwhelming, as zinc is excreted in less than milligram amounts and sodium in gram quantities.

Standards ranging in concentration from 0 to 300 μg per 100 ml should be carried through the entire analytical procedure. Matrices of standards and unknown samples must be similar.

REFERENCES

1. H. C. Hoover and L. H. Hoover, trans. of G. Agricola, *De Re Metallica*, Dover, New York, 1950.
2. A. S. Prasad, *Zinc Metabolism*, Charles C. Thomas, Springfield, 1966.
3. L. S. Goodman and A. Gelman, *The Pharmacological Basis of Therapeutics*, MacMillan, New York, 1955.
4. J. Raulin, Ann. *Sci. Nat. Bot. Biol. Vegetale*, **11**, 93 (1869) cited in E. J. Underwood, *Trace Metals in Human and Animal Nutrition*, Academic Press, New York, 1971.
5. A. L. Somner and C. B. Lipman, *Plant Physiol.*, **1**, 231 (1926).
6. W. R. Todd, C. A. Elvehjem and E. B. Hart, *Amer. J. Physiol.*, **107**, 146 (1934).
7. H. F. Tucker and W. O. Salmon, *Proc. Soc. Exptl. Biol. Med.*, **88**, 613 (1955).
8. B. L. O'Dell, P. M. Newberne and J. E. Savage, *J. Nutr.*, **65**, 503 (1958).
9. S. P. Legg and L. Sears, *Nature (London)*, **186**, 1061 (1960).
10. P. Dynna and G. N. Havre, *Acta Vet. Scand.*, **4**, 197 (1963).
11. A. S. Prasad, J. A. Halsted and M. Nadini, *Amer. J. Med.*, **31**, 532 (1961).
12. A. S. Prasad, A. Miale, Z. Farid, H. H. Sandstead, A. Schulert and W. J. Darby, *Arch. Intern. Med.*, **111**, 407 (1963).
13. B. Vallee, *Physiol. Rev.*, **39**, 443 (1959).
14. E. M. Widdowson, R. A. McCance and C. M. Spray, *Clin. Sci.*, **10**, 113 (1951).
15. I. H. Tipton and M. J. Cook, *Health Physics*, **9**, 103 (1963).

16. A. S. Prasad, E. B. Schoomaker, J. Ortega, G. J. Brewer, D. Oberleas and F. J. Oelshlegel, Jr., in *Progress in Clinical and Biological Research, Vol. 1: Erythrocyte Structure and Function*, (ed. G. J. Brewer) Alan R. Liss, New York, 1975.

17. B. Foley, S. A. Johnson, B. Hackely, J. C. Smith and J. A. Halsted, *Proc. Soc. Exptl. Biol. Med.*, **128**, 265 (1968).

18. H. Wolff, *Klin. Wochenschr.*, **34**, 409 (1956).

19. E. Hove, C. A. Elvehjem and E. B. Hart, *Amer. J. Physiol.*, **119**, 768 (1937).

20. B. L. Vallee, F. L. Hoch, S. J. Adelstein and W. E. C. Wacker, *J. Amer. Chem. Soc.*, **78**, 5879 (1956).

21. G. Weitzel and A. M. Fretzdorff, *Hoppe-Seyler's Z. Physiol. Chem.*, **292**, 221 (1953).

22. J. M. Bowness and R. A. Morton, *Biochem. J.*, **51**, 530 (1950).

23. J. M. Bowness and R. A. Morton, *Biochem. J.*, **53**, 620 (1953).

24. I. H. Tipton, H. A. Schroeder, H. M. Perry, Jr., and M. J. Cook, *Health Physics*, **11**, 403 (1965).

25. R. E. Lutz, *J. Ind. Hyg.*, **8**, 177 (1926).

26. K. M. Hambidge, C. Hambidge, M. Jacobs and J. D. Baum, *Pediat. Res.*, **6**, 868 (1972).

27. H. Smith, *Forensic Sci. Soc.*, **7**, 97 (1967).

28. W. H. Strain, L. T. Steadman, C. A. Lankau, W. P. Beiliner and W. J. Pories, *J. Lab. Clin. Med.*, **68**, 244 (1966).

29. W. J. Pories and W. H. Strain, in *Zinc Metabolism*, (ed. A. S. Prasad) Charles C. Thomas, Springfield, Illinois, 1966.

30. W. J. Pories, J. H. Henzel, C. G. Rob and W. H. Strain, *Lancet*, **2**, 121 (1967).

31. H. Spencer, *J. Nutr.*, **86**, 169 (1965).

32. B. L. O'Dell and J. E. Savage, *Proc. Soc. Exptl. Biol. Med.*, **103**, 304 (1960).

33. J. G. Reinhold, K. Nasr, A. Lahimgarzadeh and H. Hedayati, *Lancet*, **1**, 283 (1973).

34. F. H. Nielsen, M. L. Sunde and W. G. Hoekstra, *Proc. Soc. Exptl. Biol. Med.*, **124**, 1106 (1967).

35. E. J. Underwood, *Trace Elements in Human and Animal Nutrition*, Academic Press, New York, 1971.

36. D. Oberleas, R. C. White, R. S. Hadley and A. S. Prasad, *Fed. Proc. Amer. Soc. Exptl. Biol.*, **31**, 668 (1972).

37. H. Spencer, D. Osis, L. Kramer and E. Wiatrowski, in *Clinical Applications of Zinc Metabolism*, (eds W. J. Pories, W. H. Strain, J. M. Hsu and R. L. Woosely) Charles C. Thomas, Springfield, Illinois, 1974.

38. W. G. Klingberg, A. S. Prasad and D. Oberleas, in *Trace Elements in Human Health and Disease*, Vol. 1 (eds A. S. Prasad and D. Oberleas) Academic Press, New York, 1976.

39. H. M. Perry, Jr. and H. A. Schroeder, *Amer. J. Med.*, **22**, 168 (1957).

40. F. Planas-Bohne and H. Olinger, *Health Physics*, **31**, 165 (1976).

41. W. R. Beisel and R. S. Pekarek, *Int. Rev. Neurobiol. Suppl.*, **1**, 53 (1972).

42. R. A. Dieter, Jr., W. E. Neville and R. Pifarre, *J. Thorac. Cardiov. Surg.*, **59**, 168 (1970).

43. J. H. Henzel, M. S. De Weese and E. L. Lichti, *Arch. Surg.*, **100**, 349 (1970).

44. I. Vikbladh, *Scand. J. Clin. Lab. Invest. Suppl.*, **2**, 1 (1951).

45. R. D. Lindeman, R. G. Bottomley, R. L. Cornelison, Jr. and L. A. Jacobs, *J. Lab. Clin. Med.*, **79**, 452 (1972).

46. W. E. C. Wacker, D. D. Ulmer and B. L. Vallee, *New England J. Med.*, **255**, 449 (1956).

47. J. A. Halsted and J. C. Smith, Jr., *Lancet*, **1**, 322 (1970).
48. R. D. Lindeman, A. A. Yunice, D. J. Baxter, L. R. Miller and J. Nordquist, *J. Lab. Clin. Med.*, **81**, 194 (1973).
49. R. I. Henkin and F. R. Smith, *Amer. J. Med. Sci.*, **264**, 401 (1972).
50. W. I. Low and H. Ikram, *Brit. Heart J.*, **38**, 1339 (1976).
51. J. C. Smith, Jr., in *Zinc Metabolism: Current Aspects in Health and Disease*, (eds G. J. Brewer and A. S. Prasad) Alan R. Liss, New York, 1977.
52. H. H. Sanstead, K. P. Vo-Khactu and N. Solomons, in *Trace Elements in Human Health and Disease*, (eds A. S. Prasad and D. Oberleas) Academic Press, New York, 1976.
53. E. J. Moynahan, *Lancet*, **1**, 399 (1974).
54. K. M. Hambidge, P. A. Walravens and K. H. Neldner, in *Zinc Metabolism: Current Aspects in Health and Disease*, (eds G. J. Brewer and A. S. Prasad), Alan R. Liss, New York, 1977.
55. L. E. Sever, *Lancet*, **1**, 887 (1973).
56. H. E. Stokinger, in *Industrial Hygiene and Toxicology*, (ed. F. A. Patty) Wiley, New York, 1963.
57. J. P. Papp, *Postgrad. Med.*, **43**, 160 (1968).
58. R. P. Batchelor, J. W. Fehnal, R. M. Thompson and K. R. Drinker, *J. Ind. Hyg. Toxicol.*, **8**, 322 (1926).
59. A. Hamilton and H. L. Hardy, *Industrial Toxicology*, Publishing Sciences Group, Acton, Mass., 1974.
60. E. D. M. Gallery, J. Blomfield and S. R. Dixon, *Brit. Med. J.*, **4**, 331 (1972).
61. J. E. Furchener and C. B. Richmond, *Health Physics*, **8**, 35 (1962).
62. D. A. Hath, W. M. Becker and W. C. Hoekstra, *J. Nutr.*, **88**, 331 (1966).
63. S. Csata, F. Gallays and M. Toth, *Z. Urol.*, **61**, 327 (1968).
64. H. Cowling and E. J. Miller, *Ind. Eng. Chem. (Anal. Ed.)*, **13**, 145 (1941).
65. B. G. Malmstrom and D. Glick, *Anal. Chem.*, **23**, 1699 (1951).
66. G. B. Jones, *Anal. Chim. Acta*, **7**, 578 (1952).
67. T. E. Banks, R. L. F. Tupper, R. W. A. Watts and A. Wormall, *Nature (London)*, **173**, 348 (1954).
68. K. O. Raker, *Z. Anal. Chem.*, **173**, 57 (1960).
69. J. Cholak, D. M. Hubbard and R. E. Burkey, *Ind. Eng. Chem. (Anal. Ed.)*, **15**, 754 (1943).
70. A. E. Martin, *Anal. Chem.*, **25**, 1853 (1953).
71. J. Stary and J. Hladky, *Anal. Chim. Acta*, **28**, 227 (1963).
72. J. Stary, *Anal. Chim. Acta*, **28**, 132 (1963).
73. H. Bode, *Z. Anal. Chem.*, **144**, 165 (1955).
74. K. E. Kress, *Anal. Chem.*, **30**, 432 (1958).
75. J. A. Stewart and J. C. Bartlett, *Anal. Chem.*, **30**, 404 (1958).
76. J. Korble and R. Pribel, *Chem. Anal.*, **45**, 102 (1956).
77. K. Studlar and I. Janousek, *Talanta*, **8**, 203 (1961).
78. B. Tvaroha and O. Maler, *Mikrochim. Acta*, **4**, 634 (1962).
79. H. F. Liddell and S. M. Williams, *Analyst*, **83**, 111 (1958).
80. D. W. Rogers, *Anal. Chem.*, **34**, 1657 (1962).
81. L. J. Merritt, Jr., *Ind. Eng. Chem. (Anal. Ed.)*, **16**, 758 (1944).
82. S. Udenfriend, *Fluorescence Assay in Biology and Medicine*, Academic Press, New York, 1962.
83. D. F. H. Wallace and T. L. Steck, *Anal. Chem.*, **35**, 1035 (1963).
84. C. E. White and M. H. Neustadt, *Ind. Eng. Chem., (Anal. Ed.)*, **15**, 599 (1943).
85. J. A. Bishop, *Anal. Chem.*, **35**, 1035 (1963).
86. S. Watanabe, W. Franz and D. Trotter, *Anal. Biochem.*, **5**, 345 (1963).

87. J. A. Bishop, *Anal. Chim. Acta*, **29**, 172 (1963).
88. R. M. Rush and J. H. Yoe, *Anal. Chem.*, **26**, 1345 (1954).
89. T. Y. Yuan and J. G. A. Fiskell, *J. Assoc. Offic. Agr. Chemists*, **41**, 424 (1958).
90. R. H. Maier and J. S. Bullock, *Anal. Chim. Acta*, **19**, 354 (1958).
91. W. D. Duffield, *Analyst*, **83**, 503 (1958).
92. I. A. Hunter and C. C. Miller, *Analyst*, **81**, 79 (1956).
93. G. Kahle and E. Reif, *Biochem. Z.*, **325**, 380 (1954).
94. S. Yasunaga and O. Shimomura, *J. Pharm. Soc. (Japan)*, **74**, 62 (1954).
95. E. Lederer and M. Lederer, *Chromatography*, Elsevier, Princeton, 1957.
96. M. Cernikova and B. Konrad, *Biochem. Biophys. Acta*, **71**, 190 (1963).
97. A. A. Duswalt, Jr., Doctoral Dissertation, Purdue University, 1958, cited in R. W. Moshier and R. E. Sievers, *Gas Chromatography of Metal Chelates*, Pergammon Press, New York, 1965.
98. K. Yamakawa, K. Tanikawa and K. Arakawa, *Chem. Pharm. Bull. (Tokyo)*, **11**, 1405 (1963).
99. R. W. Moshier and R. E. Sievers, *Gas Chromaography of Metal Chelates*, Pergammon Press, New York, 1965.
100. N. M. Karayannis and A. H. Corwin, *J. Chromatog. Sci.*, **8**, 251 (1970).
101. M. M. Ginzburg and I. E. Elpener, *Stomatologiya*, **1**, 13 (1947).
102. R. G. Menzee and M. L. Jackson, *Anal. Chem.*, **23**, 1861 (1951).
103. O. N. Hinsvark, W. H. Houff, S. H. Wittwer and H. M. Sell, *Anal. Chem.*, **26**, 1202 (1952).
104. J. Vogel, D. Monnier and M. Haerde, *J. Electroanal. Chem.*, **3**, 321 (1962).
105. M. Brezina and P. Zunian, *Polarography in Medicine, Biochemistry, and Pharmacy*, Interscience, New York, 1958.
106. J. P. Franke and R. A. Zeeuw, *Arch. Toxicol.*, **37**, 47 (1976).
107. K. J. Koch, Jr. and E. R. Smith, *J. Clin. Endocrin.*, **16**, 123 (1956).
108. B. A. Loveridge and A. A. Smales, in *Methods of Biochemical Analysis*, Vol. 5, (ed. D. Glick) Interscience, New York, 1957.
109. W. W. Meinke, *Anal. Chem.*, **30**, 686 (1958).
110. V. P. Guinn and C. D. Wagner, *Anal. Chem.*, **32**, 317 (1960).
111. D. E. Robertson, *Anal. Chem.*, **40**, 1067 (1968).
112. D. Z. Piper and G. G. Goles, *Anal. Chim. Acta*, **47**, 560 (1969).
113. J. Scheider and R. Geisler, *Z. Anal. Chem.*, **267**, 270 (1973).
114. G. J. Clemente and G. G. Mastinu, *J. Radioanal. Chem.*, **20**, 707 (1974).
115. O. Johansen and E. Steinnes, *Talanta*, **16**, 407 (1970).
116. D. J. Mahler, A. E. Scott, J. R. Walsh and G. Hayme, *J. Nucl. Med.*, 739 (1970).
117. G. Henke, H. Mollman and H. Alfer, *Z. Neurol.*, **119**, 283 (1971).
118. K. K. Pillay, C. C. Thomas, Jr. and C. M. Hyche, *J. Radioanal. Chem.*, **20**, 597 (1974).
119. R. Becker, R. A. Veglia and E. R. Schmid, *Radiochem. Radioanal. Letters*, **19**, 343 (1974).
120. T. E. Henzler, R. J. Korda, P. A. Helmke, M. R. Anderson, M. M. Jiminez and L. A. Haskin, *J. Radioanal. Chem.*, **20**, 649 (1974).
121. B. Maziere, J. Gros and D. Comar, *J. Radioanal. Chem.*, **24**, 279 (1975).
122. A. A. Smales, in J. H. Yu and H. J. Koch, Jr., *Trace Analysis*, John Wiley, New York, 1957.
123. T. Hall, *Science*, **134**, 449 (1961).
124. A. R. MacKenzie, T. Hall and W. F. Whitmore, *Nature (London)*, **193**, 72 (1962).
125. I. Adler, in *Handbook of Analytical Chemistry*, (ed. L. Meites), McGraw-Hill, New York, 1963.

126. V. L. Grebe and F. Esser, *Fortschr. Gebiete Roentgenstrahlen*, **54**, 185 (1936).
127. S. Natelson and S. L. Bender, *Microchem. J.*, **3**, 19 (1959).
128. S. Natelson and W. R. Whitford, in *Methods of Biochemical Analyses*, Vol. 12, (ed. D. Glick), Interscience, New York, 1964.
129. S. Natelson, *Clin. Chem.*, **11**, 290 (1965).
130. B. P. Fabbi and L. F. Espos, US Geol. Survey Prof. Paper 800-B, B 146 (1972) cited in M. Pinta, *Modern Methods of Trace Element Analysis*, Ann Arbor Science, Ann Arbor, Mich., 1978.
131. M. J. Kaye, *Geochim. Cosmochim. Acta*, **29**, 139 (1965).
132. N. B. Price and G. R. Angell, *Anal. Chem.*, **40**, 660 (1968).
133. E. Murao, *Anal. Chim. Acta*, **67**, 37 (1973).
134. C. Williams, *J. Sci. Food Agric.*, **27**, 561 (1976).
135. M. Pinta, *Modern Methods for Trace Element Analysis*, Ann Arbor Science, Ann Arbor, Mich., 1978.
136. B. Holynska, *Radiochem. Radioanal. Letters*, **16**, 5 (1974).
137. C. H. Lochmueller, J. W. Galbraith and R. L. Walter, *Anal. Chem.*, **46**, 440 (1974).
138. A. W. Morris, *Anal. Chim. Acta*, **42**, 397 (1968).
139. J. V. Gilfrich, P. G. Burkhalter and L. S. Birks, *Anal. Chem.*, **42**, 1749 (1970).
140. T. G. Dzubay and R. K. Stevens, *Environ. Sci. Tech.*, **9**, 663 (1975).
141. J. H. Sheldon and H. Ramage, *Biochem. J.*, **25**, 1608 (1931).
142. W. Benoit, *Z. Ges. Exp. Med. Biol.*, **90**, S421 (1932).
143. A. Policard and A. Morel, *Bull. Histol. Appl. I*, **9**, 57 (1932).
144. A. Policard, *Protoplasma*, **19**, 602 (1933).
145. H. J. Koch, Jr., E. R. Smith, N. F. Shimp and J. Conner, *Cancer*, **9**, 499 (1956).
146. P. Monacelli, H. Tanaka and J. H. Yoe, *Clin. Chim. Acta*, **1**, 577 (1956).
147. I. H. Tipton, in *Metal Binding in Medicine*, (ed. M. J. Seven), J. B. Lippincott, Philadelphia, 1960.
148. A. Walsh, *Spectrochim. Acta*, **7**, 108 (1955).
149. J. E. Allan, *Analyst*, **86**, 530 (1961).
150. J. B. Willis, in *Methods of Biochemical Analysis*, Vol. 11 (ed. D. Glick), Interscience, New York, 1963.
151. H. E. Parker, *Atomic Absorp. Newsletter*, **2**, 23 (1963).
152. K. Fuwa, P. Pulido, R. McKay and B. L. Vallee, *Anal. Chem.*, **36**, 240 (1964).
153. S. Sprague and S. Slavin, *Atomic Absorp. Newsletter*, **4**, 228 (1965).
154. A. S. Prasad, D. Oberleas and J. A. Halsted, *J. Lab. Clin. Med.*, **66**, 508 (1965).
155. M. H. Briggs, M. Briggs and A. Wokatama, *Experentia*, **28**, 406 (1972).
156. G. Heinemann, *Z. Klin. Chem. Klin. Biochem.* **10**, 467 (1972).
157. D. Kurz, J. Roach, E. J. Eyring, *Biochem. Med.*, **6**, 274 (1972).
158. G. E. Marks, C. E. Moore, E. R. Kanabrocki, Y. T. Oester and E. Kaplan, *Appl. Spectrosc.*, **26**, 523 (1972).
159. S. Chaube, H. Nishimura and C. A. Swineyard, *Arch. Environ. Health*, **26**, 237 (1973).
160. H. T. Delves, *Atomic Abs. Newsletter*, **12**, 50 (1973).
161. G. Machata and R. Binder, *Z. Rechtsmed.*, **73**, 29 (1973).
162. J. Smeyers-Verbeke, D. L. Massart, J. Versieck and A. Speecke, *Clin. Chim. Acta*, **44**, 243 (1973).
163. J. R. Sorenson, E. G. Melby, P. J. Nord and H. G. Petering, *Arch. Environ. Health*, **27**, 36 (1973).
164. J. D. Bogden, N. P. Singh and M. M. Joselow, *Environ. Sci. Technol.*, **8**, 740 (1974).
165. A. J. Jackson, L. M. Michael and H. J. Schumacher, *Anal. Chem.*, **44**, 1064 (1974).

166. M. A. Evenson and C. T. Anderson, *Clin. Chem.*, **21**, 537 (1975).
167. B. Momcilovic, B. Belonje and B. A. Shah, *Clin. Chem.*, **21**, 588 (1975).
168. D. J. David, *Analyst*, **84**, 536 (1958).
169. C. H. McBride, *J. Assoc. Offic. Agri. Chemists*, **48**, 406 (1965).
170. H. E. Parker, *Atomic Abs. Newsletter*, **2**, 23 (1963).
171. J. R. Buchanan and T. T. Muraoka, *Atomic Absorp. Newsletter*, **3**, 79 (1964).
172. N. Baumslag and P. Klein, *Arch. Environ. Health*, **25**, 23 (1972).

CHAPTER 32

ZIRCONIUM

Although zirconium is a relatively abundant metal in the earth's crust, being more abundant than zinc for example in igneous rocks, sandstones, shales and soils, its biological significance is yet to be demonstrated. According to Bowen,[1] soils, shales, sandstones, and igneous rocks contain 300, 160, 220, and 165 ppm of zirconium compared to zinc concentrations of 50, 95, 16 and 70 ppm, respectively. However, since zirconium is a relatively insoluble and chemically resistant metal, concentrations present in waters and living organisms are rather low by comparison. It is not soluble in cold hydrochloric, sulfuric, or nitric acids, and only slightly soluble in hot hydrochloric or sulfuric acid. The metal is readily soluble in aqua regia, hydrofluoric acid, hot phosphoric acid, and various organic acids.

Vlamis and Pearson,[2] employing radioactive zirconium, were able to demonstrate that plants absorbed the metal from soils, however poorly. Presumably plant roots release organic acids which act upon the metals. Zirconium complexes with acetate, citrate, formate, glycollate, oxalate and tartrate. The metal is found in the leaves of trees, in vegetables, and in foods, possibly as chelates.

Zirconium was found in the blood of an organism *(dinoflagellate gninodinium brevis)* which caused the 'red tide' of the Florida gulf in the 1950s.[3]

Information regarding the deposition of the metal in plant, animal, and human tissues is sparse. Lounamaa's investigations of zirconium contents found in vegetation in Finland[4] have been cited repeatedly, as have the classical studies of Schroeder and Balassa on human and animal tissues,[5] and the more recent work of Hamilton et al.[6]

As with other trace elements which have been discussed, the findings reported are dependent upon differences in the population investigated as well as in the analytical techniques employed. Schroeder and Balassa,[5] using a spectrophotometric method involving measurement of the zirconium-chloranilic acid color complex, obtained higher values by a factor of more than

100 in some instances than did Hamilton *et al.*,[6] who determined zirconium by spark source mass spectrometry. For example, whole-blood zirconium concentrations reported by the former ranged between 0.4 and 1.25 μg g^{-1} (wet weight) while the latter investigator obtained values from 0.012 to 0.028 μg g^{-1}. Zirconium apparently is concentrated in the erythrocytes.

Hamilton *et al.*[6] stated that zirconium concentrations in brain, kidney, liver, lung and muscle ranged between 0.01 and 0.06 μg g^{-1} (wet weight), and levels found in lymph nodes varied between 0.3 and 0.6 μg g^{-1}. Schroeder and Balassa reported concentrations of 0.93–2.04, 0.65–5.39, 1.35–9.19, and 1.18–7.49 μg g^{-1} for brain, kidney, liver and lung, respectively. Muscle averaged 2.63 μg g^{-1}, and fat, 18.7 μg g^{-1}.

Zirconium has been found in all foods examined. Hospital diets were estimated to provide 3.49 mg of zirconium daily.[5]

Additional and quite detailed studies regarding the metabolism of zirconium in the animal body would be desirable indeed. Since the metal appears to be virtually absent from the urine, it is assumed that the intestine serves as the major excretory organ. Sites of absorption, retention and excretion of zirconium are yet to be delineated.

TOXICITY OF ZIRCONIUM

Animal studies indicate that zirconium has a low order of toxicity whether administered orally,[7,8] or injected.[8] The sulfate salt administered to mice in rather high concentration (5 ppm in drinking water) appeared to have little effect upon growth and did not prove to be carcinogenic in nature. A slight reduction in survival time was noted, however.[7]

Systemic toxicity due to contact with zirconium and its compounds in industry has not been reported. Allergic reactions to the salts incorporated into skin lotions and in deodorant preparations have been documented, however. A hypersensitivity reaction resembling that induced by beryllium has been reported to occur following use of zirconium-containing lotions prescribed for the relief of poison oak exposure.[9] Pulmonary granulomata have been induced in rabbits following exposure to zirconium lactate mists.[10] Deodorant preparations containing zirconium have been implicated as the cause of axillary granulomas of allergic origin.[11]

USES OF ZIRCONIUM

Zirconium, being highly resistant to corrosion and heat, is used as a shielding material in nuclear power reactors and in nuclear submarines. Its compounds also find application in the glass and ceramics' industries, in the production of flash bulbs, tracer bullets and detonators, and in paint manufacture. The chloride and acetate are employed as water repellant agents for textiles and papers. Zirconium sulfate is often used as a tanning agent in lieu of chromium salts.

Zircon, or zirconium silicate, in addition to being considered a semi-precious stone, is utilized in abrasives and polishers, in refractories and in foundry sands.

ANALYSIS OF ZIRCONIUM

Colorimetric–fluorimetric

As expected, colorimetric and fluorimetric methods for zirconium analysis lack specificity.

Chloranilic acid in a strongly acid medium yields a magenta-colored complex with zirconium.[12] p-Dimethylaminoazophenylarsonic acid also complexes zirconium.[13] Hafnium, uranium, thorium, tin, titanium, molybdenum, iron and copper also form colored complexes with these reagents. Incomplete removal of the latter two cations may account to a marked degree for the higher values obtained in analyses of biological materials by this method.

Quercitin in an acidic alcoholic medium yields a strong yellow color with zirconium. Antimony, chromium, tin, tungsten and molybdenum react in the same way with a sensitivity comparable to that of zirconium. Iron acts as an interference too.

Morin is a preferred reagent for fluorescence analysis of zirconium.[14] Its advantage lies in the fact that in a 2 M hydrochloric acid solution metals like arsenic, beryllium, bismuth, cadmium, copper, iron, lead, magnesium, manganese, mercury, molybdenum, nickel, thallium, selenium and zinc in concentrations as high as 8 mg per 100 ml do not fluoresce.

Aluminum, antimony and tin will form fluorescent complexes with morin in an acidic medium. However, zirconium can be estimated in their presence following the addition of sodium ethylenediaminetetra-acetate (EDTA). Zirconium fluorescence is quenched by EDTA, but that of the other metals remains unchanged. Morin has also been employed as an absorptimetric reagent.

Interference due to the brown-colored compound resulting from the reaction of morin and ferric iron can be removed by reducing the iron to the ferrous state with mercaptoacetic acid.

Alizarin and related hydroxyquinolines have been proposed as reagents for analysis.[15]

Zirconium as a thenoyltrifluoroacetonate can be extracted from a mineral acid medium by extraction with an organic solvent.[16-17] The metal is extractable in the presence of trifluoroacetyl acetone[18] and related compounds.[19-21]

Various acetylacetonates of zirconium being volatile compounds have been studied by gas chromatography.[22] To date applications of gas chromatographic analysis to real-life samples have not been described. However, the prospect of coupling a gas chromatograph with a spectroscopic instrument for such analysis does stimulate the imagination.

Activation analysis

Activation analysis has been applied to analyses of zirconium contents in various matrices.

Koch and Roesmer[23] estimating the metal contents in different meats found hafnium to be the major interference and suggested chemical separations to improve activation analysis.

Brooks[24] subsequently determined zirconium in rocks following separation by chelation–extraction. Other applications of activation analysis to the analysis of geological samples were reported more recently.[25,26]

Emission spectroscopy

Early investigators reported experiencing difficulties with estimation of zirconium by spectrographic analysis. Ahrens[27] considered zirconium the most difficult of all elements to volatize in an emission spark spectrograph and hence feared that analytical results might be too low due to incomplete volatilization. Brode[28] failed to discover zirconium in certain conditions because of the close proximity of the intense lines of iron, copper, cobalt, molybdenum, nickel and vanadium.

Lounamaa's investigation in 1956 of zirconium contents of plants in Finland were done by emission spectrography.[4]

A rather interesting approach to the analysis of zirconium in minerals and ores by DC arc spectrography was described more recently by Moroshkina and Vanaeva.[29] Samples were dissolved in hydrofluoric-sulfuric acids, treated with EDTA and passed through an anion exchange resin which was ashed subsequently, mixed with graphite, and subjected to instrumental analysis.

X-ray fluorescence spectroscopy

X-ray fluorescence spectroscopy has been applied to a limited degree to the analysis of zirconium and other metals in geological samples[30] and waters.[31] Applications to biological matrices have not been reported as yet.

Atomic absorption spectrometry

Detection limits at the 360.1 nm resonance line for zirconium by flame atomic absorption spectrometry are stated to be 5 μg ml^{-1}.[32] More favorable detection limits by emission spectrometry of 2 μg ml^{-1} and 5 ng ml^{-1} have been reported for a spark and plasma source, respectively.[33]

To date applications of atomic absorption spectrometry or plasma spectrometry to analyses of biological samples have not been described. When indicated, the problem may be approached by combining a chelation–extraction technique with flameless atomization in either the atomic emission or atomic absorption mode.

REFERENCES

1. H. J. M. Bowen, *Trace Elements in Biochemistry*, Academic Press, New York, 1966.
2. J. Vlamis and G. A. Pearson, *Science*, **111**, 112 (1950).
3. A. Collier, *Science*, **118**, 329 (1953).
4. J. Lounamaa, *Suomal. elain-ya Kasvit. Seur. van. Julk* **29** (4) (1956) cited in H. J. M. Bowen, *Trace Elements in Biochemistry*, Academic Press, New York, 1966.
5. H. A. Schroeder and J. J. Balassa, *J. Chronic Dis.*, **19**, 573 (1966).
6. E. J. Hamilton, M. J. Minski and J. J. Clearly, *Sci. Total Environ.*, **1**, 341 (1972–73).
7. H. A. Schroeder, M. Mitchener, J. J. Balassa, M. Kanisawa and A. P. Nason, *J. Nutr.*, **95**, 95 (1968).
8. J. Schubert, *Science*, **105**, 389 (1947).
9. W. L. Epstein and J. R. Allen, *J. Amer. Med. Assoc.*, **190**, 940 (1964).
10. J. T. Prior, G. P. Cronk and D. B. Ziegler, *Arch. Environ. Health*, **1**, 297 (1960).
11. D. Kleinhaus and W. Knoth, *Dermatologica*, **152**, 161 (1976).
12. B. J. Thamer and A. F. Voigt, *J. Amer. Chem. Soc.*, **73**, 3197 (1951).
13. F. S. Grimaldi and C. E. White, *Anal. Chem.*, **25**, 1886 (1953).
14. R. A. Geiger and E. B. Sandell, *Anal. Chim. Acta*, **16**, 346 (1957).
15. E. B. Sandell, *Colorimetric Determination of Traces of Metals*, Interscience Publishers, New York, 1959.
16. F. L. Moore, *Anal. Chem.*, **28**, 997 (1956).
17. K. Alcock, F. C. Bedford, W. H. Harwick and H. A. C. McKay, *J. Inorg. Nuclear Chem.*, **4**, 100 (1957).
18. K. L. Lawson and M. Kahn, *J. Inorg. Nuclear Chem.*, **5**, 87 (1957).
19. B. G. Schultz and E. M. Larsen, *J. Amer. Chem. Soc.*, **72**, 3610 (1950).
20. T. B. Pierce and P. F. Peck, *Anal. Chim. Acta*, **27**, 392 (1962).
21. J. Stary and E. Hladky, *Anal. Chim. Acta*, **28**, 227 (1963).
22. R. W. Moshier and R. E. Sievers, *Gas Chromatography of Metal Chelates*, Pergamon Press, New York, 1965.
23. R. C. Koch and J. Roesmer, *J. Food Sci.*, **27**, 309 (1962).
24. C. K. Brooks, *Radiochem. Acta*, **9**, 157 (1968).
25. H. Jaffrezic, A. Decarreau, J. P. Carbromnel and N. Deschamps, *J. Radioanal. Chem.*, **18**, 49 (1973).
26. R. Delmas, J. N. Barrandon and J. L. Debrun, *Analusis*, **4**, 339 (1976).
27. L. H. Ahrens, *Spectrochemical Analysis*, Addison-Wesley, Cambridge, Mass., 1950.
28. W. R. Brode, *Chemical Spectroscopy*, 2nd edn, John Wiley, New York, 1958.
29. I. M. Moroshkina and L. V. Vanaeva, *Zh. Anal. Khim.*, **25**, 2374 (1970) cited in M. Pinta, *Modern Methods for Trace Element Analysis*, Ann Arbor Science, Ann Arbor, Mich., 1978.
30. M. J. Kaye, *Geochim-Cosmochim. Acta*, **29**, 139 (1965).
31. A. Heres, O. Girard-Devasson, J. Gaudet and J. C. Spurg, *Analusis*, **1**, 408 (1972).
32. D. C. Manning, *Atomic Abs. Newsletter*, **5**, 127 (1966).
33. M. Pinta, *Modern Methods for Trace Element Analysis*, Ann Arbor Science, Ann Arbor, Mich., 1978.

ANALYTICAL PROCEDURES FOR ATOMIC ABSORPTION SPECTROMETRY

INTRODUCTION

The literature is replete with descriptions of proposed procedures for toxic metal analysis by atomic absorption spectrometry. Criteria for acceptable techniques have changed with increased understanding of instrumental capabilities and limitations, of the effect of matrix constituents upon absorbance measurement, and of the chemistry of the analyte in question.

Flameless sampling devices being more than ten times as efficient as the conventional flame atomizer have extended the detection capabilities of atomic absorption instrumentation. However, flameless atomization has not eliminated the need for chemical pretreatment of samples prior to instrumental analysis. The nonspecific molecular absorption generated by constituents in a matrix greatly enhances the absorbance signal due to the analyte in question. Furthermore, components in the untreated biological matrix are not removed entirely in the drying and charring cycles of flameless atomization. Background correction is only partially effective in eliminating the nonspecific molecular absorption generated during the atomization cycle by the residue remaining after charring, because trace metals are present in microgram amounts or less while matrix components, like sodium or protein are present in milligram to gram amounts.

Standards and unknowns must be in similar matrices for instrumental analyses. The relative viscosities of sample and standard solutions is a factor which must be considered. For example, blood serum being more viscous than water is aspirated more slowly into a flame atomizer. Furthermore, measuring devices such as pipets do not deliver equivalent volumes of materials of different viscosities to a flameless atomizer. Organic solvents enhance the absorbance of metallic elements in a flame. The absorbance of a standard concentration of lead or copper, for example, in methylisobutyl ketone is four times as great as that of the same concentration in water. Since solvents evaporate during the drying and charring cycles, this phenomenon of organic solvent enhancement is not noted during flameless atomization.

The chemical form of the analyte influences its behavior in an analytical system. Does lead, bound to a blood protein, atomize as readily as does lead nitrate? Comparing aliquots of blood samples, diluted with water, or a surfactant solution with aqueous lead nitrate standard is rather open to question, is it not? True, results reportedly so obtained by flameless atomization are precise, but what of their accuracy? What is actually being measured, lead or . . .?

Blood proteins can be removed by precipitation with trichloroacetic acid or else by wet or dry digestion. The choice is dependent upon the behavior of the analyte. For example cadmium, being loosely bound to blood proteins, is released by trichloroacetic acid precipitation. On the other hand, gold is tightly bound to both *in vivo* and *in vitro*. Samples must be ashed in order to recover gold. Only 5% of gold in blood is recovered following trichloroacetic acid precipitation of proteins.

Tissues, hair, nails and other solid materials should be wet or dry ashed in order to eliminate major organic constituents. Concentration of the analyte, if necessary, can then be achieved by application of a suitable chelation–extraction technique.

Direct analysis of untreated biological materials can at best be considered to be but a semiquantitative method—if that.

Irrespective of the atomizing device employed (flame or flameless) nonspecific molecular absorbance signals generated by urinary components (sodium potassium, urea, etc.) completely overwhelm absorbance signals from the metal in question. Background correction alone eliminates only a minimal portion of the signal generated by the untreated urine matrix. Employing chelation–extraction methods for the estimation of urinary trace metals contents can serve to both concentrate the analyte and to eliminate matrix effects.

The procedures described in the pages which follow are methods used routinely in the author's laboratory.

ALUMINUM

Minimal chemical pretreatment of some matrices is required before instrumental analysis. The chloride and nitrate are deliquescent. Practical detection limits by flameless atomization are favorable, being 1 pg ml^{-1}, compared with normal serum concentrations of 1 ng ml^{-1}, for example.

Maintaining a clear analytical environment represents the greatest challenge of the analysis. Aluminum is ubiquitous in a laboratory environment, being a contaminant in reagents and laboratory ware, both glass and plastic. The latter must be precleaned with 5% nitric acid and rinsed copiously in deionized water before use.

Aluminum in serum technique

1. Place 0.2 ml aliquots of deionized water into a series of clean polypropylene tubes.
2. Add 0.1 ml serum or standards (0–1 μg ml^{-1}) to each. Mix.
3. Compare 20 μl aliquots at the 309.3 nm line. The graphite furnace is programmed to dry at 100 °C for 40 s, char at 1500 °C for 120 s, and atomize at 2650 °C for 6 s.
4. Background correction is necessary.

Aluminum in cerebrospinal fluid (CSF)

Dilution is not required for CSF. 20 μl aliquots of clear spinal fluid can be compared with aqueous standards directly as described above.

Aluminum in urine

1. Into a series of precleaned borosilicate glass centrifuge tubes fitted with polypropylene screw caps, place 5 ml aliquots of urine and standards ranging between 0 and 0.2 μg ml^{-1}.
2. Adjust pH to 5–6.
3. Add 1 ml of 2% aqueous cupferron and 1 ml of methylisobutyl ketone (MIBK). Cap and shake 2 min. Centrifuge for 5 min.
4. Compare organic phases and unknowns in the graphite furnace as described above.

Aluminum in tissues

1. Wet ash 0.5–1 g of tissue with 1 ml of sulfuric acid and 20 ml or more of nitric acid. Add further nitric acid as required during digestion until the digest remaining, after the appearance of sulfur dioxide fumes, is crystal clear.

   A series of standards, ranging between 0 and 1 μg ml^{-1}, are prepared through the entire digestion procedure simultaneously.
2. Dilute digest to 25 ml with deionized water. Place 10 ml aliquots into a series of 60 ml separatory funnels.
3. Adjust pH to 5 with 2.5 N NaOH.
4. Add 1 ml of 2% aqueous cupferron and 2 ml of MIBK. Shake manually for 2 min or mechanically for 10 min.
5. Allow funnel contents to equilibrate. Remove the organic phase. Centrifuge.
6. Proceed as described in the procedure above.

Calibration procedure
Stock standard

1 ml $= 1$ mg $= 1000$ ppm $= 1000$ μg of aluminum
Prepare from aluminum sulfate—$Al_2(SO_4)_3$—and store in a Teflon bottle. 636.3 mg of aluminum sulfate are equivalent to 100 mg of aluminum, so the standard contains 6.362 g l^{-1} of $Al_2(SO_4)_3$.

Quality preparations are available commercially in the US.

Working standards
Are prepared daily by dilution of the stock standard.

ANTIMONY
Antimony in solution at pH 5.5–6.5 is chelated by sodium diethyldithiocarbamate (NDDC) and extracted into methyl isobutyl ketone (MIBK). Concentrations are determined by comparing standards and unknowns at the 206.8 nm resonance line.

Antimony in blood
1. Place 6 ml aliquots of chelating solution buffered to pH 6.5 into a series of boroscilicate tubes with polypropylene screw caps.
2. Add 2 ml of heparinized blood, or standards ranging between 0 and 1 ppm. Rinse pipet up and down. Add 2 ml of MIBK. Cap and invert contents to mix. Set aside for about 10 min. Then shake tubes for about 2 min. Centrifuge.
3. Compare organic phases of unknowns and standards at the 206.8 nm resonance line. Use 10 μl aliquots. The graphite furnace is programmed to dry at 100 °C for 20 s, char at 300 °C for 20 s, and atomize at 2400 °C for 6 s. Background correction is necessary.

Antimony in urine
1. Place 5 ml of urine, or standards ranging between 0 and 0.5 μg ml^{-1}, in a series of borosilicate tubes as described in the blood procedure.
2. Adjust pH to between 5.5 and 6.5.
3. Add 1 ml of 1% aqueous sodium diethyldithiocarbamate and 2 ml of MIBK. Cap. Invert to mix. Set aside for 10 min, shake for 2 min, then centrifuge.
4. Compare 10 μl of solvent extracts of standards and unknowns at the 206.8 nm line as described in the blood procedure.

Antimony in tissues

1. Begin wet digestion of 1–5 g of tissue with 2 ml of sulfuric acid and 20 ml of nitric acid. Add more nitric acid as required to render the digestant, remaining after appearance of sulfur dioxide fumes, crystal clear.

   A series of standards ranging between 0 and 1 μg ml^{-1} ml prepared simultaneously throughout the entire digestion procedure.

   Note: If lead is to be determined on the same digestant, perchloric acid should be substituted for sulfuric acid during the digestion process. Careful monitoring is required, however. Do not permit the flask contents to char as they may then explode. Add nitric acid as the contents begin to darken.
2. When digestion is complete, dilute digests to 25 ml with deionized water.
3. Place 10 ml aliquots into a series of 60 ml cylindrical separatory funnels.
4. Adjust pH to between 5.5–6.5 with 2.5 N NaOH. Add 1 ml of 2% NDDC and 2 ml of MIBK. Shake manually for 2 min or mechanically for 10 min.
5. Allow funnel contents to equilibrate. Separate phases. Centrifuge solvent phase.
6. Compare aliquots of the solvent phase of standards and unknowns at 206.8 nm as described previously.

Calibration procedure

Stock standard

1 ml = 1 mg = 1000 μg of antimony

Prepare from antimony potassium tartrate—$C_4H_4KO_7Sb$. 281.6 mg are equivalent to 100 mg of antimony, i.e. the standard contains 2.816 g l^{-1} of $C_4H_4KO_7Sb$.

Store in Teflon.

Working standard

Prepare standards of 0, 0.01, 0.05, 0.1, 0.2, 0.5, 1.0 ppm (0–100 μg%) from the stock standards. Make fresh daily.

Calibration standards are processed concurrently with the unknowns. Matrices of standards and unknowns are identical.

Reagents

Buffered chelating solution

Add the following together in a 1000 ml volumetric flask:
300 ml of 2% NDDC
120 ml of 1% Triton X-100
100 ml of acetate buffer pH 6.5

Dilute to 1000 ml with distilled water. Store in polyethylene. The solution is stable for two weeks. Filter or centrifuge to remove particulates which may form.

Acetate buffer pH 6.5

Place 150 ml of 2.5 mol l^{-1} sodium acetate into a 250 ml volumetric flask. Add 2.5 mol l^{-1} acetic acid until a pH of 6.5 is obtained.

To prepare 2.5 mol l^{-1} sodium acetate, dissolve 340 g of sodium acetate in distilled water and dilute to 1 l. To prepare 2.5 mol l^{-1} acetic acid, dilute 144 ml of glacial acetic acid (99–100%) to 1 l.

ARSENIC

Atomic absorption procedure

Arsenic in solution at pH 5.5–6.5 is chelated by sodium diethyldithiocarbamate (NDDC) and extracted into methylisobutyl ketone (MIBK). Concentrations are determined by comparing standards and unknowns at the 193.7 nm resonance line.

Arsenic in blood

1. Place 10 ml of 5% trichloroacetic acid (TCA) into a series of round bottom centrifuge tubes, glass or polypropylene.
2. Add 5 ml of heparinized blood or standard (0, 0.001, 0.005, 0.01, 0.02 μg ml^{-1}). Mix well by stirring with a glass rod. Set aside for approximately an hour. Stir occasionally. Centrifuge.
3. Decant the supernatants into a series of 60 ml cylindrical separatory funnels.
4. Add in addition 10 ml aliquots of 5% TCA to the precipitate in the centrifuge tubes. Mix. Centrifuge. Decant washing into appropriate funnels.
5. Adjust pH to 5.5–6.5 with 2.5 N sodium hydroxide. Add 1 ml of 2% aqueous NDDC and 2 ml of MIBK.
6. Cap and shake funnels. Permit funnel contents to equilibrate. Remove solvent layer and centrifuge.
7. Compare 20 μl aliquots of standards and unknowns at the 193.7 nm line in an atomic absorption spectrophotometer equipped with a graphite furnace. The light source must be an electrodeless discharge lamp. Background correction is necessary.
8. Programme the graphite furnace to dry at 100 °C for 40 s, char at 100 °C for 40 s, and atomize at 2500 °C for 6 s.
9. Since graphite furnace atomization of arsenic is subject to memory, possible problems with inter-sample contamination can be avoided by burning off probable residues between samplings.

Urine arsenic

1. Adjust pH of urine to between 5.5 and 6.5.
2. Place 20 ml aliquots of urine and appropriate standards into a series of separatory funnels.
3. Add 1 ml of 2% NDDC and 2 ml of MIBK.
4. Proceed as described in the blood procedure.

Arsenic in miscellaneous fluids

Proceed as with urine.

Arsenic in tissues

1. Place 1–5 g of tissue, or appropriate quantities of standard solutions, into a series of 100 ml micro-Kjeldahl digestion flasks. Include a blank.
2. Add 2 ml sulfuric acid, 20 ml of nitric acid, and 3 glass boiling beads. Mix cautiously. Allow to stand for 4 hours, or overnight, to permit tissues to soften and disintegrate somewhat.
3. Mix flask contents to permit evolution of collected gases. Begin heating flask contents over a low flame. Then increase somewhat as boiling commences.
4. As char point is reached, add more nitric acid and continue boiling. Repeat process as needed until the digest becomes crystal clear. Cool.
5. Transfer digests (quantitatively) to a series of 25 ml volumetric flasks and make up to volume with distilled water.
6. Place 10–15 ml aliquots a series of 60 ml cylindrical separatory funnels. Adjust pH to 5.5–6.5 with 2.5 N NaOH. Add 1 ml of 2% NDDC and 3 ml of MIBK.
7. Proceed as described in previous arsenic procedures.

Calibration procedure

Stock standard

1 ml = 1 mg = 1000 μg of arsenic.

Prepare from arsenic pentoxide—As_2O_5. 469.8 mg are equivalent to 100 mg of arsenic so the stock standard contains 4.698 g l^{-1} As_2O_5. Store in Teflon.

Working standards

Prepare standards of 0, 0.001, 0.005, 0.010, 0.020 ppm (0–20 μg%) from the stock standard. Make fresh daily.

BARIUM

Barium in solution at pH 6–8 is chelated by thenoyltrifluoroacetone and extracted into methylisobutyl ketone (MIBK). Concentrations are determined by comparing standards and unknowns at the 553.6 nm resonance line.

Barium in blood

About 10 ml of heparinized blood are required for the analysis usually.
1. Place 10 ml of 5% trichloroacetic acid (TCA) into a series of 20 ml round bottom centrifuge tubes. Use 2 tubes per specimen.
2. Add 5 ml aliquots of heparinized blood or standards to each. Mix well by stirring with a glass rod. Set aside for about an hour. Stir occasionally. Centrifuge.
3. Combine the supernatants from each sample into a single separatory funnel. Add 10 ml aliquots of 5% TCA to the precipitates remaining in the centrifuge tubes. Mix. Centrifuge. Decant washings into appropriate funnels.
4. Adjust pH to between 6 and 8 with 2.5 N NaOH. Add 2 ml of 2% thenoyltrifluoroacetone in methanol and 3 ml of MIBK.
5. Cap and shake funnels as suggested. Permit funnel contents to equilibrate. Remove solvent layer. Centrifuge.
6. Compare 20 μl of standards and unknowns at the 553.6 nm line in an atomic absorption spectrometer equipped with a graphite furnace. Background correction is necessary.
7. Programme the graphite furnace to dry at 100 °C for 40 s, char at 100 °C for 40 s, and atomize at 2500 °C for 6 s.

Barium in urine

1. Adjust pH of urine to between 6 and 8.
2. Place 20 ml aliquots of urine and appropriate standards into a series of separatory funnels.
3. Add 2 ml of 2% methanolic theoyltrifluoroacetone and 3 ml of MIBK.
4. Proceed as described in the blood procedure.

Barium in tissues

1. Place 5–10 g of tissue, or appropriate quantities of standard solutions into a series of 100 ml micro-Kjeldahl digestion flasks.
2. Add 30 ml of nitric acid, 2 ml of perchloric acid and 3 boiling beads. Mix cautiously. Allow to stand for at least 4 hours before beginning to heat.
3. Watch digestion process closely. Add more nitric acid as the char point is approached. Continue until digestion is complete.

5. Quantitatively transfer digests to 25 ml volumetric flasks and make up to volume with distilled water.
6. Place 15 ml aliquots into a series of 60 ml separatory funnels. Adjust pH to 6–8. Add 2 ml of 2% methanolic thenoyltrifluoroacetone and 3 ml of MIBK.
7. Proceed as described for barium in blood procedure.

Calibration procedure
Stock standard
1 ml = 1 mg = 1000 μg of barium.

Prepare from barium nitrate—$Ba(NO_3)_2$–190.3 mg which is equivalent to 100 mg of barium so that the stock standard contains 1.903 g l^{-1}. Store in Teflon.

Working standards
Standards of 0, 0.001, 0.005, 0.01, and 0.02 μg ml^{-1} (0–20 μg%) are made fresh daily.

BERYLLIUM

Beryllium in solution at pH 6–6.5 is chelated by cupferron and extracted into methylisobutyl ketone (MIBK). Unknowns and standards are compared at the 234.8 nm resonance line.

Beryllium in blood
Approximately 10 ml of heparinized blood are required for the analysis.
1. Precipitate 5 ml of heparinized blood with 10 ml of 5% trichloroacetic acid (TCA). Set up two tubes for each sample.
2. After centrifugation combine the supernatants for each sample into a single separatory funnel.
3. Wash precipitates with additional aliquots of TCA and add washings to appropriate funnels as described in the previous procedures.
4. Adjust funnel contents to pH 6–6.5 with 2.5 N NaOH. Add 1 ml of freshly prepared 2% aqueous cupferron and 3 ml of MIBK to each, cap the funnels, then shake. After funnel contents equilibrate, remove solvent phases and centrifuge.
5. Carry a series of standards ranging between 0 and 1 ng ml^{-1} (0.1 μg%) through the procedures concurrently.
6. Compare standards and unknowns (20 μl) at the 234.8 nm resonance line in an atomic absorption spectrophotometer equipped with a graphite furnace. Background correction is not necessary since the beryllium signal

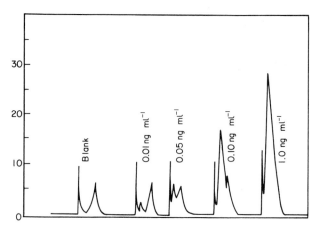

Fig. A.1. Beryllium cupferrate (flameless AA).

appears a few seconds after the non-specific molecular absorption signal (Fig. A.1).

7. Programme the graphite furnace to dry at 100 °C for 20 s, char at 300 °C for 20 s, and atomize at 2400 °C for 6 s.
8. Burning off of probable residues between samples is advisable. The graphite tube has been observed to be subject to memory effects following atomization of beryllium.

Beryllium in urine

1. Adjust pH to between 6 and 6.5
2. Place 20 ml aliquots of urine and appropriate standards (0–10 ng per 100 ml) into a series of separatory funnels.
3. Add 1 ml of 2% cupferron and 3 ml of MIBK. Proceed as described in the blood method above.

Beryllium in tissues

1. Place 1–5 g of tissue, if available, or appropriate quantities of standard solutions into a series of 100 ml micro-Kjeldahl digestion flasks.
2. Add 2 ml of concentrated sulfuric acid 30 ml of nitric acid and boiling beads. Allow to stand for 4 hours, or overnight, to dissolve tissues.
3. Mix the flask contents cautiously, but well, before applying heat.
4. When mixture approaches char point, add more nitric acid. Continue nitric acid additions until digestion is complete.
5. When digest is cool, quantitatively transfer to a 25 ml volumetric flask and dilute to volume with distilled water.
6. Place 15 ml aliquots into a series of 60 ml separatory funnels. Adjust pH to 6–6.5. Add 1 ml of 2% cupferron and 2 ml of MIBK.
7. Proceed as described above.

Calibration procedure

Stock standard

1 ml = 1 mg = 1000 μg of beryllium.

Prepare from beryllium sulfate. 1.1659 g of $BeSO_4$ is equivalent to 100 mg of beryllium, so the stock standard contains 11.659 g l^{-1}. Store in Teflon.

Working standards

Working standards are prepared the day of use and in the range the problem(s) of the moment indicate. Standards ranging between 0 and 100 ng per 100 ml are suggested.

BISMUTH

Bismuth in solution at pH 5.5–6.5 is chelated by dithiocarbamate and extracted into methylisobutyl ketone (MIBK). Unknowns and standards are compared at the 223.1 nm resonance line.

Bismuth in blood

1. Place 10 ml aliquots of 5% trichloroacetic acid (TCA) into a series of 20 ml centrifuge tubes.
2. Add 5 ml of heparinized blood or standard solutions (0, 20, 40, 60, 80, 100 μg per 100 ml). Mix with a stirring rod or by vigorous shaking. Set aside for an hour. Stir occasionally. Centrifuge for 10 min.
3. Decant supernatants into appropriately marked 60 ml separatory funnels.
4. Add 10 ml of 5% TCA to the precipitates in the centrifuge tubes. Mix well. Centrifuge. Add washings to funnel contents.
5. Adjust pH to 5.5–6.5 with 2.5 N NaOH.
6. Add 1 ml of 2% sodium diethyldithiocarbamate (NDDC) and 2 ml of MIBK. Cap and shake funnels.
7. Permit contents to equilibrate and remove the solvent layer. Centrifuge.
8. Compare 10 μl aliquots of standard and unknown extracts at the 223.1 nm resonance line. Programme the graphite furnace to dry at 100 °C for 20 s, char at 300 °C for 20 s and atomize at 2400 °C for 6 s. Background correction is necessary.

Bismuth in urine

1. Adjust 10 ml aliquots of urine or standard solutions to pH between 5.5 and 6.5, and place into a series of screw capped round bottom centrifuge tubes.
2. Add 1 ml of 2% NDDC and 2 ml of MIBK. Shake tube contents and centrifuge.

3. Compare solvent extracts of standards and unknowns as described in the blood bismuth procedure.

Bismuth in tissue

1. Place 1–5 g of tissue or 5 ml aliquots of standard solutions ranging between 0 and 1 μg ml^{-1} bismuth into a series of 100 micro-Kjeldahl digestion flasks.
2. Add 2 ml of concentrated sulfuric acid and 20 ml of nitric acid and boiling beads. Proceed with wet digestion as described previously.
3. When digestion is complete, quantitatively transfer crystal clear digest to 25 ml volumetric flasks with distilled water and dilute to volume.
4. Place 15 ml aliquots into a series of separatory funnels. Adjust to proper pH with 2.5 N NaOH. Add 1 ml of 2% NDDC and 2 ml of MIBK. Shake funnels for 2 min manually or 10 min mechanically.
5. After funnel contents equilibrate, remove solvent layer, centrifuge and compare standard and unknown extracts as described above.

Calibration procedure

Stock standard

1 ml = 1 mg = 1000 μg of bismuth.

Weigh 232.3 mg of bismuth nitrate—$Bi(NO_3)_3 \cdot 5 \ H_2O$—into a 100 ml volumetric flask. Add 5% nitric acid to dissolve the salt and dilute the flask contents to volume with distilled water. Store in Teflon.

Working standards

Working standards of 0, 0.005, 0.01, 0.05, 0.1, 0.3, 0.5, and 1 μg ml^{-1} (0–100 μg%) are prepared from the stock solution the day of use.

CADMIUM

Proteins are removed by precipitation with trichloroacetic acid (TCA) or are destroyed by wet ashing. Cadmium in solution is chelated at pH 6.5–7 by sodium diethyldithiocarbamate (NDDC) and extracted into methylisobutyl ketone (MIBK). Concentrations are determined by comparing standards and unknowns at the 228.8 nm resonance line.

Cadmium in blood

1. Place 10 ml aliquots of 5% TCA into a series of round bottom centrifuge tubes.
2. Add 5 ml of well mixed heparinized blood. Mix well by stirring with a glass rod. Set aside for approximately an hour. Stir occasionally. Centrifuge for 10 min.

3. Set up standards of 0, 0.01, 0.03, 0.05, 0.1, 0.3, 0.5, and 1 μg ml^{-1} concurrently.
4. Decant the supernatants into appropriately marked separatory funnels.
5. Wash the precipitates with 10 ml aliquots of 5% TCA. Centrifuge. Decant washings into respective funnels.
6. Adjust the pH to 6.5–7 with 2.5 N NaOH. Add 1 ml of 2% NDDC and 3 ml of MIBK. Shake.
7. When funnel contents equilibrate, remove the solvent layers and centrifuge.
8. Compare unknowns and standards by flame atomization at the 228.8 nm line.
9. Use a single slot burner head since the optimal flame is lean and oxidizing. Air and acetylene flow rates of 5–5.5 l min^{-1} and 4–4.5 l min^{-1}, respectively are suggested. When aspirating MIBK, that portion of the flame directly above the burner head should be bluish and rather featherlike. Adjust the burner height as required to attain optimal sensitivity. With a Perkin-Elmer instrument the height is set at 1.
10. Adjust to zero while aspirating MIBK. Aspiratory rate should be 1–3 ml min^{-1}.

Cadmium in urine

1. Adjust pH of urine specimens to between 6.5 and 7. Place 10 ml aliquots into a series of 20 ml round bottom screw capped borosilicate centrifuge tubes.
2. Add 1 ml of 2% NDDC and 3 ml of MIBK.
3. Treat standards, ranging between 0 and 1 μg ml^{-1} (0–1000 μg l^{-1}), similarly.
4. Proceed as described in the blood procedure.

Cadmium in tissue

1. Place 1–5 g quantities of tissue, or 5 ml aliquots of the standard solutions described, into a series of 100 ml micro-Kjeldahl digestion flasks.
2. Add 2 ml of concentrated sulfuric acid, 20 ml of nitric acid and pyrex boiling beads. Allow to stand for 4 hours, or overnight, to dissolve tissue. Proceed with wet digestion. Add nitric acid as required.
3. When digestion is complete, quantitatively transfer the crystal clear digests to 25 ml volumetric flasks with distilled water and dilute to volume.
4. Place 5 ml aliquots into separatory funnels. Adjust pH to 6.5–7 with 2.5 N NaOH.
5. Add 1 ml of 2% NDDC and 3 ml of MIBK. Shake to extract.
6. Proceed as described in the blood procedure above.

Cadmium analysis in the graphite furnace

Analysis of extracted dithiocarbamate chelates of cadmium in the graphite furnace is not practical. One observes a steady decrease in the atomization signal with repeated samplings and firings of the extracts. This may be due to some sort of an interaction between the chelate and the graphite tube.

Such phenomena, i.e. a decrease in atomization signals, are not observed when atomizing the supernatants of TCA or nitric acid precipitation of blood samples or diluted digests of tissues. There is a caveat, however. Pipette tips are liberally contaminated with cadmium. Three to four washings with 5% nitric acid followed by at least four rinsings with deionized water are usually effective in removing the contaminant.

The following modification of the Rains' procedure for blood lead has been applied in the analysis of small samples for cadmium.

Semi-micro analysis of blood cadmium

1. Preclean pipette tips (Eppendorf) and 1.5 ml conical Teflon centrifuge tubes (Eppendorf) with 5% nitric acid. Rinse copiously with deionized water. Dry tips and tubes.
2. Place 0.4 ml water into a series of cleaned centrifuge tubes.
3. Using a 100 μl Eppendorf pipette equipped with a precleaned tip, sample 100 μl of blood or standard. Wipe tip carefully. Add contents to the deionized or distilled water in the centrifuge tubes. Carefully rinse tip. Then with the same tip add 300 μl of distilled water (100 μl $\times$ 3) to ensure complete delivery of sample. Allow to stand for 10 min.
4. Add 200μl of 25% nitric acid. Cap the tubes and vortex the contents. Allow to stand for 20 min; vortex occasionally. Centrifuge 3–5 min.
5. Compare 10 μl of the supernatants at the 228.8 nm line in the graphite furnace programmed to dry at 100 °C for 40 s, char at 300 °C for 40 s and atomize at 2400 °C for 7 s. Background correction is necessary.

Tissue cadmium in graphite furnace

Diluted digests can be compared in the graphite furnace. Background correction is necessary.

Calibration procedure

Stock standard

1 ml $=$ 1 mg $=$ 1000 μg of cadmium.

Place 185.4 mg of cadmium sulfate into a 100 ml volumetric flask. Add distilled water to dissolve. Dilute to volume. Mix. Store in Teflon.

Working standards

Working standards between 0 and 1 μg ml^{-1} (0–100 μg%) as indicated in the chelation–extraction procedure are prepared by merely diluting suitable aliquots of the stock standard with distilled water.

Working standards for the graphite furnace method are prepared by diluting aliquots of the stock standard with 0.78% sodium chloride (equivalent to 150 meq l^{-1} Na).

CHROMIUM

Depending upon the matrix, protein and other sources of specific molecular absorption are removed instrumentally, by wet digestion, or both. Standards and unknowns are compared in the graphite furnace at the 357.9 nm resonance line.

Chromium in serum

1. Dilute sera 1 : 1 with 5% trichloroacetic acid (TCA).
2. Compare 20 μl aliquots of the supernatants and standards of 0, 0.005, 0.01, 0.03, 0.05, 0.1, 0.3 μg ml^{-1}, treated similarly, in a graphite furnace programmed as follows:
   Dry at 100 °C for 60 s;
   Char at 1400 °C for 180 s;
   Atomize at 2500 °C for 7 s.
3. Background correction is necessary.

Chromium in whole blood or red cells

1. Wet ash 1 ml aliquots of blood or cells with 0.1 ml of sulfuric acid and as much nitric acid as is required to obtain a crystal clear digest.
2. Treat standards ranging between 0 and 0.3 μg ml^{-1} similarly.
3. Dilute to 2 ml with distilled water.
4. Compare 20 μl aliquots in the graphite furnace programmed as described previously.

Chromium in urine

Chromium in urine samples can be determined directly in the graphite furnace as described above. Sodium, the major interference, is lost during the 3 min charring cycle. Background correction is advisable, however.

Chromium in tissue

Wet ash 0.5–5 g of tissue with sulfuric acid and nitric acids. Dilute digestants to 10 or 25 ml depending upon the sample size and quantity of sulfuric acid employed. The final sulfuric acid concentration should be less than 20%. Treat standards of 0–0.5 μg ml^{-1} similarly, and compare as described above.

Calibration procedure
Stock standard

1 ml = 1 mg = 1000 μg chromium.

Place 527.1 mg chromous sulfate—CrSO$_4$ · 7 H$_2$O—into a 100 ml volumetric flask. Add water to dissolve. Dilute to volume. Mix. Store in Teflon.

Working standards
Working standards, prepared by diluting aliquots of stock with distilled water, are not stored but are made up as required.

COBALT

Cobalt in solution at pH 5.5–7 is chelated by sodium diethyldithiocarbamate (NDDC) and extracted into methylisobutyl ketone (MIBK). Unknowns and standards are compared at the 240.7 nm resonance line.

Cobalt in blood (flame)

1. Place 10 ml aliquot of 5% trichloroacetic acid (TCA) into a series of 20 ml round bottom centrifuge tubes.
2. Add 4 ml heparinized blood or standards of 0, 0.005, 0.01, 0.05, 0.10, 0.5 μg ml^{-1}. Mix by stirring with a glass rod. Allow to stand for approximately an hour. Stir occasionally. Centrifuge.
3. Decant supernatants into 60 ml cylindrical separatory funnels.
4. Add 10 ml of 5% TCA to each precipitate. Mix. Centrifuge. Add washings to appropriate funnels.
5. Adjust pH to 5.5–7 with 2.5 N NaOH. Add 1 ml of 2% NDDC and 2 ml of MIBK. Stopper funnels and shake to extract chelate.
6. After funnel contents equilibrate, remove solvent layers and centrifuge.
7. Compare unknowns and standards at the 240.7 nm line. Use a lean oxidizing flame as described in the cadmium procedure.

Cobalt in blood (flameless)

1. Extract 1 ml of blood or standard, processed as described, into 1 ml of MIBK.
2. Compare standards and unknowns (20 μl) in the graphite furnace programmed to dry at 100 °C for 20 s; char at 300 °C for 20 s, and atomize at 2400 °C for 6 s. Background correction is employed.

Cobalt in urine

1. Adjust the pH of the urine samples or standards between 5.5–7.
2. Measure 10 ml aliquots into a series of 20 ml round bottom borosilicate tubes equipped with screw caps. Add 1 ml of 2% NDDC and 2 ml of MIBK. Shake. Centrifuge.
3. Read either in flame or furnace.

Cobalt in tissue

1. Wet ash 1–5 g quantities of tissues and suitable standards in sulfuric and nitric acids as described previously.
2. Quantitatively transfer digestant to a 25 ml volumetric flask and make up to volume with distilled water.
3. Place a 10–15 ml aliquot in a separatory funnel and adjust to pH 5.5–7.
4. Proceed as described above.

Calibration procedure

Stock standard

1 ml = 1 mg = 1000 μg of cobalt.

Place 422.5 mg of cobaltous acetate tetrahydrate—$Co(CH_3COO)_2 \cdot 4 H_2O$—into a 100 ml volumetric flask. Add distilled water to dissolve. Dilute to volume. Mix. Store in Teflon.

Working standards

Working standards are not stored but are prepared as required.

COPPER

Copper in solution is chelated by sodium diethyldithiocarbamate (NDDC) at pH 6.5–7 and extracted into methylisobutyl ketone (MIBK). Standards and unknowns are compared at the 324.7 nm line.

Copper in serum (flame)

1. Into a series of suitably labelled 20 ml round bottom centrifuge tubes place 10 ml of 5% trichloroacetic acid (TCA).
2. Add 1 ml of serum. Mix well with a glass stirring rod. Allow to stand for an hour. Stir occasionally. Centrifuge for 10 min. Decant the supernatants into suitably labelled 60 ml separatory funnels.
3. Add additional 10 ml aliquots of TCA to the precipitates remaining in the centrifuge tubes. Mix well and centrifuge. Add supernatants to the appropriate separatory funnels.
4. Treat standards containing 0, 0.5, 1.0, 1.5, 2.0, and 3.0 μg ml^{-1} concurrently in a like manner.

5. Adjust pH of all funnels to 6.5–7 with 2.5 N NaOH.
6. Add 1 ml of 2% NDDC, 1 ml of 1% sodium versenate (EDTA), and 3 ml of MIBK.
7. Shake funnels for 2 min manually or 10 min mechanically.
8. Allow funnel contents to equilibrate. Discard the lower aqueous layer. Centrifuge the solvent layer. If an emulsion persists, stir with a glass rod and recentrifuge.
9. Compare standards and unknowns at the 324.7 nm resonance line for copper. Use an oxidizing flame as described in the cadmium procedure.

Serum copper (flameless)

1. Place 0.8 ml aliquots of 5% TCA into a series of 5 ml polypropylene tubes.
2. Add 0.2 ml of serum or standards. Mix by vortexing. Set aside for 30–60 min. Mix occasionally. Centrifuge.
3. Program graphite furnace to dry at 100 °C for 40 s, char at 300 °C for 40 s, and atomize at 2500 °C for 6 s.
4. Compare 10 μl of standards and unknowns. Burn off possible residues between samplings.
5. Background correction is not necessary since the copper signal appears after that due to nonspecific molecular absorption.

Copper in semen

Proceed as described in the flameless procedure for serum. Scale expansion of $\times 2$ or $\times 4$ is usually necessary.

Copper in urine (flame)

1. Place 30 ml of urine aliquots into 60 ml separatory funnels. Set up standards containing 0, 0.001, 0.003, 0.005, 0.01, 0.03, 0.05, and 0.1 μg ml^{-1} concurrently.
2. Adjust pH to 6.5–7. Add 1 ml of 2% NDDC, 1 ml of 1% EDTA and 3 ml of MIBK. Shake funnels.
3. Allow to equilibrate. Remove solvent layer and centrifuge.
4. Compare standards and unknowns.
5. Samples containing more than 1 μg ml^{-1} (1 mg l^{-1}) must be repeated on a smaller aliquot.

Urinary copper (flameless)

1. Adjust the pH of the samples and standards.
2. Place 10 ml aliquots into a series of round bottom borosilicate centrifuge tubes equipped with screw caps.

3. Add 1 ml of 2% NDDC, 1 ml of 1% EDTA, and 2 ml of MIBK. Cap tubes. Shake to extract. Allow to equilibrate. Centrifuge.
4. Compare 10 μl aliquots as described for the blood procedure.

Copper in tissues

Destroy organic matter by wet ashing in either sulfuric–nitric or perchloric–nitric acids. Though the former is a considerably slower process, the mixture requires little monitoring since there is no danger of explosion from overheating of the acids.

If fixed tissue is being analyzed, a sample of the stock fixative solution should also be analyzed.

The size of tissue sample taken is determined by its availability. Biopsy material of 50 mg or less can be adequate in cases of Wilson's disease. The usual sample of autopsy material taken in 1 g.

Procedure

1. Place tissue in a 100 ml micro-Kjeldahl digestion flask.
2. Add 20 ml of nitric acid and 1 ml of sulfuric acid if the sample is less than 0.5 g. Add 2 ml of sulfuric acid for 0.5–1 g tissue.
3. Add a few (e.g. 3) pyrex boiling beads. Mix cautiously and allow to stand for a few hours, or overnight.
4. Set up 1 ml aliquots of standards containing 0–3 μg ml^{-1} of copper.
5. Begin heating over a low flame. Too rapid evolution of gas may cause propulsion of the tissue from the flask and consequent loss of sample.
6. As the char point is reached, add more nitric acid and continue boiling.
7. Repeat the additions of nitric acid until the digestant is crystal clear.
8. If the tissue sample weighed more than 100 mg, quantitatively wash the digestant into a 25 ml volumetric flask, make up to volume with distilled, or deionized, water and take a 5 ml aliquot for the copper determination. If the sample weighted less than 100 mg, take a larger aliquot (10–15 ml).
9. Adjust pH to between 6.5 and 7 with 2.5 N NaOH.
10. Add 1 ml of 2% NDDC and 1 ml of 1% EDTA and 3 ml of MIBK.
11. Proceed as described in the urine and blood procedure.
12. Should the copper content in an unknown be greater than 3 μg ml^{-1} repeat the chelation–extraction step on a smaller aliquot.

Calibration procedure

Stock standard

1 ml = 1 mg = 1000 μg of copper.

Place 390.9 mg of cupric sulfate pentahydrate—$CuSO_4 \cdot 5H_2O$—into a 100 ml volumetric flask. Add water to dissolve. Dilute to the mark. Mix. Store in Teflon.

Dilute stock standard

1 ml = 100 μg of copper.

10 ml of stock is diluted volumetrically to 100 ml with distilled water. Do not store, but make up as required.

Working standards (blood and tissues)

To prepare standards of 0, 50, 100, 150, 200, and 300 μg per 100 ml, 0, 0.5, 1.0, 1.5, 2.0, 3.0 ml of dilute standard are transferred to a series of 100 ml volumetric flasks and diluted to 100 ml with distilled water and mixed.

Working standards for urine

To prepare standards of 0, 0.01, 0.03, 0.05 and 0.01, μg ml^{-1}, aliquots of the 100 μg per 100 ml standard (1 μg ml^{-1}) are transferred to a series of 100 ml volumetric flasks, diluted to volume with distilled water and mixed well. The 0.001, 0.003, and 0.005 μg ml^{-1} standards are prepared by transferring 1, 3 and 5 ml of the 0.1 μg ml^{-1} (10 μg per 100 ml) standard to a series of 100 ml volumetric flasks, diluting to volume, and mixing by repeated inversion.

GOLD

Gold in solution at pH 5–6 is chelated by sodium diethyldithiocarbamate (NDDC) and extracted into methylisobutyl ketone (MIBK). Standards and unknowns are compared at the 242.8 nm resonance line.

Gold in plasma (flame)

1. Into a 100 ml micro-Kjeldahl digestion flask, place 2 ml of plasma, 1 ml of sulfuric acid, 20 ml of nitric acid, and a few pyrex boiling beads. Mix cautiously. Allow to stand for approximately an hour before beginning the ashing process.
2. Set up standards containing 0, 0.01, 0.05, 0.1, 0.5 and 1 μg ml^{-1} (0–100 μg per 100 ml) concurrently.
3. When digestion is complete, quantitatively transfer the digestants to a 25 ml volumetric flask and make up to volume with distilled water.
4. Place 20 ml aliquots of the diluted digestants into a series of separatory funnels. Adjust the pH to 5–6 with 2.5 N NaOH.
5. Add 1 ml of 2% NDDC and 2 ml of MIBK. Shake funnel contents to extract.
6. After funnel contents have equilibrated, remove the solvent layers and centrifuge.

7. Compare standards and unknowns at the 242.8 nm resonance line for gold. Use a lean, oxidizing flame as described previously in the cadmium procedure.

Plasma gold (flameless)

1. Into a series of polypropylene test tubes, place 0.8 ml of distilled water. Add 0.2 ml of sample (plasma) or standard gold solutions. Mix.
2. Program the graphite furnace to dry at 100 °C for 40 s, char at 500 °C for 120 s, atomize at 2500 °C for 8 s.
3. Compare 10 μl aliquots of standards and unknowns.
4. Background correction is suggested.

Gold in urine (flame or flameless)

1. Adjust the pH of urine samples and gold standards containing 0, 0.1, 0.3, 0.5, 1, 1.5 and 2 μg ml^{-1} of gold to 5–6.
2. Program the graphite furnace to dry out at 100°C for 40 s, char at 500°C for 120 s, atomize at 2500°C for 8 s.
3. Add 1 ml of 2% NDDC and 2 ml of MIBK. Shake to extract. Centrifuge.
4. Compare standards and unknowns in an oxidizing flame or in a graphite furnace as described above.

Gold in tissues

1. Place tissue specimen into a 100 ml micro-Kjeldahl digestion flask.
2. Add 20 ml of nitric acid and a few pyrex boiling beads. For tissue samples smaller than 0.5 g, add 1 ml of sulfuric acid. To samples 0.5 to 1 g in size, add 2 ml sulfuric acid.
3. Process standards ranging between 0 and 2 μg ml^{-1} concurrently.
4. Proceed with wet ashing as described.
5. Quantitatively transfer the digestants to 25 ml volumetric flasks and make up to volume with distilled water.
6. Transfer 20 ml aliquots of the diluted digestants to a series of separatory funnels. Adjust pH to 5–6, add the chelating agent and organic solvent. Continue as described in the gold in blood procedures.

Calibration procedure

Stock standard

1 ml = 1 mg = 1000 μg of gold.

Place 267.2 mg of gold sodium thiosulfate—Au.NaS$_2$O$_3$—into a 100 ml volumetric flask. Add water to dissolve. Dilute to volume. Mix. Store in Teflon.

Working standards

Working standards are not stored but are prepared as required.

IRON

Iron analyses by flameless atomization are quite simple. Keeping the analytical environment free from extraneous iron can present a challenge however. All glass and plastic laboratory ware used in the analyses should be pre-cleaned with 5% nitric acid and rinsed copiously with deionized water.

Serum iron

1. Into a series of 5 ml glass or polystyrene test tubes known to be iron-free place 0.8 ml aliquots of iron-free 5% trichloroacetic acid (TCA).
2. Add 0.2 ml of sera or standards (0, 0.5, 1, 1.5, 2 and 3 μg ml^{-1}). Mix tube contents on a vortex mixer. Set aside for at least 20 min. Remix tubes a few times before centrifuging.
3. Compare 10 μl of the supernatants at the 248.3 nm line in a graphite furnace programmed to dry at 100 °C for 40 s, char at 800 °C for 40 s and atomize at 2400 °C for 6 s. Background correction is not necessary.
4. Caveat: Check the iron-free status of each pipet tip used for introducing an aliquot into the furnace.
   (a) Draw up an aliquot of 5% iron-free TCA. Discard TCA. Repeat twice and discard.
   (b) Draw up 10 μl of fresh TCA solution and place into the furnace. Start the program.
   (c) Should the recorder signal be less than eight scale divisions, use the pipette tip for sampling a standard or unknown. Discard the tip if the recorder signal is more than eight scale divisions.

Iron binding capacity

Principle

Iron as ferric ammonium citrate is added in excess to saturate the siderophilin in serum. Excess iron is removed by absorption onto a resin (Amberlite IRA 410). The bound iron remaining in the supernatant represents the iron-binding capacity.

Technique

1. Place 0.2 ml aliquots of sera into a series of 5 ml iron-free glass or polystyrene tubes.
2. Add 20 μl (0.02 ml) of 5 mg per 100 ml iron solution to each. Mix. Set aside for 10 min.

3. Add about 0.05 ml of the Amberlite resin (an amount sufficient to cover a centimeter length of a spatula one centimeter wide).
4. Add 0.8 ml of barbital buffer, pH 7.5. Mix. Set aside for 10 min. Mix twice during the interval. Centrifuge.
5. Place 0.2 ml aliquots of the supernatants into fresh test tubes. Add 0.2 ml of 5% TCA. Mix on a vortex mixer. Set aside for at least 20 min. Centrifuge.
6. Determine the iron content in 10 μl aliquots of the supernatants. Observe the same precautions suggested in the serum iron procedure.
7. Concentration curve value ×2 equals total iron-binding capacity (TIBC). Iron-binding capacity (IBC) or unsaturated iron-binding capacity (UIBC) = TIBC—serum iron.
   Note: Because of the sodium content in the buffer solution employed, the recorder signal of the TIBC will be a double peak. The second peak, clearly separated from the first, is due to the iron present; the first peak is merely due to non-specific molecular absorption.

Iron in urine

1. Dilute urine samples and standards 1:5 with deionized water.
2. Compare 10 μl aliquots in the graphite furnace as described in the serum procedure.

Iron in tissues

1. Weigh approximately 0.5 g quantities of tissue into a series of 100 ml micro-Kjeldahl digestion flasks.
2. Add 20 ml of nitric acid, 1 ml of sulfuric acid and 3 pyrex boiling beads to each.
3. Set up a series of suitable iron standards concurrently.
4. Proceed with the wet digestion as described previously.
5. Quantitatively transfer the digestants to 50 ml volumetric flasks and make up to volume with distilled water.
6. Compare standards and unknowns as described above.

Calibration procedure

Stock iron standards

1 ml = 1 mg = 1000 μg of iron.

Weigh 1 g of pure iron wire into a small beaker. Add 15 ml of deionized water and 5 ml of concentrated hydrochloric acid (reagent grade). Heat to dissolve iron wire. Using iron-free deionized water, transfer the solution, quantitatively, to a 1 l volumetric flask (precleaned). Dilute to volume. Mix. Store in Teflon.

Quality iron stock standard solutions are available commercially.

Dilute stock standard

1 ml = 10 μg or iron
  Dilute 10 ml of stock iron standard to 1000 ml with deionized water.
Make fresh daily.

Working iron standards

Transfer 0, 5, 10, 15, 20, and 30 ml of dilute stock standard to a series
of 100 ml volumetric flasks. Dilute to volume with iron-free deionized water.
Make fresh daily.

Cleaning the Amberlite IRA 410 resin

1. Suspend in 3 N hydrochloric acid overnight in order to wash the resin
   relatively free from iron and to saturate it with chloride ion.
2. Decant the acid. Wash with deionized water about 10 times. Stir well.
   Allow to settle, decant, etc.
3. Suspend the resin in barbital buffer of pH 7.5. Stir well. Permit resin
   to settle. Decant buffer.
4. Wash the resin in deionized water about 4 times.
5. Dry the resin in an oven. Store in a covered container.

LEAD

Chelation–extraction procedure

Lead in solution at pH 5.5–6.5 is chelated by sodium diethyldithiocarbamate
(NDDC) and extracted into methylisobutyl ketone (MIBK). Concentrations
are determined by comparing standards and unknowns at either the 217.0
or 283.3 nm resonance lines. The former is a more sensitive line, but the
practicality of using it is dependent upon instrumental capability and stability.

Blood lead (flame)

Heparin is the anti-coagulant of choice. Versenate (EDTA) being a more
avid chelator of lead interferes with NDDC chelation. Hence even after
addition of calcium recovery of lead is incomplete.
1. Into a series of 20 ml screw capped round bottom, borosilicate centrifuge
   tubes, place 6 ml of a chelating solution buffered to pH 6.5.*
2. With a serological pipet, add 2 ml of heparinized blood. Allow to
   drain and then rinse up and down in the chelating solution.

---

* Adaptation of chelating solution containing a surfactant and buffered to pH 6.5 evolved
during the course of involvement with the NBS and AACC study group on blood lead.
Directions for preparation can be found at the end of the antimony procedure, p. 259.

3. Set up a series of standards containing 0, 0.2, 0.4, 0.6, 0.8 and 1 μg ml^{-1} (0–100 μg%) of lead in a like manner.
4. Add 2 ml of MIBK. Cap tubes. Invert about 3 times. Set aside for 10 min.
5. Shake manually for 2 min. Centrifuge for 10 min.
6. Compare standards and unknowns at the 283.3 nm line. Flame conditions are as those described for cadmium.

Blood lead (flameless)—macro

1. Prepare standards and unknowns as described in the flame method above.
2. Compare 10 μl aliquots of standards and unknowns in a graphite furnace programmed to dry at 100 °C for 20 s, char at 300 °C for 20 s, and atomize at 2400 °C for 6 s. Background correction is necessary.

Blood lead (flameless)—micro

1. Into a series of 1.5 ml Eppendorf conical centrifuge tubes, place 0.4 ml of buffered chelating solution.
2. With a 100 μl Eppendorf pipette add 100 μl of whole blood. Rinse the pipette tip with distilled water and add washings to the centrifuge tube. Repeat twice more. Add 0.2 ml of MIBK. Cap tubes and mix by vortexing. Set aside for about 10 min. Vortex again before centrifuging.
3. Compare 10 μl aliquots in the graphite furnace programmed as in the macro procedure above.

Blood lead—nitric acid precipitation

Blood proteins are precipitated by nitric acid. Lead concentrations in the supernatants are compared to standards treated similarly.

The following procedure, an adaptation of the Rains' method, is suggested when blood samples are known to contain versenate.

Preclean pipette tip and 1.5 ml conical centrifuge tubes with 5% nitric acid. Rinse copiously with deionized water to remove traces of nitric acid. Dry tubes and pipette tips.

1. Place 0.4 ml deionized water into a series of centrifuge tubes (1.5 ml).
2. With a 100 μl Eppendorf pipet with proper tips, sample 100 μl of heparinized blood or lead standards made up in 0.78% sodium chloride. Add contents to centrifuge tube. Carefully rinse tip. Wipe. Then with the same tip, sample 100 μl of clean deionized water and add to the centrifuge tube. Repeat twice, again with the same tip. Let tube contents stand for 10 min. Add 200 μl of 25% nitric acid. Cap and vortex contents to mix. Stand for 20 min. Vortex at intervals. Centrifuge for 3 min.
3. Compare 100 μl aliquots of the respective supernatants at the 217 nm line if the atomic absorption instrument is sufficiently stable. (The concentration curve has been found to be linear between 0 and 1 μg ml^{-1}).

Should the 283.3 nm line be employed, compare 20 μl aliquots of the respective supernatants.

Programme graphite furnace to dry at 100 °C for 40 s, char at 300 °C for 40 s, and atomize at 2400 °C for 6 s. Background correction is necessary.

Urine lead (flame)

1. Adjust pH of urine and standards of 0–1 μg ml^{-1} to pH 5.5–6.5
2. Place 10 ml aliquots into a screw capped borosilicate centrifuge tube.
3. Add 1 ml of 2% NDDC and 2 ml of MIBK.
4. Invert to mix. Let stand for 10 min. Then shake for 2 min and centrifuge.
5. Proceed as with blood lead (flame method).
6. If lead concentration is more than 1 μg ml^{-1}, repeat on a smaller aliquot.

Urine lead (flameless)

Proceed as described in the blood lead (flameless) procedure.

Lead in tissue

1. Begin wet ashing 1 g quantities of tissue or appropriate standards in 2 ml of perchloric and 30 ml of nitric acid. Add more nitric acid as required until digestion is complete.
2. Quantitatively transfer the crystal clear digestant to a 25 ml volumetric flask and make up to volume with distilled water.
3. Transfer 5 ml aliquots of the diluted digestants to 60 ml separatory funnels. Adjust the pH to 5.5–6.5 with 2.5 N NaOH.
4. Add 1 ml of 2% NDDC and 2 ml of MIBK. Cap funnels. Mix. Stand for 10 min, then shake to extract.
5. After funnel contents equilibrate, remove solvent phase and centrifuge.
6. Compare standards and unknowns. Use either flame or flameless atomization.

Calibration procedure

Stock standard

1 ml = 1 mg = 1000 μg of lead.

Place 159.8 mg of lead nitrate—$Pb(NO_3)_2$—into a 100 ml volumetric flask. Add 0.1 N HNO_3 to dissolve. Dilute to volume. Mix by repeated inversion. Store in Teflon.

Dilute stock standard

1 ml = 10 μg of lead.

Dilute 10 ml of stock standard to 100 ml with deionized water. Measurements must be volumetric. Make fresh daily.

Working standards

Prepare working standards of 0, 0.2, 0.4, 0.6, 0.8, and 1 μg ml^{-1} by diluting 0, 2, 4, 6, 8, and 10 ml of the dilute stock standard to 100 ml with deionized water.

Make up working standards for the nitric acid blood procedure in 0.78% sodium chloride.

LITHIUM

Lithium is determined by the direct analysis of diluted serum, plasma, urine, or tissue digestant at the 670.8 nm resonance line.

Serum

1. Prepare a 1 : 25 dilution of serum or plasma in distilled water.
2. Prepare dilute standards containing 0, 0.5, 1, 2, 3, and 5 mg per 100 ml of lithium.
3. Compare standards and unknowns at the 670.8 nm line. An oxidizing flame, not as lean as that for lead, is employed. Use gas flow rates for air and acetylene of 7 and 5 l min^{-1}, respectively.
4. Aspirate at the rate of 2–3 ml min^{-1}.
5. To convert mg% to meq l^{-1}, divide by 0.7. Example: 1.0 mg% = 1.3 meq l^{-1}.

Urine

Dilutions are prepared as required and compared with standards.

Tissue

1. Wet ash 1–5 g quantities in sulfuric–nitric acid as described.
2. Treat standards similarly.
3. Quantitatively transfer digests to a 25 ml flask and make up to volume with distilled water.
4. Compare standards and unknown as described above in the blood procedure.
5. Dilute digestants further as required.

Calibration procedure

Stock standard

1 ml = 1 mg = 1000 μg of lithium.

Place 523.3 mg of lithium carbonate—$LiCO_3$—into a 100 ml volumetric flask. Add 50 ml of distilled water and 1 ml of concentrated hydrochloric acid. Shake to dissolve. Dilute to volume. Mix by repeated inversion. Store in Teflon or polyethylene.

Working standards

To prepare standards of 0, 0.5, 1, 2, 3, and 5 mg per 100 ml, transfer 0, 0.5, 1, 2, 3, and 5 ml of the stock standard to a series of 100 ml volumetric flasks and dilute to volume with distilled water. Mix by repeated inversion.

MANGANESE

Manganese in solution at pH 6–9 is chelated by sodium diethyldithiocarbamate (NDDC) and extracted into methylisobutyl ketone (MIBK). Standards and unknowns are compared at the 279.5 nm resonance line.

Manganese in blood

1. Into a series of round bottom centrifuge tubes place 7 ml of chelating solution containing NDDC buffered to pH 6.5.
2. Add 3 ml of serum blood or standard (0, 1, 3, 5, 10 and 20 μg l^{-1}).
3. Add 2 ml of MIBK. Cap and mix by repeated inversion. Set aside for 10 min before shaking to extract the complex.
4. After centrifugation, compare the 20 μl aliquots of supernatants at the 279.5 nm resonance line in a graphite furnace programmed to dry at 100 °C for 20 s, char at 300 °C for 40 s, atomize at 2400 °C for 6 s. Background correction is necessary.

Manganese in urine

1. Adjust the pH or urine samples and standards to 6–9.
2. Place 20 ml aliquots into a series of 60 ml separatory funnels. Add 1 ml of 2% NDDC and 3 ml of MIBK. Mix. Set aside for 10 min. Shake to extract.
3. Allow funnel contents to equilibrate. Remove solvent layer and centrifuge.
4. Compare standards and unknowns in the graphite furnace programmed as described in the blood procedure.

Manganese in tissue

1. Wet ash 0.5–5 g amounts of tissue in sulfuric–nitric acid as described previously.
2. Treat appropriate standards in a like manner.
3. Quantitatively transfer the crystal clear digestants to 25 ml volumetric flasks and make up to volume with distilled water.
4. Transfer 20 ml of the dilutions to separatory funnels. Adjust pH to 6–7.

5. Add 1 ml of 2% NDDC and 3 ml of MIBK. Mix. Set aside 10 min before shaking to extract the chelate.
6. Compare aliquots of the standards and unknowns in the graphite furnace.

Calibration procedure

Stock standard

1 ml = 1 mg = 1000 μg of manganese.

Place 446.1 mg manganese acetate tetra-hydrate—$Mn(CH_3COO)_2 \cdot 4$ H_2O—in a 100 ml volumetric flask. Add water to dissolve. Dilute to volume. Mix. Store in Teflon.

Working standards

Prepare as required.

MERCURY

Mercury in solution is chelated by ammonium pyrollidine dithiocarbamate (APDC) at pH 3–4 and extracted into methylisobutyl ketone (MIBK). Standards and unknowns are compared at the 253.6 nm resonance line.

Mercury in blood

1. Into 20 ml round bottom centrifuge tubes, glass or polystyrene, place 10 ml aliquots of 5% trichloroacetic acid (TCA).
2. Add 5 ml of heparinized blood or standards containing 0, 0.001, 0.003, 0.005, 0.01, 0.02, and 0.05 μg ml^{-1} of mercury. Mix with stirring rods. Set aside for an hour. Mix occasionally. Centrifuge.
3. Decant supernatants into 60 ml separatory funnels.
4. Wash residues in centrifuge tubes with 10 ml of 5% TCA. Centrifuge and add washings to appropriate funnels.
5. Adjust pH to 3–4 with 0.5 N NaOH. Add 1 ml of 1% APDC and 2 ml of MIBK. Stopper. Mix. Set aside for 10 min and shake.
6. Allow funnel contents to equilibrate. Remove solvent phase and centrifuge.
7. Compare 20 μl aliquots of standards and unknowns at the 253.6 nm line in a graphite furnace programmed to dry at 75 °C for 40 s, char at 75 °C for 40 s and atomize at 2000 °C for 6 s. Background correction is necessary.

Mercury in urine

1. Adjust the pH of urine and standards (0–500 μg l^{-1}) to 3–4.
2. Place 10 ml aliquots into a series of 20 ml borosilicate centrifuge tubes.
3. Add 1 ml of 1% APDC and 2 ml of MIBK.
4. Proceed as described in the blood procedure.

Mercury in tissues

1. Homogenize 5–10 g of tissue in 50 ml of 5% TCA. Pour into beakers. Cover with saran (polyvinylidene chloride) and set aside for about 8 hours, or overnight.
2. Distribute beaker contents into test tubes and centrifuge. Combine the supernatants. Place in 60 ml separatory funnels.
3. Adjust pH to 3–4. Add 1 ml of 1% APDC and 2 ml of MIBK. Extract.
4. Treat aliquots of standards diluted with 5% TCA similarly.
5. Compare unknowns and standards in the graphite furnace.

Calibration procedure

Stock mercury standard

1 ml = 1 mg = 1000 μg of mercury.

Place 135.3 mg of mercury bichloride—$HgCl_2$—into a 100 ml volumetric flask. Add distilled water to dissolve. Dilute to volume. Mix by repeated inversion. Store in Teflon.

Working standards

Prepare fresh when required.

SILVER

Silver in solution is chelated by sodium diethyldithiocarbamate (NDDC) at pH 5–7 and extracted into methylisobutyl ketone (MIBK). Standards and unknowns are compared at the 328.1 nm resonance line.

Silver in blood

1. Wet ash 5 ml (of g) of blood and appropriate standards (0–50 μg l^{-1}) in 2 ml of sulfuric acid and as much nitric acid as is necessary to complete digestion.
2. Quantitatively transfer the crystal clear digestants to 25 ml volumetric flasks and make up to volume with distilled water.
3. Place 20 ml of the diluted digestants into separatory funnels. Adjust pH to 5–7 with 2.5 N NaOH.
4. Add 1 ml of 2% NDDC and 2 ml of MIBK. Mix. Extract.
5. Remove solvent layer and centrifuge.
6. Compare supernatants at the 328.1 nm resonance line in the graphite furnace programmed to dry at 100 °C for 20 s, char at 300 °C for 40 s and atomize at 2500 °C for 7 s.

Silver in tissue

Wet ash 0.5–5 g of tissue and proceed as described in the blood procedure.

Silver in urine

1. Adjust urine samples and standards (0–100 μg l^{-1}) to pH 5–7.
2. Place 20 ml aliquots into a series of 60 ml separatory funnels. Add 1 ml of 2% NDDC and 2 ml of MIBK. Mix. Extract.
3. Proceed as described above.

Calibration procedure

Stock standard

1 ml = 1 mg = 1000 μg of silver.

Place 157.5 mg of silver nitrate—$AgNO_3$—in a 100 ml volumetric flask. Add water to dissolve. Dilute to volume. Mix. Store in Teflon.

Working standards

Working standards are not stored but are prepared as required.

THALLIUM

Proteins are removed by either trichloroacetic acid (TCA) precipitation or by wet digestion.

Thallium in solution is chelated by sodium diethyldithiocarbamate (NDDC) at pH 6.5–7 and extracted by methylisobutyl ketone (MIBK). Concentrations are determined by comparing standards and unknowns at the 276.8 nm resonance line.

Thallium in blood (flame)

1. Precipitate proteins in 15 ml of heparinized blood by 5% TCA. (10 ml of TCA per 5 ml of blood). After standing for an hour, centrifuge.
2. Treat 15 ml of standards (0, 0.01, 0.03, 0.05, 0.1, 0.2, and 0.5 μg ml^{-1}) similarly.
3. Transfer supernatants to a series of 60 ml separatory funnels.
4. Wash the residues in the centrifuge tubes with additional 10 ml aliquots of 5% TCA, and add the washings to the appropriate funnels.
5. Adjust pH to 6.5–7 with 2.5 N NaOH. Add 1 ml of 2% NDDC and 3 ml of MIBK. Mix. Set aside for about 10 min. Then extract. Allow the funnel contents to equilibrate. Remove solvent layers and centrifuge.
6. Compare standards and unknowns at the 276.8 nm resonance line. Use the flame conditions as described for lead and cadmium.

Thallium in blood (flameless)

1. Treat 5 ml aliquots of heparanized blood and standards as described in the blood procedure and extract with 2 ml of MIBK.
2. Compare standards and unknowns in a graphite furnace programmed to dry at 100 °C for 20 s, char at 300 °C for 20 s, and atomize at 2400 °C for 6 s.

Urine thallium

1. Place 30 ml of urine or standards into separatory funnels. Adjust pH to 6.5–7.
2. Proceed as with analysis of blood.
   (10 ml aliquots are adequate when flameless atomization is employed.)

Thallium in tissue, hair, etc.

1. Wet ash 1–5 g of tissue in 2 ml of sulfuric acid and as much nitric acid as is necessary to complete the digestion.
2. Quantitatively transfer the crystal clear digest into a 25 ml volumetric flask and make up to volume with distilled water.
3. Process the standards similarly.
4. Transfer 15 ml aliquots to 60 ml separatory funnels. Adjust pH to 6.5–7 with 2.5 N NaOH.
5. Proceed as described in the blood procedure.

Calibration procedure

Stock standard

1 ml$=$1 mg$=$1000 μg of thallium.

Place 130.3 mg of thallous nitrate—$TlNO_3$—into a 100 ml volumetric flask. Add distilled water to dissolve. Dilute to volume. Store in Teflon.

Working standards

Prepare working standards from the stock solution when required.

ZINC

Zinc is determined by direct analysis on dilute serum, urine, semen, tissue digestant, etc. at the 213.8 nm resonance line.

Zinc in serum

1. Prepare a 1:5 dilution of serum using 5% trichloroacetic acid (TCA).
2. Treat standards containing 0, 50, 100, 150, and 200 μg per 100 ml (1–2 μg ml^{-1}) similarly.

3. Compare standards and unknowns at the 213.8 nm line in an oxidizing flame, using flow rates of 5 l min^{-1} for acetylene and 7 l min^{-1} for air.

Zinc in urine
Dilute urine specimens as required and compare with standards.

Zinc in tissues
1. Wet ash 0.5–5 g of tissue in sulfuric–nitric acid.
2. Carry appropriate standards through the digestion process also.
3. Quantitatively transfer crystal clear digest to 25 ml volumetric flasks and make up to volume with distilled water.
4. Compare standards and unknowns.
5. Dilute unknown samples further if the zinc concentration exceeds $2\mu g$ ml^{-1}.

Calibration procedure
Stock standard
1 ml = 1 mg = 1000 μg of zinc.

Place 246.9 mg of zinc sulfate ($ZnSO_4 \cdot H_2O$) into a 100 ml volumetric flask. Add distilled water to dissolve. Dilute to volume. Store in Teflon.

Working standards
Working standards are stable for months when stored in polyethylene.

Since copper and zinc determinations are usually companion assays, it is timesaving and proper to prepare working standards containing 0, 50, 100, 150, 200, and 300 μg each of copper and zinc per 100 ml.

INDEX